International Conference on Architectural Education in Asia (eduARCHsia 2017)

SHS Web of Conferences Volume 41 (2018)

Yogyakarta, Indonesia
9 November 2017

Editors:

Revianto Budi Santosa
Noor Cholis Idham
Nensi Golda Yuli
Putu Ayu P. Agustiananda

ISBN: 978-1-5108-5577-9

Printed from e-media with permission by:

Curran Associates, Inc.
57 Morehouse Lane
Red Hook, NY 12571

Some format issues inherent in the e-media version may also appear in this print version.

Printed by Curran Associates, Inc. (2018)

For additional information, please contact EDP Sciences – Web of Conferences at the address below.

EDP Sciences – Web of Conferences
17, Avenue du Hoggar
Parc d'Activité de Courtabœuf
BP 112
F-91944 Les Ulis Cedex A
France

Phone: +33 (0) 1 69 18 75 75
Fax: +33 (0) 1 69 28 84 91

contact-edps@webofconferences.org

Additional copies of this publication are available from:

Curran Associates, Inc.
57 Morehouse Lane
Red Hook, NY 12571 USA
Phone: 845-758-0400
Fax: 845-758-2633
Email: curran@proceedings.com
Web: www.proceedings.com

TABLE OF CONTENTS

The International Conference on Architectural Education in Asia (EDUARCHSIA 2017)

Being a knowledge-based practice, architecture strives to re-invent its relevance with its circumstances by developing a system of knowledge. For millenia, the knowledge of architecture was developed based on certainty. Vitruvian triad began with firmitas connoting the physical permanence of a building, followed by utilitas relying on the stability of function conducted in such building, while the last, its notion of venustas implies eternal beauty giving delight to its spectators.

In current situations, uncertainty is the norm, and certainty is the exception. Architecture exists in a world where so many different cultures are intermingle at the same time, each with their own value system. The knowledge of architecture faces an enormous complexity of values, interpretations, and readings.

Social un-stability, as well as massive mobility and migration prevail. Place attachment of people and their sense of identity and affiliation are constantly re-shaped in this situations. The absence of place and multiplicity of place might occur in an age-old place.

Enormous flows of information due to the advancement of digital technology alters our life fundamentally. In the meantime, information technology itself is in a perpetual transformation. From the tools being the extension of our hands to make things, it becomes the extension of our mind to vision things that we are going to produce. It opens possibilities and enables us to imagine what was previously unimaginable, realize what was unrealizable, as well as to conceive what was unconceivable. However, know, such tools became the extension or even foundation of our existence. Do we need to stay and gather on a certain place, or shall we move and wander depend on the digital existence of ourselves. Our needs, urgency and behavior are mediated, driven, or in some extents determined by this new apparatus.

Architecture in Asia with its immense diversity in complexity currently experience even more fundamental transformations in social relations, technological advancement, environmental management, and cultural appropriation. The existing configuration of knowledge is no longer adequate to comprehend and develop architecture in this region.

Former architectural inquiries on what is the shape of our buildings and how do we shape our building to suit and adapt with contemporary circumstances are still valid, but new challenges like do we still need buildings and why do we construct buildings.

This Conference calls for vision, insight, ideas and theories to re-chart the knowledge of architecture in order to inform the improvement of architectural education and practice in Asia particularly in the themes of:

- **Humanity and its challenges**

  One of the most noble intentions of architecture is to dignify human existence on earth. Humanity, however, currently faces serious challenges due to the local and worldwide conflicts, the propagation and advancement of machine and appliances, and the proliferation of poverty. In their thinking and practice, architects should propose new ways to regain its raison d'etre to support humanity in its dignified state.

- **Multitude of cultural complexity**

  Culture and history are not always closely related with geographical entity. Regions are fluids and overlapping. Nation states loosen their boudaries as people and capitals move from one part of the world to another massively. Vernacularism, tradition and identity as manifested themselves in architecture should be re-defined and re-invented in order to maintain their relevances with current circumstances.

- **Digital reality and digital technology**

  More than a tool to extend our hands, information and communication technology has been developed into extend our ways of conceiving the world and our ways of changing the world. Digital technology becomes digital reality as virtuality is transformed into augmented reality. Form and function—being the fundamentals of architecture—are no longer conceived as in the previous architectural thinking. Fluidity, transience and instability of forms and functions become the norm to cope with uncertainty of the world.

- **Environmentally appropriate technology**

  Environment is in the state of peril due to the massive industrialization and urbanization. This situation causes in increasing consumption of resources and proliferating negative impacts to the environment. Architecture should propose a breakthrough in the formulaition of its knowledge in order to lessen the depletion of nature.

- **Uncertainty of urbanism and territoriality**

  Since 2007 more people live in urban are than in other places. The urban areas where most of us live are not the same kind of place as before. Urbanism and architecture previously relies on the stability of space and other geographic entities. Centers virtually disappear and distributed network is more prevalent than centralized hierarchy.

https://doi.org/10.1051/shsconf/20184100001

Boundaries are dissolved and spatial entities are alwys in the state of change. Architecture needs a new way of thinking to understand of the ever-changing space and its challenges, problems and opprtunities in order to be able propose improvements.

Statement of Peer review

In submitting conference proceedings to *Web of Conferences*, the editors of the proceedings certify to the Publisher that

1. They adhere to its **Policy on Publishing Integrity** in order to safeguard good scientific practice in publishing.
2. All articles have been subjected to peer review administered by the proceedings editors.
3. Reviews have been conducted by expert referees, who have been requested to provide unbiased and constructive comments aimed, whenever possible, at improving the work.
4. Proceedings editors have taken all reasonable steps to ensure the quality of the materials they publish and their decision to accept or reject a paper for publication has been based only on the merits of the work and the relevance to the journal.

Title, date and place of the conference

International Conference on Architectural Education in Asia (eduARCHsia)
November 9, 2017
Yogyakarta, Indonesia

Proceedings editor(s):

Dr. Revianto Budi Santosa, M. Arch.
Noor Cholis Idham, Ph.D., IAI
Nensi Golda Yuli, Dr.-Ing.
Putu Ayu P Agustiananda, Dr.-Ing.

Date and editor's signature

December 14, 2017

Quality Assurance in Architectural Education in Asia – On the Perspective of 'Design' Based Architectural Education and its Holistic Assessment

Junsuk Lee [1,*]

[1]Professor of Myongji University, Korea

Abstract. This paper starts by describing the importance of 'Architecture' to our society to our everyday life. It dominates our living environment, inevitably forms visual memories of life experience. Undoubtedly, a building is a 'container' to hold one's everyday living, the grouping of architecture an urban environment is a 'container.' It is to hold one's entire life. If then, there is paramount importance in realizing how we should properly educate our future 'architects.' First, it is important that the essential of architecture education must be based on proper 'Design Studio' format. Much of reasons are self described by following quotes such as 'like art, architecture cannot be taught, but it can and should be cultivated,' 'Essentials of architecture can be learned through actual doing,' 'architecture education is most perfect liberal arts education format invented.' Secondly, it is important to realize the vast meaning of architecture and architect to our society. Luckily, there is the Charter for UNESCO/UIA for Architectural Education which serves as not only a fundamental background to this discussion but also as 'prior principle' in articulating proper education for architects. Thirdly, considering the Charter as a principle, then it is worth discussing actual methodology as a working system in delivering 'accredited/validated education.' The KAAB's Conditions is compared with the Charter. For an indepth look into the most important set of Conditions of the KAAB the Student Performance Criteria, the paper describes the origin, historic background, and evolvement of the 'Student Performance Criteria,' which is borne by the NAAB in 1980's. As it weighs much importance of the whole accrediting process, it must be carefully written to reflect society's needs of an architect. Also, it must not be written too specific nor should include quantitative measure, to avoid all school programs eventually evolve to become uniform without a character. At the same time, it must possess enough specificity to guide panel reviewers in judging competency level of the student learning at the actual site. As a conclusion, there is paramount importance in architecture education. It is the UNESCO/UIA Charter that asks all civilized architectural education to be self-assessed and peer-reviewed for continuing improvement. Considering current era, there is no doubt that there are many opportunities waiting ahead among us. Familiarizing and building sympathy to the voice of the 'Charter for Architectural Education.' will be the most proper place to start.

Our Belief in Architecture

I would like to start by raising a fundamental question of what we all believe in 'What Architecture is'. In all consciousness - past or current - our visual memories have always existed in dimension of constant spatial dimension in background, either small/large, vivid/vague, pleasant/displeasing, and etc. In fact, we have never lived our cognitive-selves outside of 'space' or 'place' or of our tactile dimension of our surroundings.

What do you remember as your early childhood memories of city-scape scenery where you were growing up? Or time to time, when there was an occasion of family outing to a 'big place' at city center day or night, what fragments of scenic views signaled you to realize that you were physically at the middle of that 'special' outing? Or throughout one's life time, when there were chances to travel other parts of the world of renowned places, the scenic impressions large or small took place in our mind as something impossible to erase from our visual memory. Often times, those visual imprints stayed with overall impression of the whole travel. Yes, we all are constantly surrounded by city scape of some sort, which has always been a conglomeration of individual buildings – works of Architecture.

Either impressed by the majestic view off from top roof of the Rockefeller Center building looking toward Central Park of Manhattan, or dazzling view toward La Défense at the top of the Arc de Triumph, it is easy to presume that one's definition of man-made scenic view of the 'life time' surely has no limit on 'what to expect'.

* Corresponding author: junsuk21@naver.com

But at the same time, I insist that all impressive civic scenery of 'life-time' are careful collections of individual, architectural efforts. There is no doubt that all buildings of architectural works were borne by 'Architect' among us, all in which reflecting our utmost fundamental function of 'living' conditions either grand or humble.

Take an example for in Hong Kong, a view of Victoria Harbor, the impressive view toward the canal, the both sides are marked by over 100 stories tall state-of-the-art multi-purpose towers. Perhaps another scenic view of the 'life time', no matter how ignorant of one's knowledge of architecture or urban planning, the view speaks itself about image of 'The Gateway to Asia' to all passerby of the area. This poses very interesting point that whenever there is impressive scenic view of the city, it is hard to escape from noticing the fact of efforts of architectural design works of the buildings are also collectively conveying the results of 'urban image' which stems from overall planning efforts of the city.

Then it is natural to assume that no matter how small the scale of individual buildings is, the designer of the building must also consider the overall cityscape, leaving its impact toward a scenic view hoping to become better neighborhood to the place. As simple as this example is, it is easy to capture the role and mission of an architect to our place and the society, which is unmeasurably meaningful and important. At the Tokyo Forum project, life of an everyday city dwellers and workers surrounding the places pass through the 'Lobby', using like their easy back-door to where ever nearby destinations – within the city block - would be. The Lobby of the project not only plays as the 'Grand Space' of the grand-scale functions, but also for the people of Tokyo, there is no doubt that this incredibly dazzling lobby has already taken as humble background for people's everyday life. It surely is such a memorable example of what a great public space should be to a greater citizen. On the other side of the globe, we all are aware of the incredible beauty of the Sidney Opera House. When I was there at a first glance, it came across how nice scenic gestures manicured well with formal compositions and array of details, although it didn't take long for me to realize the real reason why the opera house was true beauty to all visitors. It was not only the beauty we already admit by the building itself, but the real beauty was existing at the heart of the visitors by the 'place' of the opera house. The indescribably beautiful 'place' was born by the well-appointed settings of public plaza and promenades surrounding the monumental architectural work, especially by the fact that they were working as a backdrop of mind boggling scenic beauty of harbor, greeting people with places to stay. The architect and urban planner were in synchronization in conjuring up a public scene of a 'heaven', the 'place' was so perfect for anyone who wish to make life-long promises or wish to leave heart felt memories while at the 'opera house'.

Meaningful role of 'Architecture' in our life doesn't always have to be involved with urban mission. At an impressive well-lit skylight space of a modern cathedral designed by Moneo, not only toward religious minds but any visitors who can feel and reason itself about the visual presence of wondrous dialogue between proportioned materials and natural light from above can be deeply marked with meaningful definition of what grateful 'space' can be. It is even meaningful when we consider its mission toward intended function of the building. Sometimes as playful and light-weight as Walt Disney Orchestra Hall in Los Angeles or as tacit and reserved as Kahn's Salk Institute's poetic approach in formal dialogue, we all know how variety of great pieces of architecture have flourished our life with much fruitful experiences toward realms that might have never existed to us before otherwise. As architecture is even referred as a notion of 'poetics of craft construction' by Frampton (1995), it is natural for us to wonder who would be the 'qualifiers' to create one.

Design Studio in Architecture Education

Stepping in toward our topic today – 'Architectural Education' the area that needs to be addressed with what we all deeply believe in. As a starter, it is noteworthy to point out what is mentioned at one of the 'Objectives of Architectural Education' of Charter for UNESCO/UIA for Architectural Education: *"That architecture is a discipline which draws knowledge from the humanities, the social and the physical sciences, technology, environmental sciences, the creative arts and the liberal arts."* We can easily see that educating architecture involves with constant challenges of multidisciplinary knowledge and a comprehensive attitude toward offering knowledge and skills of architecture. But at the same time, it requires us (providers of architectural education) to foster learners of the architecture to become 'professional level' by execution of teaching by a proper qualification of an institution. *"That education leading to formal qualifications and permitting professionals to practice in the field of architecture has to be guaranteed to be at university/tertiary level with the discipline of architecture as the main subject and be available at universities, polytechnics and academies. This education must maintain a balance between theory and practice."* Above statements nicely sums up how comprehensive the scope of architectural education is, and it must be treated as an education at tertiary level under proper settings and qualification it deserves.

To tackle the issue of 'Quality Assurance in Architectural Education', I would like to first focus on importance of 'Design' education as a main embodiment for overall achievement. As all of us might know already, teaching ability of 'Design' is never achievable by relying on series of lecture courses nor relying on digesting tons of knowledge of 'planning methodologies'. It requires a setting of 'Design Studio' and there are numerous points & reasons why it should.

As J.B. Friedman has noted in Creation of Space (1989), *"Vision is cultivated in Studio, which is both physical place and a state of mind.... The first space an architect must make is the design studio itself, a home for the spirit of creation."* I am convinced that 'studio' is synonym for 'setting for unknown exploration'. Friedman further notes; *"As a state of mind Studio is the capacity to experiment with options, to reify the result and develop them to increasingly vivid levels of expression.... An Experimenting artist is devoted to perfecting each study not to any predictable outcome, but rather because of a basic curiosity to find what may emerge... In making architecture, as in any creative act, the end product is unknown at the beginning.... Creativity exists on the edge of the unknown. It takes effort to enter a new room.... Essentials of architecture can be learned through actual doing."*

When we agree the fact that the main embodiment of ability in architecture is skills and ability of 'problem solving', it becomes clearer that much attention is needed to realize where the 'design education' should borne from. Slutzky's point (1989) directs us clearly to kind of attitude that we all must be aware of in teaching architecture: *"Architecture, like art, cannot be taught. But it can and should be cultivated.... Born from aesthetic sensibility, architecture can only be realized by the mastery of its formal means of expression. These means can be taught.... Model making, with its attendant flexibility to research space and structure, and particularly drawing, these architectonics primer celebrates the eye and hand as it extols the language of architecture."* Although there are further points that will help navigate this passage much smoothly. Wilson & Jennings (2000) addresses that *"Different from the lecture class, in which the student's goal is likely to accept knowledge that the instructor transfers through concrete instruction, design studio centers on student's active learning and hands-on activities."* It becomes clearer that when it comes to educating 'architecture', it must have borne from somewhat difference perspective. Not only the mind set of teaching, but also the instructor's character definition deserves unorthodox approach, as per Anthony (1991) and others as follows. *"In design studio... Primary knowledge is delivered through student-teacher interaction surrounding the student's design processes." "Student-teacher interaction is a rich source of knowledge in design studio (Goldschmidt, 2010)".* Further, Austerlitz (2002) even points out that design critique is the combination of criticism and 'intimacy', putting emphasis on having close relationship in between instructor and student. Goldschmidt (2010) even takes further as follows: *"With regard to the traditional perspective of the studio instructor as master, an alternate and more current perspective is that of the studio instructor as COACH or FACILITATOR."*

As we can see, all of above findings converge into coherent attitude toward what is needed in educating architecture. At the same time, we must bear in mind of paramount importance of 'Design' education in architecture with what Schön, (1984) has noted: *"Architecture studio should be an example of education for artistry... emphasizing technical rationality in professional education decreased the tendency to educate students of artistry and increased the tendency to educate them as technicians."* Also, it must be noted that at the 'Conditions and Requirements of an Accredited School' of Charter for UNESCO/UIA for Architectural Education asks for *"studio teaching as major part of the learning"*, and it further describes *"direct teacher/student dialogue should be the basis of the learning period..."* clearly emphasizing the important role of 'design studio' in educating architecture.

In summary, I stress that the essential of architecture education must be based on properly established 'Design Studio' format. With this, mind of relentless exploration and hands-on experiences must be the prime tool in cultivating design ability for the students and it will ultimately foster overall ability of design judgement as well. Other words, training for architecture means searching for unknown answers by means of creativity and exploration of self judgement – perhaps this is why Boyer & Mitgang (1996) has stated that the architecture education is most perfect liberal arts education format invented ever since western collegiate educational system started.

Considering Quality Assurance: General Characteristics of Educational Criteria from UNESCO/UIA vs. KAAB (As one of signatory of Canberra Accord)

Despite the fact that we all are very familiar with giving a criticism or review, announcing awards for the 'building' - the end product of architecture – frequently praising only the tangible results, but not so many of us are acquainted with the fact that 'architect' must be educated under certain qualifying set of conditions/requirements with agreeable goals. Specially the fact that if a society is permitting architect of certain class as a 'licensed profession' to plan, design, and to build, then the criteria for qualified education of it should become critical concerns for all. For this reason, it is important to note that at 'General Conditions' of the Charter for UNESCO/UIA for Architectural Education, it notes the need of pursuing agreeable conditions – across the globe – as the education is critical for the qualified profession as follows: *"That the increasing mobility of architects between different countries calls for mutual recognition or validation of individual degrees, diplomas, certificates and other evidence of formal qualification. That the mutual recognition of degrees, diplomas, certificates or other evidence of formal qualification to practise in the field of architecture has to be based on objective criteria, guaranteeing that holders of such qualifications have received and continue to maintain the kind of education and training called for in this Charter."*

UNESCO/UIA Charter for Architectural Education

Looking at the Charter for UNESCO/UIA for Architectural Education must be the correct place to start our discussion and takes important position for our topic.

It is composed of categories with following important points:

1. Preamble: Stepping further into much formal aspects of what 'architect' should be to our society, here are some respectfully summarized points. First, it points that the basic goal of education is to develop the architect as a *"generalist"*. It simply means that we must prepare future architects as 'general problem solver'. Also it points out that *"While architects working in developing context, where the architects could accept the role of and "enabler", rather than that of "provider", and where the profession is open to new challenges."* This leads to the point that when a society would need an answer to any problems, properly educated architect should have ability to play a role of the leader in taking the challenges and offering resolutions to them.
2. Aims: *"The aims of this Charter are that it be used, in the first instance, for the creation of a global network of architectural education within which individual achievements can be shared by all and that it will enhance the understanding that architectural education constitutes some of the most significant environmental and professional challenges of the contemporary world."*
3. General Conditions: This describes general characteristics, assumed scope and task of architect's role in the society, and preliminary conditions that need to be considered in architectural education. It also holds emphasis on 'attitude toward general public & society' as follows. *"the vision of the future world, cultivated in architecture schools, should include: decent quality of life for all human settlements..., ecologically balanced and sustainable development of build environments..., an architecture which is valued as the property and responsibility of everyone..., architectural education should never be considered a closed process but one in which life-long learning occurs... etc."* By the 2017 addendum, followings two items are added: *"architectural heritage education is essential to: understanding sustainability, the social context and sense of place in building design, and transforming the professional architectural mentality so that its creative methods are part of a continuous and harmonious cultural process. That cultural diversity, which is as necessary for human kind as biodiversity is for nature, is the common heritage of all humanity, and should be recognized and understood, for the benefit of present and future generations."*
4. Objectives of Architectural Education: This category lists key elements that need to be taught and to be performed by students. This also lists the scope of subject areas for architecture education. It is composed of '16 fundamental objectives', and categories such as 'design', 'knowledge', 'skill'. And 'quantitative indicators' lists minimum required years of study at tertiary level of education and expected minimum professional training to be qualified for registration/licensing/certification to practice as an architect.
5. Conditions and Requirements of an Accredited School: It outlines key elements that school need to equip to perform architecture education. A range of physical resources are identified and it asks for *"studio teaching as major part of the learning"*. As it further describes the importance of *"direct teacher/student dialogue in learning period"*, appropriate space and other physical resources for architectural education are asked to be furnished. Also, it mentions appropriate exposure to practitioners for students is required by stating *"...continuous interaction between the practice and teaching must be encouraged..."* Also, it is required that school program must be based on *"self-evaluation and peer review conducted at regular intervals..."* by review panels approved by UNESCO-UIA Validation System or recognized equivalent system. This clause justifies the services rendered by legitimate boards/agencies such as KAAB and other Canberra Accord Signatories with coherent goals.

The KAAB Conditions for Accreditation

The KAAB (Korea Architectural Accrediting Board) is an inaugural signatory member of the Canberra Accord (www.canberraaccord.org), has Conditions of Accreditation with main components as follows:

1. KAAB Perspectives on Architectural Education: It points out that school program must address its educational perspectives for the 5 main 'agents' of the accreditation process (Institution, Registration Board, Professional Association, Students, General Public)
2. Educational Program and Resources: It identifies requirements for the resources to run school program such as educational curriculum, self-assessment system, human, physical, financial, and information resources etc.
3. Educational Outcome: The Student Performance Criteria of 30 items identifies criteria for evaluating student work outcome by an educational program.

It is safe to assume that the Conditions for accreditation of the KAAB encompasses key contents and spirit of the Charter of UNESCO/UIA. The considerations for 'general public & welfare of human settlements' identified at 'General Conditions' of the Charter are ethical responsibilities of rather tall orders, which covers wide range of architect's societal duty for contemporary world. The similar attributes can be found at the Conditions of the KAAB's 'Introduction' and 'Perspectives on Architectural Education'. In addition, the Charter's point of *"life-long learning..."* aspects are covered at the 'five perspectives', which requires schools to possess proper attitude toward 'life-long learning' of the field and its profession. *(Listed at the perspectives of "Architecture Education and Qualification to Practice, Registration", "Architecture Education and the Profession")*

The KAAB's 'Introduction' and 'Perspectives on Architectural Education' amply suggests the required virtues such as a notion of leadership role of architects in the society, and wider horizon for latitudinous view on matters of the world with sympathy and understanding. It also suggests that they all are important foundation to architecture education. Therefore, it is expected by the condition that all of above are administered to students by diverse learning opportunities within a school program. Since the KAAB's accrediting review process always values 'holistic assessment' of a school program, offering of variety of extra-curricular activities (such as travel-abroad programs, community volunteering, and etc.) means a lot to a school program.

In addition, there are "Procedures" which has similar weight to Conditions by the KAAB. The 'Procedures' lists overall process of accreditation, exact procedural requirements, scope of activities of process and required forms of document used. For objectivity and transparency of the whole process, it is required as a condition for accreditation that all accreditation process must occur within the frame of published 'KAAB Procedures'.

Student Performance Criteria of the KAAB

The 'Objectives of Architectural Education' of the UNESCO/UIA Charter outlines area of what professional architectural studies ought to be, identifying scope of study areas and what level of performances from students are expected. On the counterpart at KAAB to this will be the Student Performance Criteria (SPC), which is the 10th Condition of the set. It is in fact considered the most direct and important conditions for accreditation for school programs. The schools must respond to it with sufficient effort and rigor, since a school program's primary task is to educate students and running curricular program which entails student work as outcome. All accredited school program must run educational program with subsequent educational outcome which satisfies SPC as 'minimum' standard. As noted by Ryu & Lee (2009) that the SPC even acts as a primary criterion for setting up school program's educational curriculums. Therefore, actual contents with imbedded directive of how the SPC is written will practically influence all school programs of a territory, and therefore all school programs are undoubtedly at stake by the SPC. The SPC could be named or referred differently by accrediting/validating organizations, but the general characteristic of it will be similar for any accrediting/validation board or agencies around the globe.

As important as how and what is there for the SPC, it is also important to know that there are inherent difficulties in writing of 'working' SPC. The two reasons are as follows. First, the SPC must not ask in quantitative measure or overly specific in describing what needs to be taught. If it does, then it will be easy to imagine that all territory's school curriculums and study programs will eventually become identical. It is quite natural to expect that a school program always wants to find short cut in satisfying the SPC, if it is written very specifically in measurable manner, all school programs will have tendencies to become equal. As we all know that one of serious worrisome drawbacks for having robust accreditation/validation system in general is that school programs are prone become similar and losing its uniqueness. Therefore, the SPC must be written in the way that somewhat open interpretation is allowed. Secondly, the SPC at the same time needs to be written 'specific' enough that it actually is practical in 'evaluating' student work of specific knowledge/ability. At the actual site visit of a school program, panel reviewers have to rely on evaluating school's performance by examining student works based on the SPC. Therefore, there is practical reason why it has to be appropriately directive and specific to some degree. Considering all above, the SPC needs to be written in careful balance between above two contrasting aspects.

The SPC of the KAAB were first published in year 2006 with 41 criteria. As accreditation conditions & procedures have gone through regular updates, the criteria have revised to total of 37 in year 2010. The last update has occurred in 2013 revising it to 30. By beginning of year 2018, 3rd revision is expected to occur, currently a special committee within the KAAB is formed and it is working on proposed revision. So far the revision has shown a pattern of reducing down the number of criteria, but it has not been an intention to simply reduce the teaching load or narrowing the scope of education. Rather, there were much important shift which were occurring in architectural education on global perspective.

When establishing the first KAAB Conditions & Procedures for accreditation, the KAAB has closely observed on how the NAAB (National Architectural Accrediting Board)'s accrediting system is operated. And, the NAAB system was used as a model. At the time, since the Korean collegiate system in general is widely based on that of the United States', and the NAAB's Conditions & Procedures were considered the most objective and transparent in nature to be adopted by a newly forming accreditation system like KAAB (Lee, J. 2015). In fact, the term 'Student Performance Criteria' was used by the NAAB since 1980's. The NAAB's conditions for accreditation were already formed in 1940's, but it took nearly 40 years to adopt the concept of 'SPC'. It was in 1960 to 1980's that the massive expansion of knowledge in all fields occurred, and it was something that human being practically has never seen before. Especially for architecture and related fields, there were great pressure for covering expanded knowledge by the professional architecture school programs, and there were already dynamic developments and branching off of professional knowledge and specialty fields in the community. For instance, by Geddes (1967), there was a study of developing educational program by 'modulating' categories of knowledge into a system to create study fields of 'environmental design' (Fig.1). As this pioneering study shows, academia at the time started struggling to organize expanded knowledge with specialty fields to form certain study discipline or developing educational tracks. Naturally, by the 1980's, there were so much to be aware of for architects as well and important facets of knowledge that needed to be dealt by professional degree programs of architecture. It was then the concept of SPC were introduced to be used as a kind of 'check-list' of teaching. That was the reason why when it was first introduced, there were more than 70 criteria, mostly of 'units or segments of knowledge/abilities'. The list was long but they were referring simpler concepts. But things start to change by the mid 1990's, a big shift occurred in directing 'what professional architecture education' is. At the time, there were accrued dissatisfaction on actual 'performance' of graduates of professional programs at practice. There were wide spread active discussions and studies, and also there were debates on education vs practice sides of architecture, leading to general agreement that architectural education at that time was distanced away from what is actually needed at practice. This was the reason why the revised SPC starting mid 1990's was reduced in number and started emphasizing what 'actual ability at practice' was. The criteria on 'design' started to gained importance and started to act as 'consolidated learning outcome' of all knowledge/ability of a student. More and more segments of knowledge were asked to be tied as a 'design' outcome, and performance criterion started to take form of 'comprehensive' characteristic rather than 'segments/units' of knowledge. The SPC of the NAAB at time has reduced down to 53 by year 1995. This movement continued and resulted further down the number of criterion to 37 in 1998, and 34 by year 2004. It was this period that the KAAB's SPC were first adopted with similar philosophy behind as the NAAB's at the time[1].

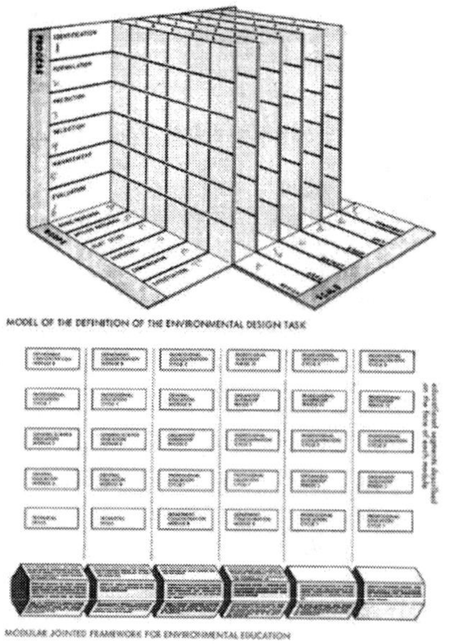

Fig. 1. Robert L. Geddes. (1967). A Study of Education for Environmental Design, Princeton U.

The KAAB's SPC is not at all obligated to follow the footsteps of how SPC of the NAAB's has evolved. But, the NAAB's past experience and seeing where it is heading next provides quite useful discussion material in preparing for next updates of SPC for the KAAB. The NAAB's SPC is moving toward giving more weight on 'comprehensive' knowledge/ability with emphasis on 'critical thinking' and 'problem solving' abilities since early 2000's. This movement of shift occurred starting in mid-1990, but the idea is shaping into much advanced term. For instance, the SPC's sub categories were set to represent 'kind' of study fields (for instance, Design, Knowledge, Professional etc.) in mid-1990's, but now their categories are according to advancements of architectural 'ability' (NAAB's SPC of 2014 is composed of 4 Realms: Critical Thinking and Representation, Building Practices, Technical Skills, and Knowledge, Integrated Architectural Solutions, Professional Practice).

Relating to issues of above, urgently suggests critical discussion we all need to involve with. Nowadays, all of us and perhaps the whole world are getting much more acquainted with the notion of "The 4th Industrial Revolution" coming, for the issues of trying to predict future job fields that are expected to flourish, sustain and disappear. Fray & Osborne (2013) even have told us that 47% of jobs nowadays will disappear in next 20 years. Amongst all, one thing for sure is that 'comprehensive problem solving skills based on creativity' is becoming such an important virtue for the future job fields in general, that we all need to work harder to correctly revise the 'criteria (SPC)' for our future generation of architects.

Conclusion

In total, as the scope for meaning of 'architecture' could cover much larger and meaningful than anything that one can think of, therefore the learning of 'architecture' clearly is a life-long mission by itself. There shouldn't be a short cut answer to it. We can only start a discussion of it by noting "Design" based architecture education is an embodiment of true learning of architecture, which involves variety of traits, unique approach with focused attitude (Lee, J. 2014), and with the notion of re-charting of human and communal value in an effort of building better, sustainable society. In all, the virtue of 'creativity' and 'self-realization' are the venues where ultimate form of architecture can be realized. And undoubtedly, building legitimacy in architectural education is critical passageway to it. Perhaps that is why Boyer & Mitgang (1996)'s appraisal[2] on architecture education is not so unfamiliar.

Luckily, the Charter for UNESCO/UIA for Architectural Education has served us as guiding light in navigation of searching for answers on 'how to' in architecture education. The charter always has served as philosophical background in efforts of erecting and

Institute of Korea) initiated researches to establish the KAAB's conditions for accreditation (SPC).

[2] Architecture education is most perfect liberal arts education format invented ever since western collegiate educational system started (Boyer & Mitgang, 1996)

articulating answers to it, perhaps to all architectural accrediting/validating boards/agencies around the globe, and of course to the Rules and Procedures of the Canberra Accord likewise.

For the global community, it is the UNESCO/UIA Charter that asks all civilized architectural education to be self-assessed and peer-reviewed for continuing improvement. For further advancement of architectural education of current era, there is no doubt that there are much opportunities waiting ahead. Such dynamic and vibrant developments currently occur in South Asian region will be clearly one of them. By familiarizing and building sympathy to the voice of the 'Charter for Architectural Education' will be the most proper place of starting point. And to understand how the spirit of the Charter can be fructified in reality by an actual system is vitally important. I believe this paper is intended to highlight this importance. Moreover, closely observing and actual participation to one of the 'live' system of accreditation (such as KAAB) will be a practical choice of being on-board with the global mission for architectural education.

On the perspective of practical side, the establishment and evolvement to become a credible system of accreditation for more than 60 legitimate school programs over past short period 10 years, It is my belief that the asset of experiences accrued by the KAAB must be first shared with neighboring region to further disseminate the 'spirit of the Charter'.

References

Anthony, K. H. (1991). Design juries on trial: The renaissance of the design studio. New York, NY: Van Nostrand Reinhold.

Austerlitz, N., Aravot, I., & Ben-Ze'ev, A. (2002). Emotional phenomena and the student–instructor relationships. Landscape and Urban Planning, 60(2), 105–115.

Boyer, E. L., & Mitgang, L. D. (1996). Building community: A new future for architectural education and practice. Princeton, NJ: Carnegie Foundation for the Advancement of Teaching.

Frampton, (1995). Studies in tectonic culture - the poetics of construction in nineteenth & twentieth century architecture, MIT Press

Frey & Osborne, (2013). The Future of Employment: How Susceptible Are Jobs to Computerisation?, Published by Oxford Martin Programme on Technology and Employment

Geddes, R. (1967). A Study of Education for Environmental Design, Princeton U., p 26, 31

Goldschmidt, G., Hochman, H., & Dafni, I. (2010). The design studio "crit": Teacher–student communication. Artificial Intelligence for Engineering Design, Analysis and Manufacturing, 24(3), 285–302.

KAAB, (2006, 2010, 2013). Conditions & Procedures for Accreditation

Lee, J. (2010). A study on Demands of Study Courses for 5 Year Professional Architectural Degree Programs on the Bases of Student Performance Criteria Matrix - Based on Examining the Professional Degree Programs in the United States, 26(10), 11-20

Lee, J. (2014). A study on the Relationship between Self-Efficacy, Personal Variables and Academic Achievement in Design Studios of Professional Architectural Degree Program Students, Journal of the Architectural Institute of Korea, 30(8), 117-124

Lee, J. (2015). A Study on Survey Results by the Site Visit Team Members for the Student Performance Criteria of the Conditions and Procedures of the Korea Architectural Accrediting Board, Journal of the Architectural Institute of Korea, 31(6), 55-63.

Lee, S. & Lee, J. (2005). A Study of Inter-relationship between Architectural Educational Needs and the Educational Contents in the United States, Journal of the Architectural Institute of Korea, 21(5), 17-26.

NAAB, (1977, 1989, 1991, 1995, 1998, 2004, 2009, 2014). Conditions for Accreditation

Ryu, J. & Lee, J. (2009). A Research on the Assessment Process of Academic Curriculum and its Management System in Accreditation Procedures of Architectural Programs in Korea, 29(10), 176.

Schöon, D. A. (1984). The architectural studio as an exemplar of education for reflection-in-action. Journal of Architectural Education, 38(1), 2–9.

UNESCO/UIA Validation Council for Architectural Education, (2014-2017). Charter of UNESCO/UIA for Architectural Education (2017 Edition)

Wilson, J., & Jennings, W. (2000). Studio courses: How information technology is changing the way we teach, on campus and off. Proceedings of the IEEE, 88, 72–80.

Multitude in Cultural Complexities in Asia

Johannes Widodo[1,*]

[1] National University of Singapore, Singapore

Abstract. The genesis of human settlements is a continuous process of production and layering of patterns, forms and spaces in different scale levels across historical periods. Our urban morphology is the product of the cosmopolitan communities, the articulation of the multi layered tangible and intangible urban traditions and modernization processes. Diversity, eclecticism, fusion, acculturation, adaptation, are the nature of our architecture and urbanism.

However, at present, we are in an urgent need to find resolutions to address serious problems posed by the climate change, ideological conflicts, economic greediness, depletion of resources, and social justice. One of an essential elements in humanity is empathy, and this empathy has been anesthetized or lost due to ignorance and greed in almost all aspects of our relationships with others and nature. The design and planning profession and education should reflect on the mistakes that have been created which have caused cultural, social, and environmental issues. We need to reconsider our present practices, i.e., to reflect on, to interrogate and perhaps to present alternatives to our existing pedagogical paradigm.

Keywords: cosmopolitan identity, contemporary crisis, ethics, architectural education

Contemporary Crisis in Cosmopolitan Identities

For centuries sustainable cosmopolitan settlements have been developing around the "Mediterranean Sea of Asia", (referring to the coastal regions around the South China Sea, Java Sea, and Malacca Strait) since the establishment of the inter-insular and inter-continental maritime trading, called "Bumantara" and "Nusantara", located at the center of international economic networks, and in-between maritime corridors for human migration and cultural exchanges. Here the cosmopolitan culture has been developing through complex layering processes of various cultures, ideologies, economies, and ecosystems sustained over a long historical period from the past to the present and will continue.

At present we are in an urgent need to find resolutions to address serious problems posed by the climate change, ideological conflicts, economic greed, depletion of resources, and the threat to social justice.

Southeast Asia has been disturbed by Rohingya refugee crisis, violent separatist threat in Marawi, the rise of religious and racial intolerance in Indonesia and Malaysia, continuous conflicts in Southern Thailand and Myanmar, conflicting territorial claims in South China Sea, environmental damage due to excessive reclamations in Melaka, Johor, and the Jakarta, persistent flooding in various cities and places, gaps between the rich and the poor, etc.

These challenges are not only becoming the concern of our generation, but seriously affecting our future generations as well. Without radical change in our attitude towards other human being, built environment, and natural environment, the future of human civilization and the survival of the planet earth are very discouraging. Therefore, Ethics becomes fundamentally important as the focus in education of the next generation, to change our behavior and perspective in saving and caring of our common planet and living habitat.

Until today the pedagogy of many Architectural education in different countries in Asia have not change much and still continuing the colonial legacies or the old school of thought, based on "teaching" paradigm and "project based" study.

There are widening gaps between academic learning and the reality outside campus, between theory and practice, between global and local, between specialization and broad-based education, between the old and the young. Our architectural education needs to develop a design-research culture to deal with social and environmental changes. Pedagogical reform is desperately needed at this moment of crisis.

Old ways of teaching Architecture is not effective anymore to deal with the multitude and the scale of the contemporary crisis. The challenge today is how to enable our students to be more resilient, critical, creative, innovative, and pro-active. Also on how to develop both research- and design capacity through research by design and design by research.

Another challenge is to re-connect the education with the profession, the academia with the common people, theory with practice, and behavior with ethics. In short, the biggest need today is to restore the original ideals of

* Corresponding author: jwidodo@nus.edu.sg

holistic Architectural education, so the new generation of Architects will be able to become a problem-solver and "vanguard" of our civilization and common planet.

Return to the Fundamentals of Architectural Education

The fundamental principles of Architectural education was clearly stated by Marcus Vitruvius Pollio who lived in the 1st century BCE in his "Ten Books of Architecture", Chapter I, "The Education of the Architect".

According to Vitruvius, the architect should be equipped with multi-disciplinary knowledge, as the ground for practice and theory. Practice is learned through continuous exercise of manual work based on the design drawings. Theory is the ability to demonstrate and explain the design productions. He stressed the complementarity of theory and practice to gain credibility as an architect.

"It follows, therefore, that architects who have aimed at acquiring manual skill without scholarship have never been able to reach a position of authority to correspond to their pains, while those who relied only upon theories and scholarship were obviously hunting the shadow, not the substance. But those who have a thorough knowledge of both, like men armed at all points, have the sooner attained their object and carried authority with them."

Vitruvius further explained that a good architect is made of two inseparable factors: natural ability or talent plus training.

"Let him be educated, skilful with the pencil, instructed in geometry, know much history, have followed the philosophers with attention, understand music, have some knowledge of medicine, know the opinions of the jurists, and be acquainted with astronomy and the theory of the heavens."

An architect shall be educated and knowledgeable in drawing and sketches to show the appearance of the work that he proposes. For this, he must master geometry, optics, shadow, perspective, and building costing. He must learn history to understand the ideas and concepts behind forms and decorations.

Philosophy is needed to make an architect "high-minded and not self-assuming, but rather renders him courteous, just, and honest without avariciousness.". Vitruvius further stressed:

"This is very important, for no work can be rightly done without honesty and incorruptibility. Let him not be grasping nor have his mind preoccupied with the idea of receiving perquisites, but let him with dignity keep up his position by cherishing a good reputation. "

He must also understand the principles of law, so that, before he begins to build, he may be careful not to leave any possible dispute with the client after the works are finished, and so that in drawing up contracts the interests of both employer and contractor may be wisely safe-guarded.

Another basic knowledge of physics to deal with natural and environmental elements like water and wind. The fundamental principles of physics shall be learned from philosophy to prevent damages to the building. Architect must also understand music, so that he may have knowledge of the canonical and mathematical theory to deal with acoustical issues and properties.

"From astronomy we find the east, west, south, and north, as well as the theory of the heavens, the equinox, solstice, and courses of the stars. If one has no knowledge of these matters, he will not be able to have any comprehension of the theory of sundials."

An architect shall also have a knowledge medicine, so he can account on the questions of climates, air and water quality, and the healthiness or unhealthiness of sites. Without these considerations the healthiness of a dwelling cannot be assured.

Mastering Architecture takes time and cannot be rushed, because the learning is very vast in extend. Although it seemingly impossible for a human to comprehend and to memorize such a considerable number of studies, but because of many commonality and overlapping of one and another knowledge, it can be achieved easily. Nevertheless, as a human architect, he will not be able to achieve perfection. Therefore, make a good architect, even if someone has a great talent, training and capacity building in all subjects are undeniably needed.

"I think that men have no right to profess themselves architects hastily, without having climbed from boyhood the steps of these studies and thus, nursed by the knowledge of many arts and sciences, having reached the heights of the holy ground of architecture."

Restoring Architectural Pedagogy: from Teaching to Learning

At present, Architectural education has been strayed too far from the ideals as stated by Vitruvius. Reformation of the pedagogy in Architectural education from teaching to learning in developing the students' affective, cognitive, and psychomotor holistically is not optional anymore, but has become imperative.

Research, education, and practice in architecture should be based on the reality on the ground and not just based on alien theories. Design should be based on the understanding of the real contextual problems, to formulate sensible, sensitive, and holistic solutions.

Ethics and empathy must be reinstated in our students' heart and mind to lead them in the problem solving and designing for the future by reiterating the inherent relationships of ethics, art, science, and technology in Architecture.

Learning shall be aimed to the understanding of self by knowing the local history, paradigms, and practices in relation to the general history of humanity, to replace the narrow-exclusive-chauvinistic paradigms with a more open-inclusive-reflective paradigm.

Discourses and debates on the local identity in relation to the universal aesthetical and ethical values need to be revived through the rediscovering process of links, layers, and continuity of nature, culture, and architecture across national boundaries. This can be achieved by building a comprehensive database and critical assessments of global and Asian architectural legacies and heritages.

The design studio problem-based learning and training shall be directed to the production of culturally sensitive, socially responsible, environmentally sustainable design solutions. Design shall be restored as the core of Architectural education, and our research – especially in Asia - shall be aimed to the production of new knowledge and not just recycling old knowledge as consumer, in the spirit of openness and sharing.

By these efforts, the multitude of cultural complexities in Asia will not be seen as liability or threat, but as the strength of Asian identity. Unity in Diversity is the key to harmonious relationship between different ideological, cultural, and social differences based on respect and camaraderie.

Architect, guided by his restored ethic and sense of responsibility, will be able to take a better care of our common home and environment.

Considering the urgency of the multitude and seriousness of the current crisis, the pedagogical reform in the Architectural education cannot be delayed anymore.

References

Ismail, R., Shaw, B.J., Ooi, G.L. (editors). (2009). *Southeast Asian Culture and Heritage in Globalising World – Diverging Identities in a Dynamic Region*. Surrey: Ashgate.

Widodo, J. (2002). *Restoring Studio Culture and Architecture Professional Education.* In Laurens, J.M. The Design Studio. Surabaya: Petra Christian University. 85-94.

Widodo, J. (2007), *Towards the Renaissance of Architectural Education in Asia.* Keynote paper presented in the International Conference "Challenges and Experiences in Developing Architectural Education in Asia". Yogyakarta: UII.

Widodo, J. (2010). Current State of Modern Asian Architecture Discourse and Networking. *Journal of Architectural Education*, 63(2), 79-81.

Intercultural Hands on Projects – Experiences in Architectural Education in Asian and European Context

Marcus Hackel[1,*], *Andrea* Gaube [1], and , MA *Sabrina* Lampe[1]

[1]Wismar University, Germany

Abstract. The duties of German architects include the indepth design process as well as a thorough quality supervision during the construction process with the goal of the "build success". They are reflected in the "Hands on Projects" organized by German Universities. The best results and broadest findings come out of international and interdisciplinary cooperation and projects with participants coming from the diverse cultural background and even integrating refugees into these projects.
Students get in touch with different philosophies, attitudes, values, and approaches. They learn about intercultural communication and develop unique solutions. Different social and cultural background leads to different behavior. Not being aware of the cultural differences may lead to misunderstanding and irritation. Analysing the cause of these misunderstandings and getting knowledge about the cultural influence on architectural planning, communication and problem solving is one of the mayors tasks of these intercultural and interdisciplinary projects.
Two case studies from Thailand and Germany published in this paper show different experiences with intercultural and interdisciplinary "Hands on Projects".

Keywords: German architecture, architectural education, Hands on Projects, intercultural communication, international cooperation, interdisciplinary cooperation.

Intercultural Hands on Projects

In global comparison German Architecture has an image of high quality combined with care and preservation of the architectural culture and heritage. The complexity of the German architectural duties from design to detail design and working drawings and the depth of quality supervision especially on site go beyond similar duties of architects in many countries of Asia or in other continents. The German „Werkvertragsrecht" in the Bürgerliches Gesetzbuch (Palandt 2017), Law of Contract for the Final Result, obligates architects in Germany not just to plan but successfully conclude their work, e.g. an approved building application or a completed useable building as ordered. Buildings are seen as long term investments that demand highest quality in design and build result. The long term effects of urban design and architecture are recognized by most people as an important factor for the social and cultural well-being of the society.

All this leads to a reputation of realistic, reliable and well-engineered architecture.

Today numerous regions of the world experience a growth in economy combined with a booming building sector. This creation of value within the real estate field must be seen alongside with its possible negative consequences in the matters of ecology, social life and culture. This in turn leads to an increasing responsibility of architects towards sustainability.

The duties of German architects including the in depth design process and quality supervision during the construction process with the goal of the „build success" are reflected in the "Hands on Projects" organized by German Universities. They focus on all fields of sustainability: ecology, economy and socio- cultural sustainability. Nevertheless the best results and broadest findings come out of international and interdisciplinary cooperation and of projects with participants coming from diverse cultural background and even integrating refugees into these projects.

Students get in touch with different philosophies, attitudes, values and approaches. They learn about intercultural communication and develop unique solutions. On the one side German students tend to be straight forward and project related in their initial approach. Their communication is "Low Context". Asian students on the other hand use a different more indirect approach to communication and problem solving. Very often these different behaviours lead to misunderstanding and irritation. Analysing the cause of these misunderstandings and getting knowledge about the cultural influence on planning, communication and problem solving is one of the mayor tasks of these intercultural and interdisciplinary projects. In most of the cases at the end of this learning process all project

* Corresponding author: marcus.hackel@hs-wismar.de

partners are able to collectively solve misunderstandings and project problems. In the wake of the "Hand on Projects" technological skills as well as intercultural understanding and "soft skills" are developed. The cooperation on site leads to deeper understanding between the international participants and very often to long lasting friendships.

Hands On Project Case Studies

Hands On Project Case Study 1
Sustainable "Sala" Pavilions for the Water Village Amphawa, Samut Songkhram Province, Thailand

Introduction

As part of the sustainable development process for the Thai village of Amphawa, located at the river Mae Klong the Chulalongkorn University of Bangkok, the Tianjin University and the Technical University of Berlin designed and built two Sala pavilions meant as gathering places for the villagers.

Sala pavilions are defined by open structures combined with large overhanging roofs for sun or rain protection. This type of building is traditionally Thai and can be found all over Thailand, as "Salawats" in temple compounds, on the country side as "Sala Asai" for travelers, along roads as "Sala Rim Thanon" or like the one in Amphawa at a river´s bench or at channels as "Sala Thanam".

There are numerous functions for Salas, as the Thai society is oriented on community and confraternity. At this the Salas serve as meeting point for the neighborhood or families and as recreation spots outside the individual home. Additionally they are used as quays for the river traffic.

This open shady space is perfectly adjusted to the local climate and works without energy consuming air conditioning through its design for transverse ventilation. Therefor the typology of the Thai Salas is exemplary for a sustainable and ecological architectural concept.

Recycling used building materials, in our case old wooden beams, can help to downsize the ongoing deforestation of the last primeval forests in South East Asia. Compared to the recycling methods we know from Europe, where wood is shredded, glue added as a composite material and finally formed into chipboards, the recycling of old wooden beams in whole has a long tradition in Thailand. The traditional Thai timber house consisted of prefabricated structures easily to be dismantled and reconstructed at a new location without greater efforts. Groups of various neutral modules could be assembled around a central patio and contained a potential for flexibility and adaption if needed.

Unfortunately the majority of these buildings have vanished over the last decades due to the dramatic upheaval in economy and changes in the Thai society. The future viability of traditional Thai living clusters is more and more questioned, however it is of major importance to fight the loss of traditional building types, as they are unique and an intelligent way to sustainability, as the "Thailand Cultural Environment Project" gives proof.

Nevertheless it is necessary to not only preserve the existing architecture but find future oriented ways for development. The Thai – Chinese- German Sala designs attended to this task in their different approaches.

There are different methods of recycling visible in the two Sala designs built. The Thai design work focusses on the usage of old pillars, beams, rafters, stringers and laths in their former function creating a complete new design out of recycled components. The German Sala design goes beyond this stage and combines slim profiles to replace the hard to get larger and therefor expensive massive wood profiles. With this new method of joining parts it is possible to build in larger measurements than those of the former building that serves as material source.

This class is part of the curriculum of the Master program at the Technical University of Berlin within a series of design and practice projects of "Studierende bauen im Ausland / Students built Abroad". The contents include urban planning, architectural design and a complete detailing for the building process, which needs an international project management to be organized and scheduled. With the final building activities this class conveys holistic knowledge of the designing and building procedures.

The persons of charge were Prof. Dr. Zhang Yukun in cooperation with Prof. Dr. Paolo Genovese for Tianjin University, Prof. Dr. Budit Chulasai and Prof. Dr. Siriwan Silapacharanan cooperating with Terdsak Tachikatkachorn for Chulalongkorn University of Bangkok and for the Technical University of Berlin Dr.-Ing. Marcus Hackel and Prof. Rainer Mertes in cooperation with Prof. Dr. Klaus Rückert.

The project itself focused on two major topics, each with a number of goals to be achieved.

Project Topics

Project Topic No.1 : Conveyance of Profession-Specific Knowledge

1. Experience the problems and specific needs of intercultural communication, development of problem solving approaches with the goal to understand the mechanism of optimized international project communication, analysis of the existing social and urban structures
2. Urban and architectural design within extra-European context
3. Learning of building methods under the aspects of self-help, sustainability and financial feasibility
4. Implementation within multinational teams
5. Comparative studies of German, Thai and Chinese design concepts, working methods, presentation

techniques, group behavior, dispute resolution policies and courses of negotiation

6. Preparation of all participants for future international projects
7. Conveyance of knowledge in international project management
8. Detail planning and contemplation during the realization process

Project Topic No 2: Encounter with Foreign Students and Scientists

The planning process took place simultaneously at all three Faculties of Architecture in Berlin, Tianjin and Bangkok. Meetings were held with representatives of the Amphawa government and villagers to learn about their points of view and expectations. Finally the participating locals voted for one Thai and one German design to be built at two allocated lots at the river promenade.

Following phase 1 and 2 the students travelled to Bangkok and Amphawa to exchange their experiences in analyzing techniques and design approaches. During this visit the two designs were built by the participating students of all three universities. Besides the practical work the students learned about different working methods, communication and mentality of their Thai, Chinese and German project partners and the impact of these on the project result all along.

Project Phases

Project Phase No 1 Analysis and Preliminary Design

The first step was to draw up regional surveys of the geographic, social, cultural and economic structures as well as the technical and constructive analysis of the existing building stock. This knowledge was used to create urban planning concepts and architectural designs concurrently at Berlin, Tianjin and Bangkok.

A workshop with all participating professors and students of the three universities was held in Berlin to analyze the results and choose the most suitable designs to be continued working on during a joint Summer School. Here the approaches and results got evaluated again and principles for further processing rephrased together.

The revised preliminary designs were presented by the Thai students to the Amphawa public to choose by a questionnaire.

Project Phase No 2 Design Selection, Working Drawings, Logistics

Based on the results of the questionnaires one Thai and one German design were selected to be detailed and afterwards built. These design works needed to be optimized in concept and economy, the supporting framework and all details had to be updated ready for building. At the same time the logistics, schedules and building management had to be conceptualized.

Project Phase No 3 Building Process

During two months of presence at Amphawa village the students of the three participating universities built the structures of the two chosen designs. Within this realization process experiences were gathered, shared and exchanged regarding the goals of project topic No.1 in the aspects of building practice, sustainability, expenses and savings, and most of all the subjects of characteristic mentality, communication forms and methodology of Thai, Chinese and German partners including their effects on the final results.

Project Phase No 4 Evaluation

In the aftermath about one year after the realization process a review visit to Amphawa was held to validate the use, the users and the frequencies of use of the two Salas. The survey showed a use of 24/7, every day at every possible time. Locals as much as visitors from e.g. Bangkok went there for recreation, fishing or dining. Even the most popular daily soap of Thai TV chose the German Sala as a location for one of its episodes.

Further inquiries with the Mayor of Amphawa and the building department gave a positive feedback on the two projects. The intended function as a gathering place was achieved.

The pavilions got an inspection to look for possible damages or traces of deterioration and the building department agreed on regular check-ups especially on the stiffening bracings and eventually necessary repairs for the years to come to guarantee the durability of the structures.

An evaluation among students of the PR of China about the differences of teaching and designing concepts in Germany and Asia resulted in the following summarized insight:

Final Thoughts

"In retrospect, the whole course of the cooperation project, it`s worth thinking about the collision on the ideas and difference behavior between the Chinese and Ger-mans. In the design phase, Chinese students paid more attention to the idea or con-cept, to the form, and to the excellent drawings but lack technic and details, while German students did more work on material and conformation. They are more ra-tional and used to calculation, but lack a little creativity in the design. In the construction phase, Germans showed to be more experienced. One of the reasons maybe due to the German architecture education system dating back to Bauhaus which lays more importance on technique while aesthetics is at higher position in Chinese traditional architecture education. Through this self-construction practice, from sawing one piece of wood to hammering a nail, we have redefined architecture design for ourselves. Maybe we can`t denial the difference between design and construction, but how wide the difference will be is in our architects control. Without going through the whole constructing process by yourself, you can not realize the difference, not to mention control. Attention to construction, please. It`s the high time." (Chen Yu, Fu Lei, Hao Guanmin, Liu Yujie, Yang Shen, Zhou Ting in Mertes, Hackel 2008)

Hands On Project Case Study 2 Studierendengärten Wismar / Student Gardens Wismar

Hochschulgarten „Hinter dem Mühlenteich"/ University Garden "Behind Mühlenteich"

Since 2012, a university garden has been developed and run by the Wismar University in the allotment colony "Hinter dem Mühlenteich", which incorporated different objectives in teaching. The annual changing and accompanying themes included

1. Design of landscape gardens and their inclusion in urban development
2. History of allotment culture in Germany
3. International principles of garden design
4. Constructions and buildings for temporary uses and living in the garden
5. Ecological building methods, recycling and upcycling
6. Possibilities of ecological gardening
7. Self-sufficiency as alternative forms of life

Furthermore there is always the intention that students in the architecture program get a relation to agriculturally usable ground and nature, which will be converted by their future professional activities, often irretrievably.

The university garden has been subjected to enormous changes during the years, as such a kind of allotment generally determines changes.

After an initial use for the cultivation and harvesting of fruits and vegetables, students wished to design and build a garden house by themselves. A one-year design and building class was initiated, including the selection of preferred variants, the working and detail drawings and finally the building process.

All this was only possible to be realized through sponsoring. There had been a number of tasks, which the participants had to pursue autonomously and self-organized. Thus was the demolition of an old garden shed, including the orderly disposal or sorting of the material for recycling and the organization of materials for the new building – from sand to wood, cement, metal parts made in an apprentice workshop, alternative sealing materials made of recycled material for the roof etc. This was made possible by a lot of communication and support by friends, garden neighbors, and local companies. The construction work could not be completed in one semester, but a constant change and further development took place fostered by the students who continued to work in this project during the following year.

With the refugee flows in 2015, an urgent world-wide topic was added. Thru this garden project we have been able to bring refugees in contact with the German culture and language, to learn from each other and from different cultural values as well as to experience ways of cooperation in a student project. Thus, the university garden lead to international exchange, involving students from different countries, study courses and refugees. The continuing enormous interest of the following semesters in this project led to two additional gardens taken in use with different key aspects put in.

The first garden, which has been pursued for a number of years now, was prior subject for further building activities, including materials and techniques from other countries such as mudbrick building. The change and further development of the garden house design according to the ideas of the new users was a good experience in dealing with existing substance. Additionally the cultivation of vegetables as well as the signposting of the various plants in German, English and Arabic as well as the joint processing of the products in a self-made clay oven promoted the joint experience in the project.

Communication and cooperation with the members of the garden association became an important aspect in this project, too. For this purpose students from Canada, who were not able to speak German fluently, participated in a work assignment of the garden association, which was unusual for the other gardeners. The young men helped with the maintenance of the premises and thereby promoted the acceptance of the project. The unusual, but now daily presence of many foreigners in the colony became a matter of course to the members of the garden association.

Fellow students as well as guests are invited to garden festivals and the annual exhibition at the end of each summer semester. At these events the results of the semester are presented and visitors may savor international dishes made from the fruits of the garden. The collective practical work and the experience of the directly visible results, which can also be consumed confirmed the success of the project and will form the base for the years to come.

Garden at Filmbüro (film office) Mecklenburg-West Pomerania in Wismar

In 2016, the Filmbüro M-V in Wismar provided an approx. 80 sq. m. of unused space for the garden project on its site in the immediate vicinity of the campus. Eleven students (including two refugees who were interested in studying and were enrolled as a guest auditor) from five nations and three different courses (architecture, interior design and business) developed a garden concept and realized it in the 1st year. Besides, the students contributed their own different experiences and ideas. Due to the different cultural origins, the students have been able to fall back on very different previous knowledge: Thus, Yazan, a guest student who came as a refugee from Syria to Germany in 2015, and Waleed, an architecture student from Cairo/Egypt, had never seen an earthworm before. Also the cultivation of a compost was for many student's new territory as well as the winning of seed. Salim, a 15-year-old refugee from Afghanistan, was glad when he heard that zucchini was growing here in Germany and that he was able to find a little piece of native country here in the garden. This was also an important aspect of the project – the discussion about different garden concepts and the awareness of the origin of fruit and vegetables, which we regard here as self-evident and local in Germany, although their origin is in quite different countries.

Students get seeds and plants, as well as building materials, through donations by DIY stores, garden centers and plant flea markets. The donation collection has given the international students an enormous self-confidence as they are often hindered by missing of language skills. Within a semester the participants grow on fruits and vegetables, put on a bar-footpath filled with materials from the garden (clay, pine cones, pebbles and bark) and built a terrace and seating furniture's as well as two large patches and a bar. As a material served, primarily, recycling materials and discarded palettes. In particular, the different approach of the single students should be emphasized here: While the architecture students planned everything in detail, the students from business directly worked on the jointly developed ideas after a short discussion, because a lot of improvisation and spontaneity was necessary by realization. The communication between each other worked astonishingly well despite some language barriers – while the one during the seminar expanded their knowledge of English by not so common vocabulary, the international students learned amazingly fast and well German.

The first harvest was donated to the "Mittagstisch für Leib und Seele" (lunch for the destitute) of St. Nikolai church in Wismar. None of the students had heard of the Church's offer so far, and this action was something special for all. Together with volunteers, they cooked in the small kitchen of the church, prepared the dining room and give the food to destitute. Yazan and Feras were very grateful to be able to make a contribution for the society as well. So they were able to give back some of the support they have received since 2015. Also for Mohammad, a student from Iran, was this kind of help and support of the volunteers and the church quite a new experience. Thus it was possible to make a social contribution even beyond "gardening".

The project goes on – in the 2nd year, the existing concept, the community gardening was continued and extended: Smaller repairs were made, a bird pot of concrete was poured and an herbal spiral was laid. Also common activities with the Filmbüro M-V took place. The joint launch in the garden season was held together with the film studio, interested guests and many children. Thematic film events, barbecues and much more have been performed. The publicly accessible garden and the cooperation with the filmbüro offer a whole series of common activities and a cultural cinematic supplement to the garden work.

The students show a growing awareness and a growing interest in the cultivation of local fruit and vegetables and the re-use of materials. Also the possibility to have the opportunity to implement and realize own ideas is also important and a challenge for many students. Therefore, the Hands On Garden projects offer students a great opportunity to get together outside of the lecture hall, learn from others due practical works, learn about different approaches and cultural influences, and deal with different ecological, social and cultural aspects.

Final Thoughts: Quotes from Students

"... The garden house was an interesting cultural experience to participate in. there's a lot of history rooted in the formation of these communities and it allowed me to gain a greater understanding of their purpose and origin." Spencer Chrisholm, Master architecture, 2nd semester.

"This elective has helped me to gain fresh insights in teamwork and coordination. I find the program to be very practical. Not only has it helped me to develop skills in gardening, but also to connect with people from different cultures and backgrounds. Through this program, I realized that Germans are actually friendly people once you get to know them. It has helped to debunk the myth that Germans are cold people. Aside from that, the sense of achievement is indescribable when you see your hard labour come to fruition. Although the wedding process may be annoying, to see plants grow from small seeds to lush bushes is truly a rewarding experience." Yu Jin Lim, Master architecture, 2nd semester.

"... Another aspect I liked was the social coexistence. You always felt that you were creating something together." Gina Weser, Master Interior Design, 2nd semester.

"Through this project, the participants have gained a deeper understanding of the importance of sustainable gardening and the aspects of maintaining such a garden. This project focused on the value of the community interactions. "Gardeners need community". As well as trend and influences, educational and accessibility issues were examined for opportunities in self-sufficiency gardening. The sheer volume of gardener exchange was apparent throughout this project. Community members could receive a large amount of information and practical knowledge by being involved in the discussed activities. On the other hand, the benefits for refugees are summarized in socializing with the students, learning languages and being integrated in different cultures." Mouaz Alghandour, Master architecture, 2nd semester.

Conclusions

In summary, the findings for future sustainable student projects in the intercultural and interdisciplinary context can be defined as follows:

1. Sustainable student “Hands on Projects” must reconcile economic, environmental, social and cultural requirements.
2. The fundamentals are the well-balanced analysis and management of economic efficiency, ecological, social and cultural aspects - considering all stakeholders and the whole lifecycle of buildings.
3. Sustainable architecture learns from history and takes into account the responsibility for future generations.
4. Hands On and Self-experience in practice should be enabled by new teaching methods.
5. Intercultural projects in architectural education include "hard skills aspects" such as analysis methodology, design methodology, technology, construction management, cost and time management as well as “soft skills aspects” such as communication, consensus and conflict resolution, design approaches and culture, tradition and change and stakeholder participation.
6. Traditional ways of design, decision-making and communication must be questioned from all sides.

Fig. 1. German Sala Amphawa.

Fig. 2. Thai Sala Amphawa.

Fig. 3. Hochschulgarten, Hinter dem Mühlenteich“, Wismar.

Fig. 4. Garden at Filmbüro, Wismar.

References

Mertes, Hackel (2008), „*Sala Amphawa“ Sustainable Low-Cost Pavilions for Thailand*, Berlin, Berlin University of Technology

Palandt (2017), *Bürgerliches Gesetzbuch*: BGB, Munich, C.H. Beck

Scholz, Wellner, Zeitner, Schramm, Hackel, Hackel (2017), *Architekturpraxis Bauökonomie*, Wiesbaden, Springer Vieweg

Renewing Theories, Methods and Design Practices: Challenges for Architectural Education

Yandi Andri Yatmo[1,*]

[1]Department of Architecture, Faculty of Engineering, Universitas Indonesia, Indonesia

Abstract. Architectural education should promote the advancement of knowledge that is necessary as the basis for the development of excellent design practice. Architectural education needs to respond appropriately to the current issues in the society. To find its way into the society in an appropriate way, architecture needs to be liquid. The ability to address the liquidity of architecture requires educational approach that promotes the ability to work with a range of design methods and approaches. There are several principles that become the basis for developing architectural education that could strengthen its position within the society: to promote knowledge-based design practice, to embrace variety of design methods and approaches; to keep a balance between design knowledge and design skills; while at the same time to aim for mastery and excellence in design. These principles should be the basis for defining and developing the curriculum and the process of design learning architectural education. Then the main challenge is on our willingness to be liquid in developing architectural education, which needs continuous renewal and update to respond to the changing context of knowledge, technology and society.

Keywords: design methods, design knowledge, liquid, society, education

Introduction: Why Do We Need to Renew?

As a discipline that addresses the inhabitation of human, architecture is very closely related to the society. The practice of design in architecture needs to respond directly to the needs and issues in the society. Design can be defined as "the human capacity to shape and make our environment in ways without precedent in nature, to serve our needs and give meaning to our lives" (Heskett, 2002, p. 5). This definition at least implies two things: the importance of innovation to find something new with no precedent, and the importance of design practice to become meaningful for the life of the society.

The demand to serve the society has inevitable lead to the challenges of architectural education to offer sufficient knowledge and skills for its students and graduates to be ready for those meaningful, innovative design practice. Architectural education sits between two requirements: to contribute to the society through appropriate design practice, and to promote excellence in design practice that could break the boundaries of innovation in architectural methods and approaches. Keeping the balance between these two requirements become a challenge. It is imperative that we consider architectural education institutions not merely as factories to produce the graduates to supply the demands of the society. Architectural education should always be the place where knowledge is cultivated, and where new things are invented.

To address this challenge, renewing architectural education becomes critical. Architectural education should become the generator for promoting renewal of theories, methods and practice, in order to achieve advancement in architectural design practice. This paper proposes five main principles as the basis for developing architectural education in responding to these challenges.

Architecture Practice Must be Liquid

The first principle suggests that we need to understand that architecture needs to be liquid so that it becomes appropriate for the society where it exists. In physics, the liquid characteristic refers to the ability of a substance to change and adjust to its container. Architecture could become liquid when it emerges as the manifestation of the design process that responds appropriately to various aspects of socio-culture, everydayness, locality and agency of the society. Architecture becomes liquid since it depends on many things that are located external of the architecture itself (Till, 2009).

The idea of liquid architecture also contradicts the common understanding of architecture as fixed and complete. Architecture should be liquid because it may evolve as a process that does not always match with initial drawings of the architect; it evolves as a process that provides opportunities for change and modification that might be necessary so that it becomes more meaningful for the society. Liquid architecture is the

* Corresponding author: yandiay@eng.ui.ac.id

opposite of the complete architecture. Architecture that is incomplete may give space for the society to complete it; it becomes as space for the society to leave traces, as a way to mark their existence as an integral part of the architecture.

Liquid architecture requires architects to understand their appropriate position in the context where they work. It requires the ability to understand the deep structure of the context, to understand the everyday of the society, to develop appropriate strategies for intervention, and to be ready to adapt and to adjust. Because architecture should be liquid, therefore it is necessary for the practice of architecture to develop appropriate design methods that could fit properly with various situation of the society. As a consequence, it is imperative that architectural education promote the ability to maneuver in between such liquid characteristics of architecture, through the renewal of its curriculum and learning approaches.

Architectural Education Should Promote Knowledge-based Design Practice

The ability to respond appropriately with the design contexts and issues requires more than just the acquisition of the design skills. It is necessary for the architectural design practice to be based strongly on knowledge. There is an important statement from Vitruvius that delivers an important message for all teachers and learners of architecture.

"Architects who have aimed at acquiring manual skill without scholarship have never been able to reach a position of authority to correspond to their pains, while those who relied only upon theories and scholarship were obviously hunting the shadow, not the substance. But those who have a thorough knowledge of both, like men armed at all points, have the sooner attained their object and carried authority with them" (Vitruvius, 1960)

Vitruvius clearly pointed out the importance of theory and scholarship as a basis for the practice of architecture. Theory has a critical role as "an intellectual tool" for the development of architecture as a discipline (Plowright, 2014, p. 58). It becomes necessary to find the scholarship of design and use it as the basis of our teaching and practice in architecture. "since knowledge is defined within a discipline, rather than externally, scholarship is also defined by discipline" (Powers, 2007, p. 18).

Without the development of knowledge, it is unlikely that the practice and method of architecture could be developed; "the advancement of architecture is inextricably linked to the acquirement of knowledge" (Till, 2009, p. 2). There has been a challenge on how theory could be used in design practice, in particular to become the generator of architectural design (Plowright, 2014). Theories are taught in most schools of architecture, however, the extent to which we could directly relate the theories taught in theory classes and the design practice in the studio remains a challenge.

So the second principle I propose here is that architectural education should promote knowledge-based design practice. The design learning process needs to prioritize thinking and inquiry throughout the process of designing. This will ensure that the process of design learning becomes the process of developing knowledge and generating ideas and not just the process of manual crafting and generating product. Architectural education institution should become the place where thinking is nurtured and developed.

Architectural Education Should Embrace Variety of Design Methods and Approaches

The first principle stands on the belief that architecture needs to be liquid so that it could address the context of the society more appropriately. The second principle emphasizes on the challenge of architectural education to teach design knowledge as the basis for design practice. Further implications of these two is the needs for architectural education to open opportunities to learn variety of design methods, in order to equip the students with the ability to manoeuvre in between the different contexts with strong basis of knowledge.

The scope of architectural education should cover a wide range of design methods and approaches and celebrate the strength of each. The objectives of architectural design studios should not be to promote mastery in certain method only, as this may limit the ability of the graduates to respond to the complexity of design issues and problems in the society.

We can take an analogy of learning to swim for learning to design. To learn to swim, students need to be introduced to and required to practice a variety of swimming stroke. The acquisition of different strokes become more important then the ability to swim the furthest distance with only one particular stroke. Each swimming stroke offer different strength, and each needs to be used in different terrain. Particular stroke might give benefit in reaching particular distance with speed, while another might be better to survive in a long distance travel. This situation is similar to design methods; each method has its own strong point when applied into different design contexts. Exposing students to different design methods and approaches could prepare them to respond to different contexts in reality. The development of the students' ability to manoeuvre in different terrains of design contexts become the main challenge for architectural education.

This analogy is particularly relevant for undergraduate education where the students should acquire holistic understanding on architectural design and not just specializing on particular method or approach only. To expose students with a variety of design methods and approaches, the curriculum of architecture should define different kinds of design knowledge for each stage of design studio. The contents of design studios at each level should not be based on types or project complexity as often developed in the curriculum in many architecture schools. It becomes necessary to define clearly the types of knowledge to be delivered and the main design approached to be acquired in each studio.

Architectural Education Should Keep the Balance Between Knowledge and Skills

While knowledge should become the heart of design learning, this does not necessarily mean the ignorance of relevant skills in design. As stated by Vitruvius, the acquisition of knowledge only will not result in completeness of being an architect, and so is the acquisition of skills only. Therefore, architectural education should keep the correct balance between knowledge and skills.

The types of skills to be trained in architectural education cannot be separated from the knowledge developed. This is what I call knowledge-based skills – that is the skills that are developed in parallel with the development of knowledge. Skills of representation need to be developed in relation to the ideas, the thinking, the methods and approaches that are developed in the design process. Each design method and approach may require different techniques of representation and thus require different skills. Particular design methods may require certain technical skills and acquisition of certain technical tools. Therefore, the content of knowledge in each design studio needs to be accompanied with relevant skills that enable the students to exercise particular design method and approach more comprehensively.

Other sets of skills that need to be developed in architectural education are the skills that are required for being liquid in the process of developing architectural ideas. It is important that the students are equipped with a set of skills that enable them to manoeuvre easily within the society – to listen, to adjust, to adapt, and to go with the flow, and to design in between this process. These skills might be in contrast with the belief that the education should allow the students to speak, to be confident, to defend the idea, and to win the argument, as the part of the studio culture developed in most architecture schools. It is time to redirect the studio culture into the constructive environment where the skills for being liquid are nurtured, to enable the students to adapt easily within the society and to be flexible in facing different situations.

Architectural Education Should Aim for Mastery and Excellence

Education at postgraduate level requires different approach. While the education at undergraduate needs to expose students with different knowledge, skills, methods and approaches, these are mainly at the basic level that opens opportunities for improvement. Education at postgraduate level should aim for mastery and excellence in design practice. The learning at postgraduate level needs to promote the ability to determine the position among possible design approaches.

Many architecture schools mixed up the mastery in design with the mastery in problem solving, or mastery in technical skills. This is an incorrect path of architectural education, as mastery should be developed as the mastery in the scholarship of design. Mastery and excellence at postgraduate level needs to be developed through the advancement of knowledge, the depth of the inquiry, the experiment in design methods and the invention in the modes of representation. All these become important ingredients to achieve the mastery and excellence in design practice that should be promoted in architectural education.

Research by design (Fraser, 2013) becomes a key aspect in the design education particularly at postgraduate level. It promotes the ability to see design as a process of inquiry, where new design knowledge may emerge and new design method may develop. Supervising the research by design is not an easy task for architectural schools, since there are a wide range of possibilities in the process of inquiry and the process of design. Nevertheless, research by design becomes an appropriate way to promote mastery and excellence in design practice.

Towards Renewal of Architectural Education

The five principles that I propose here has come out of my experience in innovating with different methods and approaches of design studio learning. These are the principles that I believe should be the basis for developing curriculum in architectural school in order to promote the ability to develop design practice that are both excellent and responsive. There are, however, some consequences that will come with these principles. The principles will require the renewal of curriculum, to distribute the content of design knowledge more appropriately, to achieve the balance between knowledge and skills, and to nurture the culture of inquiry. Others challenges may include the necessary updating of the teachers, promoting collaboration between different areas of architectural learning, and providing the infrastructure to support the process. However, the main challenge is on the willingness to be liquid in developing architectural education, which needs continuous renewal and update to respond to the changing context of knowledge, technology and society.

References

Fraser, M. (Ed.) (2013). *Design Research in Architecture: An Overview*. London: Routledge.

Heskett, J. (2002). *Design: A Very Short Introduction*. Oxford: Oxford University Press.

Plowright, P. (2014). *Revealing Architectural Design: Methods, Frameworks and Tools*. London: Routledge.

Powers, M. (2007). Toward a Discipline Dependent Scholarship. *Journal of Architectural Education*, 61, 15-18.

Till, J. (2009). *Architecture Depends*. Cambridge, MA.: MIT Press.

Vitruvius (1960). *The Ten Books on Architecture*. Trans. Morris Hicky Morgan. New York: Dover Publications.

Educating Architects in Turkish Culture, Gains and Loses: A Historical Perspective

Ibrahim Numan[1,*], and *Revianto* Budi Santosa[2]

[1]Fatih Sultan Mehmet Vakif University, Turkey
[2]Department of Architecture, Universitas Islam Indonesia, Indonesia

Abstract. Ottoman and Turkish architecture have exhibited a constancy and change in the course of history. Nevertheless these changes can only be understood by taking architectural education into account. Here it will be tried to discuss the gains and losses between the increase in the degree of architectural education/specialized knowledge and the blindness to the value judgements of the society and their reflection on the Turkish architecture. The edifice can not be separated from the architect, architect from the society; society and education from the idea. An architect is educated by society. Architect in turn is the product of the common idea of the civilisation he or she is affiliated to. Even a small diversity in the depth of belief or change in the approach in life style brings a deviation in the understanding of arts and architecture consequently architectural education also experience gains and losses along with it.

Introduction

Architecture is a social art, conceived and realized in a society. An edifice can not be separated from its architect, in turn, the architect from the society he or she belongs to; as well as the architect's education is also developed from an idea nurtured in a society. To certain extent, an architect is educated by society. An architect, in turn, is the product of the common idea of the civilisation he or she is affiliated to.

In the dynamics of a society, a diversity in the depth of belief or change in the approach in life style brings a deviation in the understanding of arts and architecture consequently architectural education also experience gains and losses along with it.

Architecture in Turkey experience a remarkable dynamics along with major changes in its society. Following the Anatolian Seljukids in 1071-1308, Ottomans in 1299-1923 (comprising Formation, Classical, Reformation, and Westernization periods), Republic of Turkey 1923- date (Modernization), Turkish culture have experienced a great range of diversity in social value and in belief systems which are profoundly reflected on its architecture.

Critical Threshold In Ottoman Architecture

Sign of change in Ottoman architecture appeared during the reign of Ahmet III (r. 1703–30). Intending to broaden the empire's political and cultural contacts, he sent ambassadors to Paris, Vienna, Moscow and Warsaw. French or Versailles in particular was primarily appealing. The ambassador, Mehmet Efendi returned from Paris with twelve engravings of Versailles, inspiring the ruler to build garden, palaces and fountain in Istanbul. The most impressive building of this period is the Fountain of Ahmet III situated in front of Bab i Humayun or the Imperial Gate of Topkapi Palace.

Further steps of Westernization were taken in the construction of Nuruosmaniye Câmii (1748-1755) next to the Grand Bazaar in Istanbul. This mosque and its religious complex (*kulliye*) was built under the order of Sultan Mahmut I and completed during the reign of Sultan Osman III, therefore, the mosque was named after him. The design and construction of this mosque was carried out by Mimar Mustafa Aga and his assistant Simon Kalfa, a non-muslim Greek.

Nuruosmaniye represents a threshold between the Classical Ottoman and the Westernization period. It was the first major Ottoman mosque Ottoman to exemplify the new baroque style introduced from Europe. It consists essentially of a square room covered by a large dome resting on four pillars on the corners and four slightly pointed arches on the walls, boldly appear in the interior as well as in the exterior.

The mihrab and the courtyard preceeding the main prayer hall have distinct shape. The mihrab is fashioned like a semi circular apse prominently appear to the qibla façade. A porch with five bays standing on the entrance side facing a semi circular courtyard enclosed by an arcade covered with nine domes.

* Corresponding author: inuman@fsm.edu.tr

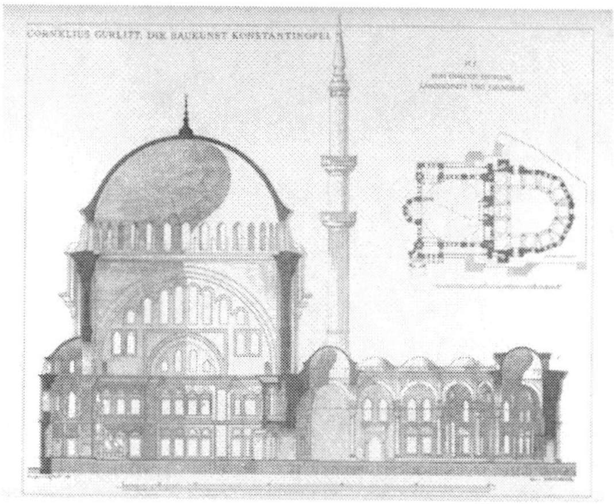

Fig. 1. Nurosmaniye Mosque in Istanbul.
Source: Gurlitt, 1912.

Baroque and Roccoco elements such as the C and S curves in the plans and architectural elements, and by rococo ornamentation with leaf forms, scallops, and volutes further emphasize the novelty of characters of this mosque. Later, this newly introduced style was applied in a number of imperial mosques in Istanbul, such as Laleli Mosque (1783) and Buyuk Selimiye Mosque (1805) in Uskudar.

Changes towards to westernization have reflected almost in every aspect of the society from the way of life to dressing to ornamentation; hence, from buildings to education in architecture.

Empire, Eclecticism Period (1808-1876)

Under the influence of the foreign architects and non-muslim builders ideas have been developed which divert Ottoman architecture towards eclecticism. These ideas were mainly had been constructed on a discourse that Ottoman architecture should be enriched by reflecting different details of different styles of the different regional areas that empire was spreading.

Early in this period, to cope with the technological advancement especially in the military area Sultanate have increase the relationships with the Europe. During the reign of Sultan Selim III (1789-1807), Yirmisekiz Çelebi Mehmet Efendi was sent to Paris on a diplomatic mission. He conveyed messages about his experiences and contacts in Europe describing opera, palaces, landscape and interior architecture beside other issues. Plans and drawings of palaces and other buildings were brought as samples of different European styles. Which consisted models for new architecture in Istanbul.

Mainly non-Muslims were active in spreading this idea. The Balyan family of Armenian origin played important roles by serving Ottoman sultans during the 18th and 19th centuries for five generations. Some prominent members of Balyan families studied architecture in Paris. Based on Neo Classical knowledge and education, they developed a unique architectural style by blending European baroque and imperial styles with Ottoman architecture and oriental ornamental style.

Sultan Mahmut II's wife, mother of Sultan Abdülaziz, Pertevniyal Valide Sultan comissioned a mosque in the district of Aksaray in İstanbul in 1869-1871. This mosque is a good example of these kind of patch-works executed by a patch-work group of non-muslims.

Fig. 2. Petrevniyal Valide Sultan Mosque, Aksaray, Istanbul.
Source of Picture: Avniyazici & Gothika, commons.wikimedia.org.

The Mosque's chief architect was the Italian Montani Pasha. Its plan was drawn by the Armenian Sarkis Balyan, while Hagop Balyan participated in the preparation, supported by the draftsman Osep. Application and site management was undertaken by naster builder Bedros Kalfa, the bricklayer Ohannes and carpenter Dimitri.

Other mosques bearing this style were Nushretiye Mosque (1826), Dolmabahçe Mosque (1855) and Buyuk Mecidiye or Ortakoy Mosque (1856). A number of bulding serving non religious purposes such as, Dolmabahçe Palace (1856), Çırağan Palace (1867), Beylerbeyi Palace (1861) and Selimiye Barracks (1828) were also employed this style.

The involvement of non-muslim European architects further promoted under the Reform Policy or known as Tanzimat initiated by Sultan Mahmut II (r. 1808-1839) and his son Sultan Abdulmecit I (r. 1839-1861) that began in 1839.

During this Tanzimat Era, some European architects like the Italian Raimondo D'Aronco (1857–1932) and the French Alexandre Vallaury (1850-1921) participated both in practice by developing certain distinct style of Oriental or Ottoman Art Noveau (Freely, 2011). They were also involved architectural education teaching muslim and non-muslim students alike.

Usul-I Mimari-I Osmani And The Rationalization Of Ottoman Architecture

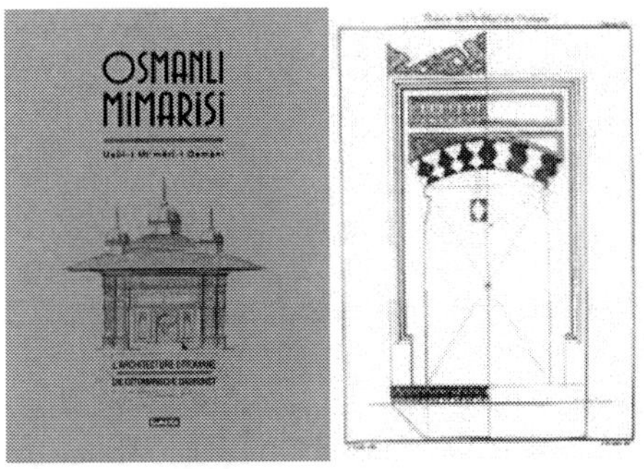

Fig. 3. Book *Usul-i Mimari-i Osmani*.

Usul-i Mimari-i Osmani or *Principles of Ottoman Architecture* was prepared by the collaboration of a group under the directorship of the Minister of Trade and Public Works, İbrâhim Edhem Paşa to represent Ottoman empire in the World Exhibition in Vienna 1873, beside two other important books. Although the purpose was to exhibit Ottoman Styles to the world it had been and still used as a one of the text books in the education of architects. The book illustrates the Ottoman Architecture with details. Technical Documents constituting a separate section entitled "The Theory of Ottoman Architecture", was provided by Pietro Montani (or Montani Efendi), an Ottoman Levantine artist of Italian origin. Montani also executed most of the drawings and color plates. Other collaborators in preparation were Adolphe Maillard, few plans by Marie de Launay and some additional renderings of decorative components by the French artist Eugène Maillard and the Ottoman Armenian painter Bogos Chachian (Nalbantoğlu, 1988).

Baydar (2004) considers the publication of *Usul-i Mimari-i Osmani* "the time of the birth of Ottoman nationalism with the aim of providing social cohesion between the various ethnic and religious groups that constituted the empire".

This monumental book marks the general shift of architectural historiography and theoretical frameworks from an Islamic to a national identity. It primarily consits of a section on historical development of Ottoman Architecture and another section on principles of architectural orders. According to the former section, Ottoman architecture undergoes three successive phases. Started from relatively obscure beginnings, Ottoman Architecture matures until the sixteenth century during which it attains perfection. The process of decline started in the seventeenth century (Baydar, 2004).

Recent Shift In Turkish Architecture

There seen a change in the approach in the recognition of an architect and architecture by the Westernization. The architect who previously had been considered as member in a guild of building activity started to be esteemed as "an entity".

A "Self"-centred system, where the individual designers are prominent figures, have emerged due to the increase in the individualistic designs (Ödekan, 2009: 47). Conjointly, there also seen a shift from the education conducted by the "Hassa Mimarlık Ocağı" or Sultan's architectural chamber and by the builders guild towards a formal architectural education disturbing the balance between the ethical-moral codes and technical learning outcomes.

In this Europeanization period Ottomans have introduced modern educational systems and have initiated military, engineering, veterinary, architectural and fine arts schools to transfer new ideologies and value judgements in philosophy, arts, architecture, decoration etc.

New institutions, such as, post offices, train stations, banks etc. and new needs also emerged a shift from the classical architectural education and organization like Hassa Mimarlık Ocağı on which hesitations on the efficiency of the classical educational manner had been aroused (Yavuz, 2009).

Here a complete shift is observed from apprentice-master relation to the academic education.

During the reign of Selim III Military Engineering schools opened. In 1884 the first civilian *Hendese-i Mülkiye Mektebi* or Engineering School with five years education had introduced taken the French school, *Ecole des Pontes et Chaussees* as a model and European teachers recruited (Karaca, 2012: 156). Architecturel courses were given in the Engineering Schoos.

Ebniye-i Hassa replaced Hassa Mimarlar Ocağı. Its Director Abdulhâlim Efendi has appeal Sultan the need for a civilian architectural school where "fenn-i mimâri" or art of architectural courses will be given and he received the grand of the sultan (Turan, 1963: 179). By the effords of Osman Hamdi Bey the Museum Director and Nazır Suphi Paşa the Trades Minister in 1881 Sanâyi-i Nefîse Mekteb-i Alisi The first Higher Fine Arts School was established (Turan, 2012: 179)

The first formally educated architects were graduated from this school.

This movement is the main and the greatest gain in terms of "Academical Architectural Education" in Turkey.

There was a clear distinction between the studio education in the Fine Arts School "Sanayi-i Nefise Mektebi". While Giulio Mongeri was teaching and designing "ideal and pretensious" projects, Vedat Bey, the first Turkish architectural studio mentor and his students were in more "realistic" side (Eldem, 1983: 11).

1. First Nationalistic Movement Architecture

One of the first graduates of this formal academical architectural education Mimar Kemalettin and its first Muslim Ottoman teacher Mimar Vedat have started a new movement towards the new interpretation of the Classical Ottoman Architecture against the widely applied European styles specially by the foreign architects and non-Muslim Ottoman subjects builders.

https://doi.org/10.1051/shsconf/20184102004

This movement is called the First National Architectural Style which find many followers still today.

In the early period of Republican Architecture this First Nationalistic vision/movement in architecture which somehow can be considered a neo classical interpretation of Traditional Turkish Architecture started during the late late Ottoman period have strenghten by the new ideology of the Republic of Turkey and constitute the new Architecture of Ankara.

There is another movement from Europe to Turkish lands during and after the first World War. The four architects took place in the architectural arena of the Republic of Turkey as instructors, designers, advisors, planners. These were mainly Clemens Holzmeister who built TBMM, Jansen (the urban planner of Ankara), Ernst Egli, Theodor Jost, Martin Wagner, Martin Elsaesser, Bruno Taut who is teaching the Fine Art School, Robert Oerley. The Contemporary movements of Europe was imported to Turkey and Turkish Architectural Education

2. Second Nationalistic Movement Architecture

Another movement which is called the second nationalistic movement during the 1950s. Because of a new attempt had been observed in interpreting the Turkish architecture with the contemporary lines depending on the roots in architectural history, mainly by Sedat Hakkı and Turgut Cansever.

The Gain In Architectural Education

Contemporary schools and contemporary education around the 1940s maybe 1930s on words still they establish of the Higher Education Council of Turkey.

There are 4 types of education programs in architecture:

1. 2-year program of vocational schools
2. 4-2 year-program of teachers school.
3. 4-5 year-program of academy departments; with arts oriented schools following the State Academy of Fine Arts in Istanbul, and technically oriented schools like the Yildiz Architecture and Engineering Academy in Istanbul.
4. 4-5 year-program of university department where there are the arts and theory oriented schools following American architecture model like art and theory oriented schools in Middle East Technical University (METU) established in 1956, and technically oriented schools like Istanbul Technical University following EcolePolytechnique model as the first architecture and engineering school.

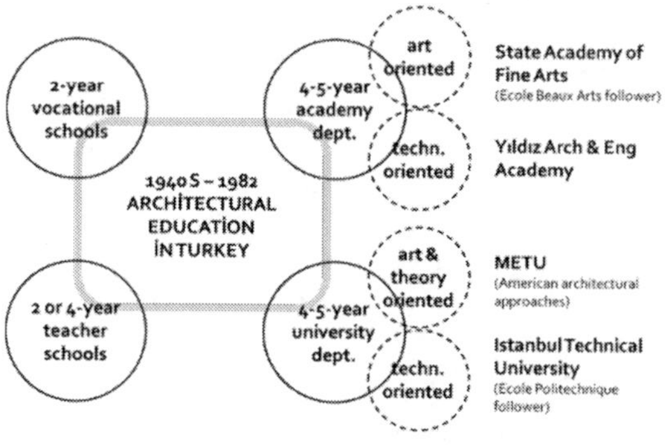

Fig. 4. Four Types of Educational Programs in Architecture.

Then come to the establishment of the higher educational council in Europe. There is abolishment of the academies in university. The 4+2 system is applied, and again there are some distinctions between the art oriented and theory universities and technically oriented schools.

These education sytems, however, changed in 2010 on works under the United European decision in an effort called the Bologna Process. Started with Bologna Declaration signed in 1999, the Bologna Process aims to increase the comparability and readability of degrees across all member state higher education systems ensuring all stakeholders' greater compatibility and comparability in the systems of higher education in all of EU member countries. Turkey participated to Bologna Process since 2001 and it is turned into a formal process for the country with the Council of Higher Education (YÖK) being the primary institution responsible from the implementation of Bologna Process.

The Bologna Process is started to be applied in all 82 architecture schools in Turkey. This broad unity in teaching methodologies learning outcomes for the university. What are the games here? The games the systematical education set of learning outcomes in parallel word UNESCO, UIA and EU integrating to the word architectural education.

The gain during this process are the provision of systematic architectural education, equipped with a set of well-defined learning outcomes. This change brings the mass education, very little architectural ethics is embedded in courses. Students are assessed only on courses. Teachers can not be parenting any ethical approach to any student or graduate. In short terms, they do not know the characteristic of students. There are coming as masses and graduated as masses. They can not nurture of the students' ethics.

The Loss In Architectural Education

Return to Ottoman architectural education to analyze what has been lost. The Turkish-Islamic architecture, especially Ottoman Architecture, is the result of a great composition that originates from the common spirit of society. But regrettably, the artist/architect, who brought this composition into being out of traditions related with an oral cultural understanding, and this oral ottoman education very few little written sources have reached today from the ottoman time. So of the books out there are Teziretul Bunyan of Sai Mustafa Celebi the narration of Mimar Sinan and Risale-I Mimariyye was about Mehmet Aga, the architect of Sultan Ahmet Mosque by Cafer Efendin in (1614) from these we clearly convinced that Mimara Haca Ocag Sultan's Architectural Chamber had great role in Architectural Education during The Ottoman era.

The talented youngsters were admitted as apprentices and being educated by the more experienced ones. The apprentice-master educational system was accepted during the Ottoman Era. Besides, the on hand education drawing, perspective-elevation, mathematics and geometry were being thought as theoretical system. The value/moral systems of the society and in depth belief education was essential in the education of these youngsters.

After a certain level of experience and many tests in morals and ethics, they were promoted to affiliated and then to full tenured masters and Chief Architects. Some of these selected among these chief architects (Turan, 1963) for a lifetime duty have been selected Ser-Mimarân-ı Hassa or the Chief of Court Architect, such as, Mimar Hayreddin, Mimar Sinan, Davud Ağa, Dalgıç Ahmed Ağa, Sedefkâr Mehmed Ağa, Kasım Ağa, and Mehmed Tahir Ağa.

On the other hand as the other guilds also in the architect built them guilds a moral ethics religious and especially Sufism studies were essential in shaping their students and members. In the proceeding in the conference ceremony of the sultan, on the left hand side the guild's member carrying a model of Suleymaniye mosque and the procession.

The book of Mimar Sinan mentions, "It can be deduced that as long as man, as the domain of heart and soul, does not build the pavilion of his existence and does not decorate the 'mosque of the heart' with the tapestry of the beauty of good morals, and make it flourish, he cannot make his environs flourish either" (Crane & Akin, 2006).

Return To The Spirit

According to Islamic thought, as intellect, the heart and the spirit determine the perfection of a human being, one advances on the endless path to perfection, acquiring knowledge, refinement and wisdom, according to depth one gains in this direction. (Numan,2017)
According to the Islamic thought like has it state in the verse 89: 28.

مَّرْضِيَّةً رَاضِيَةً رَبِّكِ إِلَى ارْجِعِي

Come back thou to thy Lord, well pleased (thyself), and well pleasing unto Him.

Table 1. Stations on the Path to Perfection.

Intellect	Heart	Spirit
Knowledge	Refined Knowledge	Wisdom
Skill	Refined Taste	Meaning

As intellect, the heart and the spirit determine the perfection of a human being, one advances on the endless path to perfection, acquiring knowledge, refinement, and wisdom, according to depth one gains in this direction. Can these stations on the path to perfection be applied to the artisan and architect, or even to edifices, in Islamic art and architecture?

The ones Who surpasses intellect, and searching for beautiful taste in the direction of the inspiration of their heart and soul, surpasses the technical knowledge and skill that can be acquired through intellect! There is another step beyond this refined taste, and that is only for an edifice that has reached the level of meaning that a sage architect, by his wisdom has seized, collecting it from the spiritual depths thatbelong to the "True Artist"

The adopted common meaning and beautiful taste expressed sometimes an idea, sometimes as a perception or concept of civilization, determines societal life comprehensively and will command society in all its parts ofits part, it's a fact. In arts too, including in architecture, which the idea of Unity that dominates Ottoman Society's thought and its deeper comprehension and the tune of this idea were reflecting the education in totality, including the architectural education, and was consequently embodied in every kind and expression of art and Architecture (Numan, 2017).

References

Baydar, G. (2004) *Journal of Architectural Education*. "The Cultural Burden of Architecture", pp. 19-27.

Crane, H. & E. Akin (2006) *Sinan's Autobiogrphies: Five sixteenth-century texts*. Leiden: Brill.

Efendi, C. (1987) *Risale-I Mi'mariyye: An early Seventeenth Century Ottoman Treatise on Architecture*, Leiden: E.J. Brill.

Eldem, S. H., 1983. *Sedad Hakkı Eldem 50 Yıllık Meslek Jübilesi*. Mimar Sinan Üniversitesi 100. Yıldönümü Armağanı, Mimar Sinan Üniversitesi Yayını, İstanbul.

Freely, J. (2011) *A History of Ottoman Architecture*. WIT Press: Southampton.

Karaca, M. ve diğ. (2012) *İstanbul Teknik Üniversitesi ve Mühendislik Tarihimiz*, İstanbul Teknik Üniversitesi, İstanbul.

Nalbantoğlu, G. (1988) "The Birth of an Aesthetic Discourse in Ottoman Architecture", *Middle East Technical University Journal of Architecture*, vol 8.2. pp 115-122.

Ödekan, A., (2009) "Zanaatan Mimarlığa Doğru/From Craft to Arcitecture", *Mimar Kemalettin ve Çağı*, TMMOB Mimarlar Odası ve Vakıflar Genel MüdürlüğüOrtak Yayını, Ankara, pp. 47-52.

Pamir, H. "Architectural Education in Turkey in its Social Context". in *Architectural Education in the Islamic World*, Ahmet Evin (ed.). Singapore: Concept Media/Aga Khan Award for Architecture.

Sey, Y. and Tapan, M (1983) "Architectural Education in Turkey: Past and Present", *Mimar* 10: pp. 69-75.

https://doi.org/10.1051/shsconf/20184102004

Turan, Ş. (1963), "Osmanli Teşkilatında Hassa Mimarlari" in *Ankara Universitesi Dil ve Tarih-Cografya Fakultesi Dergisi*, vol. I, issue 1, pp. 157-202.

Yavuz, Y. (2009). *Imparatorluktan Cumhuriyete Mimar Kemalettin 1870-1927*, TMMOB & Vakiflar GM Yay, Ankara.

Re-charting Architecture or the End of "Architecture" as We Know It?

Ilya Fadjar Maharika[1,*]

[1]Universitas Islam Indonesia, Indonesia

Abstract. As a discipline, the exchange of ideas between architecture and other disciplines or subject areas occurs as a rule rather than an exception. In an attempt to re-chart architecture, mapping the exchange may give a more complete picture as a reflection of architecture itself. In doing so, a survey of the use of the word "architecture" was carried out. Scopus.com was used extensively to explore articles using the word in their abstract, keyword, or title. The finding shows that the word "architecture" is extensively used in nearly all subject areas in Scopus stretching the meaning of architecture to its widest as the organization of things. That "Architecture" itself seems to be diminishing leads to more questions particularly regarding Architectural Education 4.0.

Keywords: Scopus, information technology, Architectural Education 4.0

Introduction

Elizabeth Grosz's writing "Architecture from the Outside" (2001) has alerted us to the position of architecture and its outside. This "outsideness" is not about designing architecture from building envelope inward or space between building in urban context, but rather from outside of architecture as a discipline. For Grosz, as a philosopher, thinking about architecture from the outside is to spark architecture to "think." For her, the outside is capable of a lot of positivity and innovation. Hence, architecture has that opportunity as well if we try to confront it from the outside, from other disciplines. From this point of view, as Grosz has understood, and also perhaps as architects and architecture scholars have agreed, architecture is commonly thought of and conceived as an art and a science of manipulation of space, and at large, a systematic and powerful model for spatial organization and modification (Grosz, 2010: 108-9).

The purpose of this paper is to re-chart architecture. In doing so, the position of architecture itself needs to be interrogated. It is precisely in this position, that what Grosz has posited architecture vis a vis its outside that may induce something new, a new chart of architecture. However, unlike Grosz who talks about architecture as an outsider, this position will differ, as the author is both an architect and architecture educator. Architecture by its nature is a porous body of knowledge. This porosity makes the flow of ideas between the inside (of architecture) and outside (as other discipline or subject areas) as the rule rather than exception. Although accepted by scholars in architectural fields as well as architects, the influence of other disciplines in the architectural body of knowledge, there is little attention paid as to how this flow of knowledge happens. There is little understanding whether architecture may influence these other subjects and then perhaps the influence is cycling back to reshape architectural knowledge.

This paper tries to give an insight into this flow of ideas that form a continuous dynamic in shaping and reshaping architecture. Perhaps this back and forth of ideas also gives an overview of the overall discourse of architecture creating somewhat a chart or re-chart of the knowledge of architecture.

Furthermore, since the context of this re-charting is in the domain of education, it is necessary to look at the newest trends and challenges in education, namely Education 4.0. Induced by the Industrial Revolution 4.0 where information becomes a wild token of change, many believe that it will also change the manufacturing process, value chain, distribution and customer service processes, as well as a wide range of the utilization of the Internet of Things (IoT). In higher education, the rise of MOOCs (Massive Open Online Course), the need for global identity and "education for you" will push forward the paradigm of student-centered learning and allow students to manufacture their interests for their future careers. The change will certainly reshape the business of higher education (Asmaa AbuMezied, n.d.).

Thus, it is also necessary, especially for architectural education, to understand how information and information technology may influence architectural knowledge and practices. Currently, we have exposed the flow of information to which ideas may transfer from one discipline to another in much more dynamic ways. Hence, it is also an interesting view into how information and computer technology have shaped architecture and vice versa. Indeed, since the term 're-charting' itself is a vague delineation of a task specifically dealing with the flow of ideas created by the porosity of architecture itself, it is necessary to limit ourselves through the question we deal with.

Thus, the question will be twofold. The first is the relation between architecture and its outside. The questions arisen

* Corresponding author: maharika@uii.ac.id

will be such as what kind of disciplines or subject areas of knowledge have an exchange of ideas with architecture? Do they change architecture or vice versa? What aspects and what kind of impact does this exchange of influence have on architecture and to architecture education? The second is on the specific relationship between architecture and information especially the information technology that shapes the current Industrial Revolution 4.0. The question may be escalated to what impact does information technology have on architecture and architecture education.

Of course, not all of our question will be or are able to be answered easily in this short paper. The proposed methodology does address the limitation of this research.

Mapping The Inquiry: Strategy

In an attempt to examine and then limit the scope in addressing these questions, the strategy of inquiry is proposed. In this paper, the author will use the term Architecture for "architecture" that is commonly understood as conventional line of thought of architecture by architects (or what is being taught in architecture schools). Hence, strategically speaking, this paper will use a literature exploration and survey to identify the use of architecture in many other disciplines represented in the articles. As Wang and Groat have believed, the literature contains historical lineage showing a reservoir of historical information. It also shows genetic link and roots with an idea in the background. Exploring literature gives an understanding of the conceptual landscape in its contemporary context, the vantage point that create nuance, and of the intellectual agenda behind (Groat, L, and Wang, D. 2002: 58). Consequently, it may situate Architecture as a discipline in front of challenges from other disciplines.

In term of tactics of this exploration, since a thorough survey is impossible; the author has limited the primary source and tools to Scopus (Scopus.com). For initial information about the extent of this survey, Scopus was searched using the keyword "architecture" in the article title, abstract, and keywords. It reaches 693,114 documents. There are 138,806 documents using the word "architecture" in their title and 321,279 documents that use it as a keyword (search date started October 4 - 20, 2017, some inconsistency of this number may appear due to the dynamic nature of the Scopus record).

Indeed, the choice of Scopus itself creates somewhat of a bias in favor of natural sciences, engineering and biomedical research, impairing social sciences, arts and humanities let alone architecture (Mongeon, P., & Paul-Hus, A., 2015). Scopus also declines many Architecture magazines - which to architects are vital. But since the purpose of this paper is to expand Architecture through the outside, the bias itself perhaps is an advantage that create windows of opportunity to find a novel perspective on how the ideas of architecture are being developed over the course of time, place and subjects to reveal what Grosz has stated as positivity and innovation.

Cross-disciplinary Use of "Architecture"

The word "architecture" has reached many disciplines and subject areas. For the purpose of this paper, the author starts from documents that use architecture in title, abstract, and keywords and then narrows it down to documents that use architecture in their keywords. The survey was also narrowed down to the "only article" type of documents. At this initial mapping, it showed 111,580 articles (search date 16 October 2017). Based on these numbers, the analysis available in Scopus is used to create an overview of this expanding chart of architecture.

Seen from the disciplinary use of architecture, the overall picture of the articles in Scopus can be seen from its subject areas in Fig. 1.

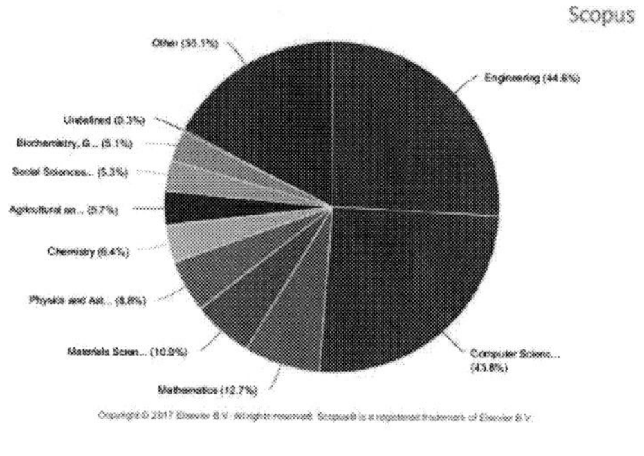

Fig. 1. Number of articles in Scopus per subject area.
Source: Scopus.com, date of searching 16 October 2017.

Currently, from the chart of Fig. 1, we see that the fields using the word architecture in their keyword extensively is primarily engineering (44.6%) and then computer science (43.8%). Surprisingly only 3128 articles in "art and humanities" which are generally the "host" of Architecture is in the chart included in the "other" (30.1%).

Seeing the source of this Scopus data, most of the journals come from computer science. There are three top sources, namely *Lecture Notes in Computer Science* (Springer Verlag, 1885 articles), *IEEE Transactions on Computers* (EEE Computer Society, 1123 articles), and *IEEE Transactions on Very Large-Scale Integration (VLSI) Systems* (Institute of Electrical and Electronics Engineers Inc. with 746 articles). The architectural journals listed are *ARQ* (Pontificia Universidad Catolica de Chile, 191 articles) and *Architectural Record* (McGraw-Hill Construction Dodge with the contribution of 125 articles).

Seen from the "historical" use of architecture in Scopus, Fig. 2 shows a dramatic increase in the last 20 years only.

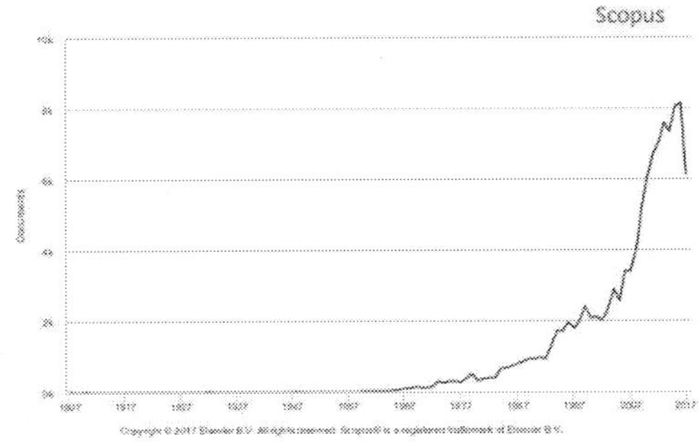

Fig. 2. Number of article in Scopus per year.
Source: Scopus.com, search date 16 October 2017.

In fact, the use of architecture in Scopus seems to have expanded only very recently. The numbers in the chart (Fig. 2) shows that the first record is in the year 1907 with only a small number of papers and then increases in the 1970s and a rapid increase after the year 2000. From 1907 to 1950 there were only five articles covering medicine (3 articles), arts and humanities, and psychology as well as an undefined area regarding animal behavior (Mowrer, 1932). The total amount of papers from the earliest recorded until 1969 was only 104 or 7.4 articles per year. After 1970, the number increased reaching 24,214, or an average of 807 articles per year. After 2000, the number increases dramatically as it reaches 87,249 articles with an average of 4,847 articles per year marking the widest use of the term "architecture" in the scientific world.

Which discipline dominates the contribution use of architecture can be seen from the composition of subject areas. As a 'timeline' of architecture, from 1951 to 1970, the composition of subject areas that used architecture in their keywords changed radically. There were 586 articles, 80.2% of them from the field of medicine, followed by engineering (7.2%) and social science (4.4%), and then by dentistry (3.8%) and health professions (3,6%), computer science reached only 1.4%, ranked 10th on the list. A Picture of 1971 to 1980 shows a dramatic change as well. From 2,759 articles, the five largest compositions of subjects are engineering (69.8%), and then medicine (15.7%), computer science (11.9%), mathematics (12.7%) and nursing (1.8%). The rise of computer science in this era is evident.

In the decade 1981-1990, the field of engineering is still in the forefront as it reaches 64.4%, followed by computer science 38.0% and mathematics (12.5%). Rather surprising, physics and astronomy comes in fourth place (7.2%), followed by the ever-existent subject of medicine (6.0%).

There is no significant change in the following decade, 1991-2000, as the big three were still engineering (51.8%), computer science (50.4%), and mathematics (12%). A swap appears between medicine (5.1%) in the fourth and physics and astronomy (5.1%) comes in fifth place. However, the years 2001-2010 mark the first-time computer science (47.8%) took a lead replacing engineering (40.5%), which placed second and medicine lost its position in the top five. It dropped to the 11th position. Then material science (8.3%) entered the list of the big five as well as physics and astronomy (7.6%). Since then, in the last decade of 2011 to the present (2017), material science is in third place after computer science (41.8%), engineering (41.3%) and material science (14.5%) as the big three followed by mathematics (11.8%) and physics and astronomy (11.6%).

Ideas Exchanged Through The Use of Architecture

While it would be too difficult in this paper to reveal all ideas behind the use of architecture in all Scopus indexed articles within the range of time in detail, what has been pinpointed is the "interesting" use of architecture that may give indication of new discourse for Architecture. Again, using the tactic of making a historical time line of thematic use of architecture it perhaps indicates the idea exchanged between the field of Architecture and other disciplines that use architecture and how it has evolved.

The oldest article in Scopus was in 1907 in the field of psychology addressing an experiment on the aesthetic value of architecture and design. Commenting on his original paper, Rowland explains how observers had difference introspections and ways of enjoying decorative design characterized by opposing demands of art as well as of life (Rowland, 1907).

The second oldest was in the field of health care. Ziemer (1910) addressed a history of the medical research group, the Mayo Clinic. The first time the word "information" appears in the keyword to mention the information center at the Mayo Clinic. Since the document shows only limited information, it seems that the document lists one of the oldest clinics in the United States of America, which has now become a global center of "hope and healing" (Mayo Clinic History & Heritage, n.d.).

After over twenty years, the third oldest article appeared in 1932, in the field of psychology. Instead of writing about architecture in our current sense, it addresses the "animal architecture" of a rooster. The architecture here points rather to the shape of animal kingdom (Mowrer, 1932).

The term architecture appears in a journal of medicine entitled School Architecture (1945). It is not entirely clear in what way the term was used since there is only very limited information, but interestingly, also in the year 1945, the notion of "biophilic architecture" appears, potentially for the first time. However, it is not from the field of what we now familiar with, the biophylic form of Architecture for instance, but from the field of medicine addressing the molecular architecture of cells (Leake, 1948). From this point, we see that the idea from other field may reoccur in Architecture if a supporting factor is "ready." In the case of the biophylic form of Architecture, today's expression of architecture of cells as real building is possible due to computer technology and parametric programming.

In the 1960s, many publications that use architecture are addressed on the topic of health. It is expressed extensively on the principles of design of housing that appeared in the medicine journal *Royal Society of Health Journal* (Hutchings, 1960). Furthermore, in 1961, three papers appeared all addressing architecture and its relationship with health. Gottlieb (1961) outlines the history of western medicine, especially the temples and hospital sites in the city-state of Epidaurus where the shrine Asklepios was devoted to the gods of medicine. In the current situation, the temple is regarded as a world heritage site due to its valuable insight into the healing cults of the Greek and Roman eras (UNESCO, n.d.). The relationship between material and health was addressed by Kolmovskoi (1961) mentioning glass as a hygienic construction material. A deeper understanding of the relationship between human health and its environment was also outlined in the context of the office environment in the *Journal of Occupational Medicine*. In the paper, it is mentioned, perhaps, for the first time, the terms human-engineering and bioengineering (Gad, 1961). In 1962, there were six articles all in the context of sanitation, medicine, and health including attention to hospitals and space perception in the ocular field. In 1963, there were 19 documents and in 1964, 16 documents about medical and public health facilities such as *Space Planning Principles for Biomedical Research Laboratories* (Snow 1963). In 1964, the name of Frank Lloyd Write appeared in a psychoanalytic interpretation of his designs (Globus, G. G., & Gilbert, J., 1964). In 1965, 49 documents were published also in the medicine field, especially on oral and dental hygiene. In 1966 there were 60 documents also published in medicine including from the *Journal of Archaeology* addressing *The Hospice of the Knights of St. John in Akko* (Goldmann, Z. 1966). A further 92 documents were published in 1967.

Although health and medicine become a central theme of publication in the early records using architecture, the central theme of these early years was sanitation and hygiene. However, between 1965 and 1970 many of the publications in that domain shifted to specific factions, namely in dentistry, such as surgery space, dental offices, and the like.

From a general point of view, in this initial exchange of ideas, it seems that the use of architecture is primarily speaking of buildings that facilitate medical needs. There is also the use of architecture in terms of its metaphorical meaning such as the "architecture of behavior" in psychology although it was still very limited. The notion of biophilic architecture was also being introduced, albeit for cells, which today is known as the embryonic ideas of current trends.

The survey of the early years of Scopus finds small numbers of articles making it possible to inspect their content rather in detail. After the 1970s, we are unable to investigate in more detail in each article as the numbers grow significantly. Hence, we expose only the field or subject areas and pinpoint certain specific articles containing novelty in the use of architecture, or which have considerable citations or addressing specific issue or interest.

In the early 1970s, we also encounter the word of architecture used for the first time in computer science. New concepts arrived, as shown by 124 documents published containing words such as 'operating system architecture' or 'new architecture for mini computers.' In the abstract, the word architecture was used to mention the context of the new science of computers such as the 'architecture of the system.' There was also mention of the role of the architect in the field of site planning for power stations to enhance the landscape setting (Holford & Shepherd, 1970). It's also shown at this point of time that the field was starting to shift from medical and health to computer science although some publications on dental office still appeared.

In 'Architecture' as early as the 1970s, there appeared a discussion on the issue of large complex building design in *Nature* suggesting programming and schematic design with computer-aided techniques (England, 1971). Computers began to enter the world of architecture and architects such as a paper entitled "Computer Programs for Architects and Layout Planners" (Lee 1971). This is a clear example of a call for the use of computer science as a tool for Architecture design. The object study, however, was still in the domain of medicine, namely the large complexes of hospitals.

Photogrammetry enters the domain of architecture describing the latest models of equipment used in architectural representation in the 1970s. Parameters of photogrammetric and stereometric cameras, plotting instruments, stereocomparators and image rectifiers are briefly discussed. There was also discussion on architecture in elderly, disabled and modern hospital design. On this type of building, numerous publications appeared addressing hospital, rehabilitation, airport medical facility, and dental office design including the topic of programming with systematic control and system approaches, management of planning, safety design, and ergonomics. There were also introductions to planning and design of tall buildings, the issue of environmental design and concerns towards ecology and nature in cities. Also within this range, there were articles that have high citations on the subject of acoustics in the *Journal of the Acoustical Society of America*, perceptions of space in the *Scandinavian Journal of Psychology* and the *British Journal of Psychology* as well as formulas of determining population from floor area ratio in the *Journal of World Archaeology*.

There was extensive use of coupling word such as software architecture, computer architecture, network architecture, as well as system architecture throughout the 1970s to 1990s. In the field of engineering, the *Journal of Artificial Intelligence* earned extensive attention showing huge numbers of citations. Many of them were addressing cognitive architecture. Special attention can be made to the work of Laird, J. E., Newell, A., & Rosenbloom, P. S. (1987) that earned 1201 citations showing a system capable of general intelligent behavior. Special attention can be addressed to the *Design Studies* journal especially the work of Newland, P., Powell, J. A., & Creed, C. (1987) delivering the idea of information handling and transfer for designers. They showed how designers perceive and learn about their world which is influenced by learning styles, perception, and culture. The paper earned 22 citations and perhaps marked the rise of how design and information are closely interrelated.

In the 2000s, computer science dominated in the use of architecture. In the middle of this new interaction, some significant keywords need attention. The first is the rising significance of the building information modeling (BIM) (138 articles). The big five of journals addressing BIM are the *Automation in Construction* (Elsevier), *Journal of Construction Engineering* (American Society of Civil Engineers), *(Journal of Information Technology* (Kungliga Tenkniska Hoegskolan/Royal Institute of Technology), Architectural Design (Wiley-Blackwell) and *Journal of Computing in Civil Engineering* (American Society of Civil Engineers (ASCE). From subject area composition, BIM covers the engineering side (78%), computer science (32.6%) and business management and accounting (13.0%) with some new notions coming to the surface of discussions such as automation, collaboration platforms, supply chain integration, and performance modeling. The pace of this new approach reflects its significance as integration models of planning, design, construction, and operation as well as management of the facility in architecture, engineering, and construction (AEC) industry as reflected by Azhar's article (Azhar, 2011) that earns the 267 citations in this area. The notion of integration is even pushed forward into new integration through an autonomous assembly of collective production. This may replace old generic construction as indicated by Lu (2017) in *Architectural Design*. There is also concern on how the emerging subject areas of BIM and sustainability are being adopted in education including in distance learning, multidisciplinary collaboration, and industry collaborations as surveyed by Becerik-Gerber, B., Gerber, D. J., & Ku, K. (2011).

An article that also earns attention in the potential of integration of BIM is gaming technology as outlined by Yan, W., Culp, C., & Graf, R. (2011). The integration will open an entirely new form of Architecture as it merges places and buildings as physical dynamics with virtual environments and user activities in real-time interaction. Here the integration will be pushed forward into simulation and visualization, character modeling and animation, collision detection, navigation, path planning, materials lighting, and so on.

Another venue discussion is the rising notion of smart cities. There are over 660 articles addressing this new technology, rising drastically in the last five years (from 23 documents or less prior to 2012 to 149 in 2017). An article that earns attention is Zanella, A., Bui, N., Castellani, A., Vangelista, L., & Zorzi, M. (2014) addressing the use of the Internet of things for smart cities. Smart city refers to a broad term of communication technologies supporting services for both the administration of a city and the citizens. Here the Internet of things covers various devices from the sensory level, network infrastructure to data management. This technology may transform the existing physical urban system to a hybrid cyber-physical new system as also outlined by Jin, J., Gubbi, J., Marusic, S., & Palaniswami, M. (2014).

An entirely new way of using architecture is shown in the word nanoarchitecture that has developed in biochemistry and advanced material. The word addresses a way to build artificial nanostructures that can mimic biological systems and organs as outlined by Lin, C., Liu, Y., & Yan, H. (2009). This article also earns considerable attention with 193 citations. Another use of architecture is on the advanced bio-inspired or bio-mimetic composite as Studart, A. R. (2012) has outlined. Another article that shows the emerging scientific discipline of bio-mimetics in the field of Architecture is explored by Knippers, J., & Speck, T. (2012) that earns 65 citations and appears in *Bioinspiration and Biomimetics* journal from IOP Publishing.

Architech-in-formation: An Epilogue

The above conversation between architecture and Architecture through literature exploration opens up a discussion on the ontological level of the current reality architectural discourse as Grosz has urged to pay attention to the outsideness of architecture. The above new realities push us – architecture scholar, educator, and architect – to not think of Architecture in the Pvesnerian or Vitruvian view anymore. The Pvesner duality of architecture as the conceptual value of Lincoln Cathedral as a piece of Architecture versus a bicycle shed as a building is no longer the standard. Indeed, our Vitruvian Architecture today is still very much alive but soon it will no longer be enough to contain the new wave of information technology marking the Industrial Revolution 4.0. What we are experiencing now is a radical disruption of Architecture. The concept of *firmitas*, durability of materials, has been pushed forward into biological forms that grow as well as self-assembly polymers for future building materials. The concept of function, *utilitas*, has also been pushed forward into new frontiers of the game. There will be no 'fixed' functional programs as the function itself will only be the result of continuous interaction between the cyber and the real. Convenience is not designed but programmed. The concept of beauty and delight, *venustas*, has also been transformed into stimulatory and transient qualities of beauty. Through this, we are not only filling a leftover space of Vitruvius's *Ten Books of Architecture*. It has become necessary to write another ten books of architectures for the modern era. Today, architecture is architecture-technology information as well as architecture information.

Regarding architectural education, what we face in Architectural Education 4.0 is not merely about the disruption of education created by massive open and online courses or the like, but rather the very essence of architecture itself. Architecture in formation will be about its materiality, its relational formation of human (and nonhuman) interactions with both its permanent and transient environment as well as its continuous manipulation of neural architecture for delight.

References

Asmaa AbuMezied. (n.d.). "The role of Higher Education in The Fourth Industrial Revolution." Retrieved October 18, 2017, from https://www.weforum.org/agenda/2016/01/what-role-will-education-play-in-the-fourth-industrial-revolution.

Azhar, S. (2011). "Building information modeling (BIM): Trends, benefits, risks, and challenges for the AEC industry." *Leadership and Management in Engineering*, 11(3), 241-252. doi:10.1061/(ASCE)LM.1943-5630.0000127.

Becerik-Gerber, B., Gerber, D. J., & Ku, K. (2011). "The pace of technological innovation in architecture, engineering, and construction education: Integrating recent trends into the curricula." *Electronic Journal of Information Technology in Construction*, 16, 411-432. Retrieved from www.scopus.com.

England, R. (1971). "Planning complex building systems." *Nature*, 229 (5282), 233-236. doi:10.1038/229233a0.

Gad, R.K. 1961. "IV. The Office Environment—fact or Fancy." *Journal of Occupational Medicine* 3 (6).

Globus, G. G., & Gilbert, J. (1964). "A Metapsychological Approach to the Architecture of Frank Lloyd Wright." *Psychoanalytic Review, 51*, 285-297. Retrieved from www.scopus.com.

Goldmann, Z. (1966). "The hospice of the Knights of St. John in Akko." *Archaeology, 19*(3), 182-189. Retrieved from www.scopus.com.

Gottlieb, B.J. 1961. "Greek Gods of Medicine and Their Temples. II. Epidaurus: Impressions of a Study Trip to Hellas." *Folia Clínica Internacional* 11.

Grosz, E. (2001). *Architecture from the Outside, Essays on Virtual and Real Space*. Massachusetts London, England: The MIT Press Cambridge.

Holford WG, & Shepheard HH. (1970). "Architect and power engineering." *Electronics and Power*, 16, 449-453. Retrieved from www.scopus.com.

Hutchings, A. (1960). "The Design of Housing." *Royal Society of Health Journal* 80.

Jin, J., Gubbi, J., Marusic, S., & Palaniswami, M. (2014). "An information framework for creating a smart city through the Internet of things." *IEEE Internet of Things Journal,* 1(2), 112-121. doi:10.1109/JIOT.2013.2296516.

Knippers, J., & Speck, T. (2012). "Design and construction principles in nature and architecture." *Bioinspiration and Biomimetics,* 7(1) doi:10.1088/1748-3182/7/1/015002.

Kolmovskoi, A.A. 1961. "Glass as a Hygienic Construction Material." *Gigiena I Sanitariia* 26.

Laird, J. E., Newell, A., & Rosenbloom, P. S. (1987). SOAR: An architecture for general intelligence. *Artificial Intelligence,* 33(1), 1-64. doi:10.1016/0004-3702(87)90050-6.

Leake, C.D. 1948. "Molecular Architecture of Cells." *Federation Proceedings* 7 (1).

Lee K. (1971). "Computer programs for architects and layout planners." 139-147. Retrieved from www.scopus.com.

Lin, C., Liu, Y., & Yan, H. (2009). "Designer DNA nanoarchitectures." *Biochemistry,* 48(8), 1663-1674. doi:10.1021/bi802324w.

Lu, A. (2017). "Autonomous assembly as the fourth approach to generic construction." *Architectural Design*, 87(4), 128-133. doi:10.1002/ad.2205.

Mayo Clinic History & Heritage (n.d.). Retrieved October 18, 2017, from http://history.mayoclinic.org/timelines/history-timeline.php and http://history.mayoclinic.org/tours-events/w-bruce-fye-center-for-the-history-of-medicine.php.

Mongeon, P., & Paul-Hus, A. (2015). "The Journal Coverage of Web of Science and Scopus: a Comparative Analysis." *Https://arxiv.org/.* Retrieved from https://arxiv.org/ftp/arxiv/papers/1511/1511.08096.pdf.

Mowrer, O.H. 1932. "A Note on the Effect of Re-peated Hypnotic Stimulation." *Journal of Abnormal and Social Psychology* 27 (1). doi:10.1037/h0071988.

Newland, P., Powell, J. A., & Creed, C. (1987). "Understanding architectural designers' selective information handling." *Design Studies, 8*(1), 2-16. doi:10.1016/0142-694X(87)90026-3.

Rowland, E.H. 1907. "A Further Application of a Result Obtained in Experimental Aesthetics." *Psy-chological Review* 14 (4). doi:10.1037/h0068206.

School Architecture. 1948. *The Journal of the Royal Institute of Public Health and Hygiene* 11 (7).

Snow, D. L. (1963). "Space Planning Principles for Biomedical Research Laboratories." *Public Health Monograph,* 71, 1-52. Retrieved from www.scopus.com.

Studart, A. R. (2012). "Towards high-performance bioinspired composites." *Advanced Materials,* 24(37), 5024-5044. doi:10.1002/adma.201201471.

UNESCO (n.d.). Sanctuary of Asklepios at Epidaurus. Retrieved October 18, 2017, from http://whc.unesco.org/en/list/491.

Wang, D., & Groat, L. (2002). *Architectural Research Methods.* John Wiley & Sons, Inc.

Yan, W., Culp, C., & Graf, R. (2011). "Integrating BIM and gaming for real-time interactive architectural visualization." *Automation in Construction,* 20(4), 446-458. doi:10.1016/j.autcon.2010.11.013.

Zanella, A., Bui, N., Castellani, A., Vangelista, L., & Zorzi, M. (2014). "Internet of things for smart cities." *IEEE Internet of Things Journal,* 1(1), 22-32. doi:10.1109/JIOT.2014.2306328.

Ziemer, R. 1910. "Profile: Mayo Clinic Historical Suite and Archives." *Watermark (Archivists and Librarians in the History of the Health Sciences)* 33 (1).

Socio-Spatial Typology In Karanrang Island

Rahmi Amin Ishak[1,*], *Slamet* Trisutomo[2], *Ria* Wikantari[1], and *Afifah* Harisah[1]

[1]Architecture Department, Hasanuddin University, Indonesia
[2] Regional and Urban Planning Department, Hasanuddin University, Indonesia

Abstract. The phenomenon of community life on the small island is influenced by the stimulating factor of harmonious social interaction system through cooperation, kinship, economic activity, children playing, transportation system, religion and other social activities. The social dynamics of small island communities appear in the layout and environment in which they live, how they manage and utilize space, both indoors and outdoors. The purpose of this paper is to describe the socio-spatial typology of settlements on Karanrang Island, including a description of the spatial pattern of communalenvironments. Research approaches through spatial similarities and differences in the classification of behavioral setting, including physical, non-physical, socio-spatial arrangements. Karanrang Island as a research focus which has an area of 7.8 Ha is one of small islands inhabited in cluster PangkajeneIslands (Pangkep) South Sulawesi, with characteristic of dense settlement, and diversity of tribe, also inhabited by 434 families. The method of this research is observation, data collection through field survey with descriptive analysis based on empirical data on meso / environment which is divided into:1) inter building space; 2) Space in the building; 3) Open space, and; 4) Environmental facilities. The results showed that classification of socio-spatial typology of communal environment is divided into four types of socio-spatial models based on the configuration of social interaction activities, namely:1) Type of Linear Centripetal, at the inter buildings space; 2) Type of Centripetal Cluster, space on the building; 3) Type of Centrifugal Cluster, at green open space/field; 4) Type of cluster Centripetal, at environmental facilities. The socio-spatial type based on actor's activities, occupancy, and territory, can be distinguished on: 1) Type of children's activity; 2) Type of mother'sactivity; 3) Type of father's activity, and 4) Type of combination activity.

Keywords: small island settlement, socio-spatial typology, social settings, spatial settings

Background

The settlement environment in small islands is influenced by social aspects of society, adaptation to island conditions, and settlement growth. Based on the theory of cultural value orientation by Kluchkohn (1961), the essence of the meaning of community life on the island can affect the spatial order of the settlement and social environment of society, thus binding the community to live. Community settlement and social interaction are also influenced by a series of adaptations to the physical constraints of the island environment, spatial, social, and physical aspects of the environment that affect each other in the formation of the settlement environment.

The growth of settlements on small islands is characterized by the community's attachment to the island, partly due to strong family ties, rural social interactions, hereditary land ownership, and the aspect of a value orientation that is entirely dependent on marine resources. It distinguishes the socio-spatial characteristics of small islands which tend to be solid and homogeneous, with other lands tend to be structured and segregated. In addition, remoteness and isolation factors make the island community more survive and independent in the fulfillment of daily needs. The occurrence of social interaction is characterized by the existence of space encounters both fixed, semi, and temporary, so that social spaces play an important role on the sustainability of social order. The architectural forms of the island environment also affect the activity and behavior of society and vice versa, so that both are in a dynamic cycle, and produce different socio-spatial phenomena at all times.

In order to support the sustainability of the social order and to increase the carrying capacity of the environment of the settlement and conservation of small islands in Pangkep regency, socio-spatial research is needed that focuses on social interaction, in particular the behavior setting on Karanrang Island. The analysis of the socio-spatial typology of small island environments is based on the social behavior setting of the community and the spatial physical condition of the island, within the meso sphere of the settlement environment (inter-building space, space in buildings, open spaces, and environmental facilities).

Literature Review

Social Space

Hall (1984) in the 'The Hidden Dimension', states that social space is an abstract space that is not real physical boundaries. Along with this, Lefebvre in 'Production of Space' (1991)

* Corresponding author: rahmiamin@unhas.ac.id

argued that space is a social product, resulting from social constructs based on values and meanings that are produced socially as well, so that space influences the perception of the spatial humans that produce it. If it is assumed that space is a product of the inhabiting process of space, then the result of the process lies in each part of space and becomes the contents of space. The content of space is everything that exists in the space either produced naturally or by society. All that is meant, are living creatures that live in them, objects, work products including signs and symbols located in space.
Space is created from the way of living that is influenced by physical aspects (natural conditions), cultural aspects and social relations. Lefebvre (2008) explains that space is created by the way we live in social life (lived space), which in the reality of life intersects with the physical material aspects of the perceived space and the non-material (mental) aspects of space conceived in our minds (conceived space). Therefore, an emerging space is the product of adaptation to the environment and socio-culture.

In a study of environmental and behavioral architecture, Rapoport (1977) classifies the boundaries of public and private space in three main typologies:

1. Type of occupancy: 1) personal occupancy (limited to the scope of the family); 2) community occupancy (limited to one group member or one particular social group); 3) society occupancy (free for the whole community); 4) free occupancy (no restrictions).
2. Space type: 1) urban public (open to all); 2) urban semi-public space (limited users); 3) public group (free for some groups of city dwellers); 4) group private (limited to one family only); 5) family private (limited to one family); 6) individual private (only for individuals).
3. Type of territory: 1) public territory (free for everyone, but must comply with the norms set in the area); 2) interactional territory (an area to communicate between certain social groups); 3) home territory (controlled by one family); 4) body territory (personal space that is individual).

There are two early spatial metaphors used to describe social space, the "ecological zone" model by Baacke and the "island model" by Zeiher. In the ecological zone model Baacke (1999), explains that the social development of children takes place within the local social space. This model illustrates the social attachment of children in concentric circles gradually.

1. The "ecological center" describes the family and home environment, the place where children find firsthand the main personal reference, and where it spends most of its time.
2. "Nearer ecological environment" describes a closer environment that allows children to find first-line relationships outside the family in the neighborhood, in the local area or in the village.
3. "Ecological sectors" describe public places such as schools, playgrounds, shops or swimming pools. They require certain role behaviors and therefore demand new skills from the child.
4. "Ecological periphery" is the field of contact outside of the usual routines such as vacations, trips and contacts to other spaces that are outside of everyday experience.

Further studies by Zeiher (2005) found that ecological zones in concentric layout arrangements are no longer adjacent, but are located in separate zones fed as islands connected. These islands are in a larger space that is only crossed but not all are experienced. Children realize that the island where they live as a center and journey on their way to school, friends and family through other social spaces without feeling connected.

Types of Social Groups

According to Soekanto (2014) social groups can be divided into two types, namely regular and irregular social groups. Regular types are: 1) In-group and out-group; 2) Primary and secondary groups; 3) Paguyuban and patembayan; 4) formal group and informal group; 5) Membership group and reference group; 6) Occupational and volunteer groups.
Types of irregular social groups, namely:

1. Crowds are individuals who come together by chance at a place at the same time, consisting of casual crowd and lawless crowd
2. The public is a group that is not a unity. Interaction occurs indirectly through communication tools, and tends to be more dominated by individual interests.
3. Communities are people who reside in a region (geographical) with certain territorial boundaries, the dominant interaction between members within their territory.

Criteria for the classification of society are:

- Population;
- The area, resources, and population density of the inland areas;
- Special functions of the local community towards the whole community;
- The local community organization concerned.

Research Methods

The research using Deinet (2002) approach, to analyze the social space based on the context of qualitative research, are: narrative interviews with "key person" who have a special impression on social space and can share their experiences. In addition, structured interviews were conducted using questionnaires with open questions, to gain a deeper understanding of community activities. Interviews on institutions, with using structured questionnaires, community and institutional staff were asked to provide their assessments of local community and social infrastructure such as schools, day centers, social services, health institutions, and so on.

Identification of socio-spatial typology using spatial ground analysis techniques, overlay mapping and formal descriptive analysis using sketch / drawing techniques. The types of data required include primary and secondary data. Physical aspects of the settlement environment include the unit of analysis: spatial function (fixed, semi, temporary), identity, sign, spatial boundary, center and spatial orientation, frequency of population of social space that often arise. Social aspects of society, including unit analysis: behavioral arrangements (activities, behaviors, roles by age and sex), time and activity frequency. Samples of society as social actors, community leaders, village officials as respondents. Primary data sources: island communities as respondents, community leaders and village heads as informants, and the physical situation of the island, housing / buildings and settlements on Karanrang Island. Secondary data are data collected from other sources, such as location maps and statistical data, population demographics. Secondary data sources: village office, statistical data centers, relevant regulations and policies of local agencies, referral literature, and internet search.

Karanrang Island as the focus of the study is part of a group of islands of Pangkep regency, South Sulawesi, Indonesia, spread over the sea area of 11,464.44 km^2, there are 114 islands, and 94 uninhabited islands. In the meso / environment scope is divided into: 1) inter-building space; 2) space on the building; 3) open space, and; 4) environmental facilities.

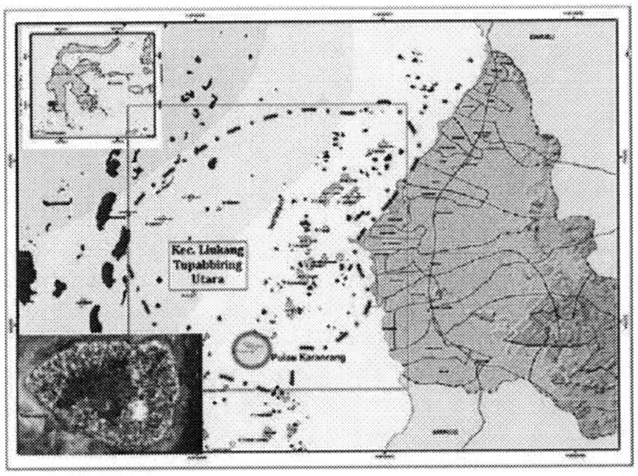

Fig. 1. Locate of Karanrang Island on Pangkep Regency.

Results And Discussion

Spatial Pattern

Karanrang Island as an inhabited island, has an elongated settlement pattern and surrounds along the coast of the island, with a generally irregular settlement arrangement. In the central part of the island there are grave area, and green open spaces as groundwater binders, fields, and environmental facilities (Fig. 2). Karanrang Island has high growth settlements and tends to expand to open spaces in the middle of the island and graveyard. The problem of environmental density and land constraints on Karanrang Island is due to the high population growth and high enough housing each year.

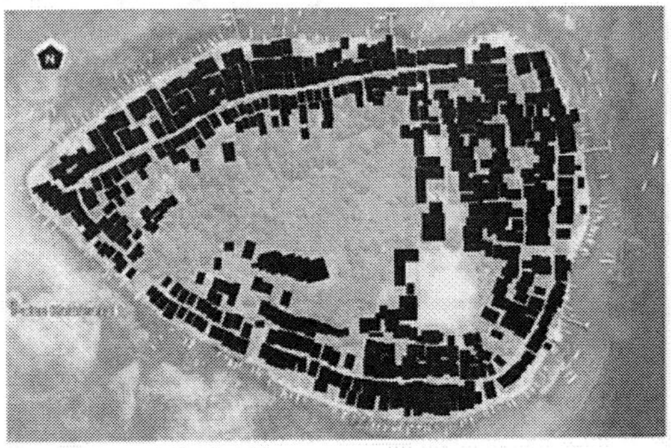

Fig. 2. Figure Ground of Karanrang Island.

Population growth tends to be in line with the growth of houses, in the last year there are 3 - 5 new house built by the population. The current trend of people prefer to build a stone house compared to the stage house. Stone houses are generally built with a fence between the neighboring house, with a more orderly circulation of the road.

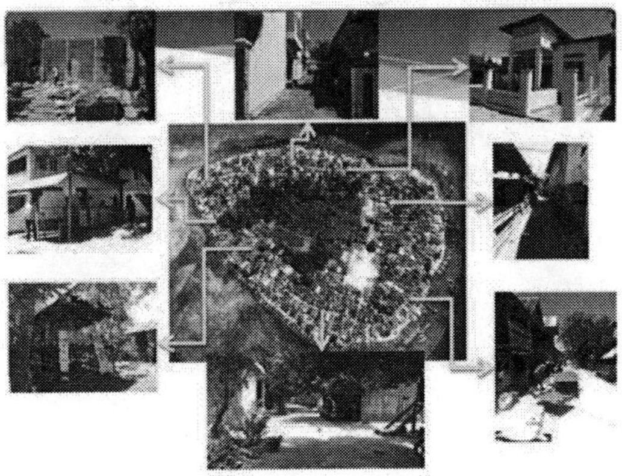

Fig. 3. Spatial Existing Condition of Karanrang Island.

Spatial Function

Karanrang Island have facilities such as elementary, junior and senior high schools, village health centers (PUSKESDES), mosque and mushallah, and a field/open space located in the center of the island, which is generally close to the settlement (50 - 100 m). The location of the mosque/mushallah from the resident's house is generally; 46% near (50 - 100 m); 39% far enough (200 - 300 m). The location of daily shopping is generally; 69% very far (over 500 m), the location of education is generally; 77% near (50 - 100 m), the location of the gathering with the neighbors are generally; near 92% (50 - 100 m).

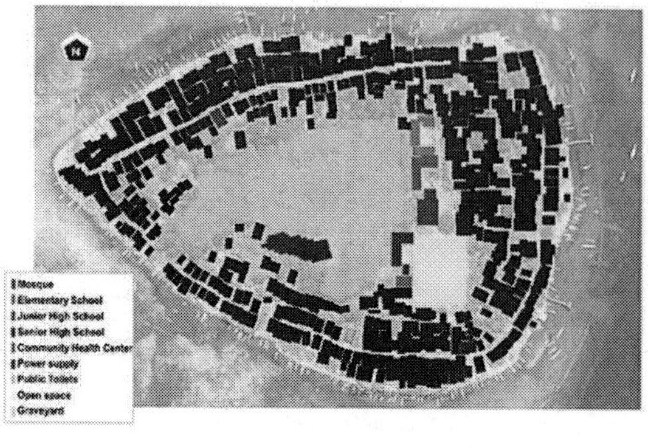

Fig. 4. Spatial Function of Karanrang Island.

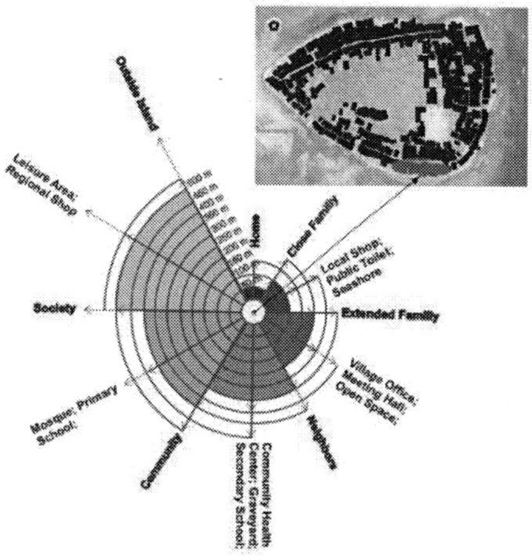

Fig. 5. Distance of facility to residential-edge of the island.

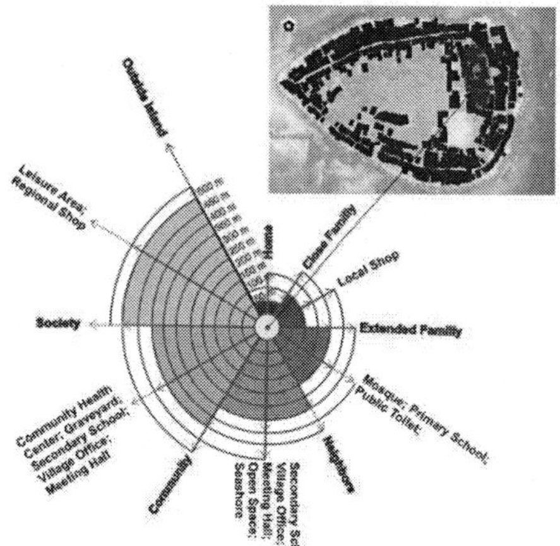

Fig. 6. Distance of facilities to residential-middle of the island.

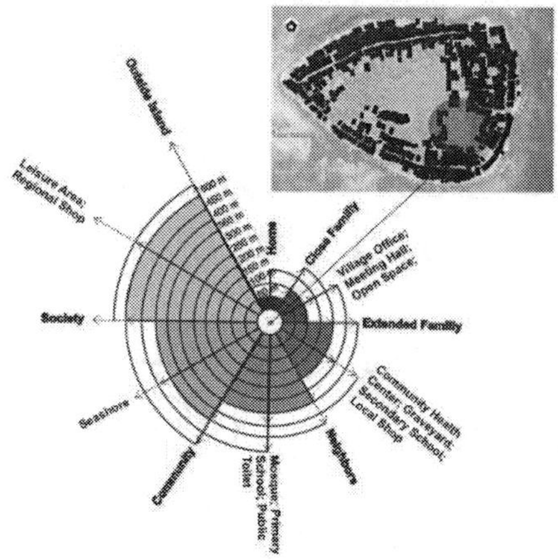

Fig. 7. Distance of facilities to residential-center of the island.

Behavior Setting

Karanrang Island is dominated by 77% indigenous people (46% Bugis-Makassar mixed tribe) and 23% of migrants. In everyday life, the sea provides a great influence on the population (85%), as well as a source of livelihood and transportation. The highest education population is generally at the primary level (69%), and junior high (23%), the livelihood of head of household is generally fisherman with average income every month generally 1 million - 1.5 million rupiah.

Community social relations in the form of gathering with family or neighbors generally tend around the house (77%), either in the yard, or in front of the road home. Other activities related to social relations are marriage receptions (46%), aqiqah (46%), religious events (8%), 78% of the community doing activities in front of the house or in the neighborhood, 23% in open spaces near the house. or field in the middle of the island. Most of the population participates in any social activity (69%).

Social interactions such as communication with dominant neighbors take place around the dwelling (77%). The relationship between communicating activities, reception, visits, or recreational activities with neighbors / families and gathering places, indicates that social interaction activities require space to gather with neighbors / family. Similarly, the relationship between communicating activities with neighbors and house layout, the tendency of communication intensity often occurs in settlements with irregular building layouts (31%). While the activities of receiving neighbors / guests tend to take place around the house (77%).

Based on the settlement layout, visiting activities tend to occur frequently in irregular home layouts. In addition, visiting activities are also linked to the distance between homes. A total of 77% of houses with a building distance of 1-2 meters tended to have higher interaction with neighbors, compared to houses with a building distance of 3-4 meters. It also appears on the pattern of circulation path and visiting activity, the tendency of such activity occurs in irregular circulation patterns (46%).

Visiting neighbors tends to take place in circulation paths and irregular settlement layouts (54%). Related distance between buildings, visiting activities tend to take place in buildings with a distance of 1-2 meters (69%). This activity is also associated with a gathering place around the house (77%). Based on the circulation pattern, people tend to visit neighbors in the settlement with irregular circulation patterns (46%).

Typology of Socio-Spatial

To achieve full socio-spatial dynamics, social space analysis needs to focus on interactive and relational interactions between individuals and social structures. The concept of social space can be used as a metaphor for this perspective and helps analyze the process of social networking and its interrelationships.

The following table 1 shows some behavior setting that happened on: 1) inter buildings space; 2) space on the building; 3) open space, and; 4) environmental infrastructure. Configuration and character of the activity tend to be affected by the behavioral setting of the activity, activity type, and user activity. Communal play activities by children in open spaces tend to be dynamic-spreading, whereas individual play activities under trees, on boats, or in the hallway tend to be static-centered. Social interactions that occur in inter-building space tend to occur incidentally and temporally, where the distance between buildings and circulation patterns also affect the frequency and intensity of interactions that occur.

Table 1. Configuration of Character and Activity on Karanrang Island.

	Place		Activity character	Activity Configuration	
1	Inter building space		**Semi-static** (Circulation, chatting / relaxing, hanging clothes)	Linier – Centripetal 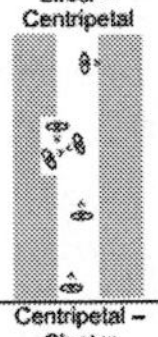	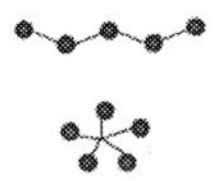
2	Space of building		**Static** (keeping children, selling, talking / relaxing, fixing tools, preparing work)	Centripetal – Cluster 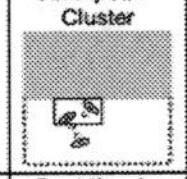	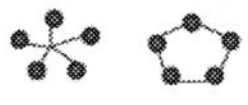
3	Open Space		**Dynamic** (individual-communal play, sunning fish, chatting / relaxing)	Centrifugal – Cluster 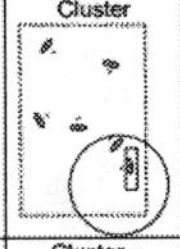	
4	Environment Infrastructure		**Semi-dynamic** (chatting / relaxing, village meetings, washing, praying, studying)	Cluster - Centripetal 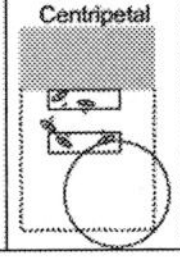	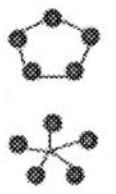

Based on the type of actor's activity, occupancy and spatial territory, the socio-spatial type on Karanrang Island can be distinguished on: 1) Type of children's activity; 2) Type of mother's activity; 3) Type of father's activity, and 4) Type of combination activity.

Children's social space moves from home to major environmental facilities (schools, mosques, green open spaces), then to beachside playgrounds, and on holiday out of the island. Children's daily activities on the island tend to spread in the middle and edge of the island, while the center of the island or village is a field and graveyard, which tend to be rarely visited by children, especially during the day. Children prefer to play in a shady and sheltered environment (Fig. 8).

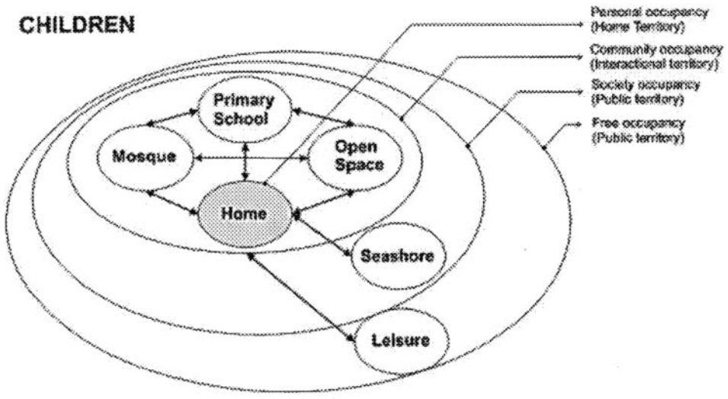

Fig. 8. Type of socio-spatial activity of children on Karanrang Island.

Compared with the daily activities of children, the daily activities of mothers tend to be more varied. The main social interaction space of mothers moves from house to family home / neighbor - public toilets - and stalls. While the weekly secondary activity is a recitation activity in the mosque - checking health at the health center - a celebration event - chatting / relaxing on the beach. Monthly shopping activities tend to be conducted in regional markets outside the island (Fig. 9).

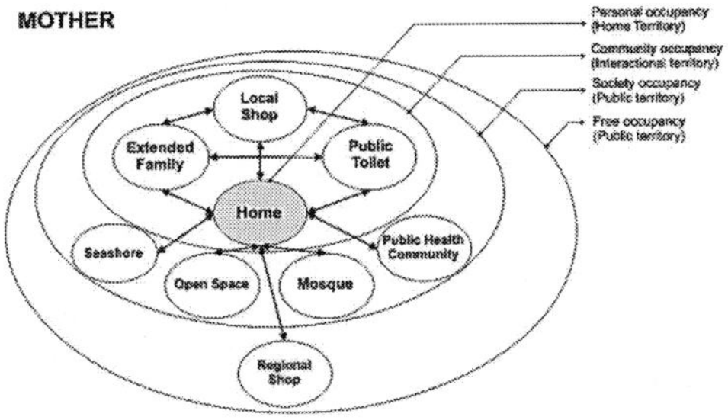

Fig. 9. Type of socio-spatial activity of mothers on Karanrang Island.

The activities of fathers, who work as fishermen tend to spend more time on the seas and beaches. Generally the father's social space moves from home - family / neighbor - beach - green open space. Weekly / monthly meeting activities tend to be in the Village Office - Mosque / Mushallah - Village Health Center - Village Hall. The furthest activities outside the island are often conducted outside the island to go to the sea, trade or shop for basic necessities (Fig. 10).

https://doi.org/10.1051/shsconf/20184103001

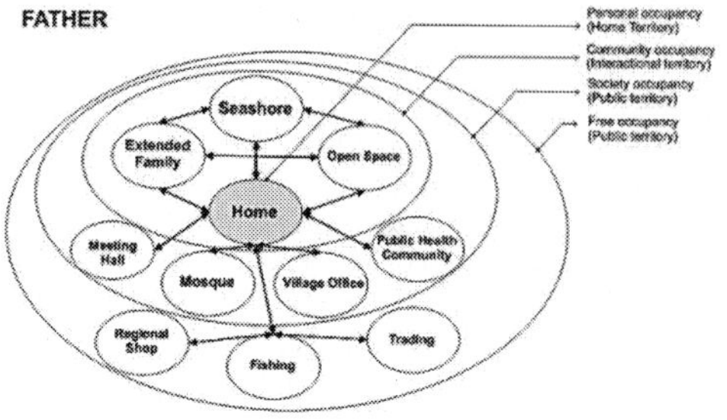

Fig. 10. Type of socio-spatial activity of fathers on Karanrang Island.

The combination type of children's – mother's – father's activities, tend to spend more time on 3 zones. Generally the combination's type of social space moves from home - family / neighbor - green open space. Weekly / monthly meeting activities tend to be in Mosque / Mushallah – Public Health Community – Seashore. The furthest activities outside island, often conducted to Regional Shop for primary and secondary necessities, also to have some leisure (Fig. 11).

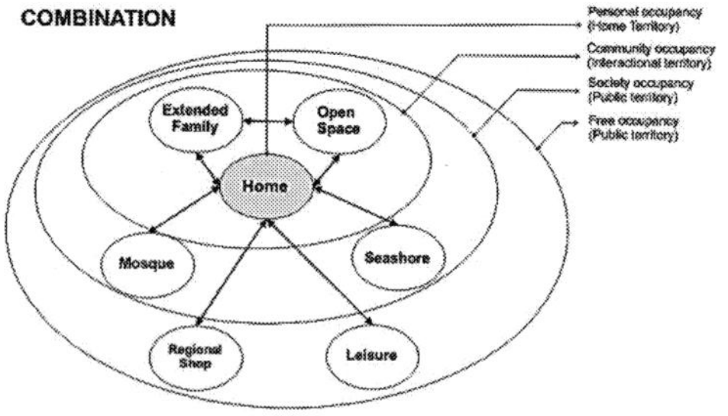

Fig. 11. Type of socio-spatial activity combination on Karanrang Island.

Social space is not merely a "container" but a relational arrangement of people and material goods in certain places that are always dynamic and fluid. Through the process of "distance" one can acquire material places, establish new relationships and create new social spaces of their own (Deinet, 2009). In this sense, space is always socially determined. In addition, this understanding allows multiple social spaces in one geographic place to be identified. Social space can be changed and also disappears as their producers leave the place. Following this idea, social space can be regarded as a small and temporary community at the local and regional levels.

To achieve full socio-spatial dynamics, social space analysis needs to focus on interactive and relational interactions between individuals and social structures. The concept of social space can be used as a metaphor for this perspective and helps analyze the process of social networking and its interrelationships.

Conclusion

The socio-spatial typology on Karanrang Island is distinguished based on the characteristics of the social interaction on behavior setting, and divided by type of social space (inter buildings space, space in buildings, open space, and environmental facilities).

In the meso / settlement environment, there are four types of socio-spatial models based on the configuration of social interaction activities, namely: 1) Type of Linear – Centripetal, at the inter buildings space; 2) Type of Centripetal – Cluster, at space on the building; 3) Type of Centrifugal – Cluster, at green open space / field; 4) Type of Cluster – Centripetal , at environmental facilities.

The socio-spatial type based on actor's activities, occupancy, and territory, can be distinguished on: 1) Type of children's activity; 2) Type of mother's activity; 3) Type of father's activity, and 4) Type of combination activity.

References

Baacke, Dieter (1999): "Die neue Medien-Generation im New Age of visual thinking: Kinder- lind Jugendkllitur in der Medienkultur". Leske + Budrich, Opladen.

Deinet, Ulrich (2009), Methodenbuch Sozialraum, VS Verlag für Sozialwissenschaften | GWV Fachverlage GmbH, Wiesbaden.

Hall, Edward T. (1984) The Hidden Dimension, Anchor Books Editions, USA.

Hengst, Heinz; Zeiher, Helga (2005) Kindheit Soziologisch, VS Verlag für Sozialwissenschaften | GWV Fachverlage GmbH, Wiesbaden.

Koentjaraningrat (1995), Pengantar Ilmu Antropologi, Aksara Baru, Jakarta.

Laurens, J. Marcella (2004), Arsitektur Perilaku Manusia, PT. Gramedia Widiasarana Indonesia, Surabaya.

Lefebvre, Henry (1991), The Production of Space, Oxford, UK & Cambridge, USA: Blackwell

Madanipour, A (1996), Design of Urban Space, in Inquiri into a Socio-Spatial Process, John Wiley & Sons, New York.

Rapoport, A. (1969). House Form & Culture. Englewood Cliffs, N.J: Prentice-Hall

Rapoport, Amos (2005), Culture, Architecture, and Design, Locke Science Publishing Company, Inc, Chicago.

Ritzer, George. Editor (2005) Encyclopedia of Social Theory, Volume 1, California: Sage Publications, Inc.

Ritzer, George. Editor (2005) Encyclopedia of Social Theory, Volume 2, California: Sage Publications, Inc.

Ritzer, George & Smart, Barry (2003) Handbook of Social Theory, California: Sage Publications, Inc.

Spatscheck & Ostermann (2009) Social Space Analysis and the Socio-Spatial Paradigm in Social Work, Working-paper serien 2009:1, ISSN 1650-8971.

Teaching Co-Design to Promote Students' Community Engagement

Wiryono Raharjo[1,*]

[1]Department of Architecture, Faculty of Civil Engineering and Planning, Universitas Islam Indonesia, Indonesia

Abstract. The current landscape of architectural education has been predominantly based on studio teaching the mode of teaching that can be traced back in the age of French L'Ecole des Beaux Arts. It was during the Beaux Arts era that studio teaching and jury system first deve-loped, and it continues today. The approach is characterized by the use of atelier (studio) for coaching, where juniors learn from seniors and competition among students are the norm. The approach also presumes that architects are the provider of design services who think that they are more competent than their clients in solving the design problem; which is a paradox to the fact that most parts of the built environment anywhere in the world are the creation of ordinary people. In other words, ordinary people can contribute significantly to the decision making within the architectural design process. They can actually be treated as a partner of architects in designing their building, which is what Co-Design all about. This paper aims to describe the introduction of Co-Design approach in architectural design studio teaching at Universitas Islam Indonesia. The research questions this paper intends to respond: to what extent could Co-Design be adopted in architectural design teaching? What lessons have the students learned from their experience in involving people in the urban design process? The case study methods employed in this research, through which author looks at the Co-Design teaching delivery in the Architectural Design Studio 7 in the past three years. While overall result of the study shows that time limitation has constrained the students' community engagement, most cases examined reveal that introduction of Co-Design approach has paved the way for students to understand how ordinary people contribute to the shaping of the urban built environment.

Keywords: Community Engagement, Urban Co-Design, Kampung

Introduction

Since the age of L'Ecole des Beaux-Arts, architectural education all over the world has been predominantly based on studio teaching. Donald Schön (1987) uses the term "coaching" to call the mode of teaching in the studio where students are guided by a coach in the gradual process of learning to design. In such mode of teaching the coach is usually a more senior person, either a lecturer or a teaching assistant, who stimulates dialogue with students to "reframe" the design problems (Schön, 1987). Often the dialogue between students and their coach (mentor) mimics the dialogue between architect and client, where student acts as an architect while mentor plays the role as a client (Raharjo, 1993).

However, studio teaching also sees the client as a person to be served by the architect. While the end results subject to client's approval, such situation assumes the superiority of architect in running the design process. It disregards the fact that most parts of the built environment are not designed by the architect. In other words, ordinary people have the capacity to contribute significantly to the making of the built environment. They are the designer by nature, whose voices shall be taken into account. These people can be treated as the partner of the architect in conducting the design process, which leads to the egalitarian relationship between architect and client. Within the broad range of design literature, such design approach is often termed Co-Design.

Co-Design frequently appears as "an umbrella term for participatory, co-creation, and open design processes" (Chisholm in http://www.designforeurope.eu/what-co-design accessed on 30 September 2017). It places the central role of clients in the decision-making within the design process.

This paper describes how Co-Design approach is adopted in the architectural design studio teaching. It intends to respond to the following questions:

1. To what extent can co-design approach be adopted in architectural design teaching?
2. What lessons have the students learned from involving the people in making design decision?

The research was conducted through empirical observations of teaching activity and the environment of the Architectural Design Studio (ADS) 7- a course delivered within the Department of Architecture, Faculty of Civil Engineering and Planning, Universitas Islam Indonesia (UII). The teaching documents observed are those delivered in the past two years (2015 and 2016), but the ongoing course (2017) is also examined to enhance the analysis.

* Corresponding author: raharjo@uii.ac.id

Co-Design: review of related literature

Co-Design is about making consensus among people in deciding the best choice of design (Day, 2003). There are two among many definitions of Co-Design this paper considers essential. The first one is by Barrie and Edwards (2003:10), who writes that Co-Design is "… the process of deliberately engaging users of the system, deliverers of services and other experts, being led by experts such as designers, to actively understand, explore and ultimately change a system together". Another definition is suggested by Hagen (2014), which says that Co-Design is the involvement of the community in the design process as active participants, to create shared value.

The root of Co-Design can be traced back in the 1960s in Britain when community participation began to emerge as a norm in design. The British Housing Act (passed in 1969) stated that people must be consulted during the preparation of the plan, not after the plan is ready (Day, 2003). Co-Design has been integrated into the development public services in Australia and New Zealand since the 1990s (Barrie & Edwards, 2013). A series of key terms that embody Co-Design are communicated, consult coordinate, and collaboration (Barrie and Edwards, 2003). Barrie and Edwards (2003:11) further mention that "Co-Design is an approach that can be utilized in a range of design disciplines and processes", while Myers (2012) writes that Co-Design is about collective thinking, a collaborative action of design between client, designer, and end user.

The approach can be regarded as the root of various other design approaches as depicted in Table 1. It has also been adopted in various fields, from automobile design to urban housing (see for example Albinsson, Lind, and Forsgren, 2007; Rodriguez and Astrand, 1996).

Table 1. The scope of Co-Design (adapted from Hagen, 2014).

No.	Co-Design related areas
1	Participatory action research
2	Participatory design
3	Social design
4	Transformation design
5	Co-production
6	Service design
7	Community consultation
8	Open innovation
9	Collaborative inquiry
10	Design thinking
11	Design for social innovation
12	Change management
13	Experience-based design

The process of Co-Design varies according to the field and complexity of the design problems, as well as the number clients, i.e., members of the community involved in the design. In the field of urban housing, as experienced by Rodriguez and Astrand (1996), the process typically begins with initial contact between designer and the future clients. The objective of initial contact is to make sure that these clients are prepared to engage in the design process. When designer and clients are ready to collaborate, the design process may follow any methods agreed by both parties.

The literature reviewed above mostly discuss the experiences of implementing Co-Design in the real design environment. None of them actually deals with how Co-Design is taught within the academic setting.

Teaching Co-Design: the methodology

As mentioned earlier, the course used as a medium of Co-Design delivery is Architectural Design Studio (ADS) 7. It is a mandatory course offered in the Department of Architecture, Universitas Islam Indonesia (UII) to the fourth year students. This course may be more appropriate to be termed "Urban Design Studio" since its content is mainly about the practice of urban design process. It is the end of a series of ADS courses prior to the commencement of Bachelor Final Project.

In the past three years, the average number of architecture students attending ADS 7 is ranging between 140 and 150 persons. This year (2017), the total number of lectures coaching ADS 7 is 10, each of them handles a maximum of 15 students. Therefore, there are 10 classes running ADS 7 simultaneously. Two of this classes, including author's class, is assigned to introduce Co-Design in the design exercise delivered to the students.

There were 15 students in author's class both in 2015 and 2016, as well as in 2017. In the first week of the semester, the author typically asked students to work in a team of maximum 5 members. The reason for making students work in a team is to accelerate the process of finding clients due to the limitation of time. In addition, it also made students more confident in communicating with people who would be chosen as their clients. However, this teamwork only valid for the first half of semester; students must work individually thereafter until the final submission of their works at the end of the semester.

This paper looks at the cases of Co-Design delivery in 2015, 2016, and the ongoing ADS 7 in 2017. It reviews the samples of work of students in ADS 7. The samples are selected using purposive sampling methods, where author selects these samples based on the quality of works as indicated by the final grade obtained by students, in addition to the characteristics of settlements used by students as their case study. Furthermore, the selected work is examined based on how students practice the Co-Design, by looking at the length of initial contact (LOIC) with the targeted clients, the length of community engagement (LOCE), and the Number of Clients (NOC).

Case study analysis and findings

The urban areas chosen by students for their case studies (Table 2) indicate two types of settlements, i.e. formal and informal (*kampung*). While formal settlements often called *kampung* as well, in this paper author defines the informal *kampung* as a settlement that emerged from the illegal subdivision of government managed land. The formal *kampung*, such as Kotagede and Kauman, occur as a result of the accumulation of individually designed on housing on unplanned plots but with legal tenure.

Table 2. Types of targeted settlements chosen by the design teams.

Year	Formal Kampung	Informal Kampung
2015	2 teams	1 team
2016 (*)	2 teams	2 teams
2017	1 team	2 teams

(*) There were 4 teams in 2016, two of them consisted of 3 and 2 members; the rest consisted of 5 members.

2015 Cases

The year of 2015 marked the beginning of Co-Design application in teaching architectural design studio at UII. As depicted in Table 2, the 2 teams have selected formal *kampung*. One of the works received A grade of this category belong to a student named NAL, which focuses on the development of urban farming in Cepit Community. As stated in the preceding section, the number of clients (NOC) she and her team found, the length of initial contact (LOIC), and the length of community engagement (LOCE) are indicators that author uses to measure the Co-Design performance. Table 3 shows the comparison between Sample 1 (NAL) and Sample 2, which belongs to a student named BS. Sample 2 represents informal settlement. In this case, BS and his team investigated Kampung Gowok, an informal settlement on Gajahwong Riverbank.

Table 3. Comparison of two cases (2015).

Student's name	NOC	LOIC (Weeks)	LOCE (Weeks)
NAL (Formal *Kampung*)	5	4	6
BS (Informal *Kampung*)	4	2	13

Note: NOC = Number of Client; LOIC = Length of Initial Contact; LOCE = Length of Community Engagement.

The above findings show that length of initial contact (LOIC) performed by BS is shorter than that of NAL. Author's observation suggests that BS could achieve such LOIC because he had developed a relationship with the community prior to the commencement of the course. He was involved in an NGO called Arkom (*Arsitek Komunitas* – Community Architect), which had been engaged with Kampung Gowok community for some time. For BS, the LOIC explains the length of time needed to consolidate the clients and his team members.

The length of community engagement (LOCE) for BS is also longer than that of NAL. This was because BS continued to get involved with Arkom in addition to working on the ADS assignment even beyond that, since he further used the above case for his Bachelor Final Project. However, NAL and her team could get 5 clients, ranging from formal neighborhood leader (RW[1] Chief) to a businessman. This shows that NAL team was able to overcome their challenge of client finding.

[1] RW stands for *Rukun Warga* (Community Association).

2016 Cases

In 2016, Co-Design continued to be adopted in the ADS 7 assignment with the same delivery method, but students are not allowed to choose the same area as that of the previous year.

Table 4 shows two samples taken for analysis. The first sample is the work of HAU, which represents the case of formal *kampung* called Prawirotaman – well-established settlement in southern part of Yogyakarta that has been experiencing gentrification in the past 3 decades as a result of the tourism industry. The second sample is the work of YSA and his team, who chose to work on informal *kampung* called Ratmakan. This *kampung* is located in the center of Yogyakarta, on the Chode riverbank.

Based on the findings depicted in Table 4, we can see that HAU and her team took a longer duration in both finding NIC and LOCE compare to that of YSA. However, she obtained smaller NOC than YSA's. Author observation during teaching suggests that HAU and her team faced difficulties in finding the appropriate clients. The longer LOCE in her case was due to the team's struggle in finding what exactly the community wants to achieve for making Prawirotaman a better *kampung*. Furthermore, team's observation informs that despite the fast pace of tourism industry within this *kampung*, the community is getting more fragmented.

Table 4. Comparison of two cases (2016).

Student's name	NOC	LOIC (Weeks)	LOCE (Weeks)
HAU (Formal *Kampung*)	3	5	10
YSA (Informal *Kampung*)	6	4	8

Note: NOC = Number of Client; LOIC = Length of Initial Contact; LOCE = Length of Community Engagement.

The work of YSA and team on Kampung Ratmakan shows a shorter LOIC and LOCE than that of HAU, however, YSA and team gained more NOC. Author observation during teaching suggests that such achievement was due to the active role of the key player in Ratmakan development, i.e. the young and energetic Head of RT 07[2]. This person had opened access to the team, which led to the gaining of 6 NOC. The access also enabled the team to attend several community gatherings to discuss how *kampung* Ratmakan mitigate the flood disaster.

[2] RT stands for Rukun Tetangga (Neighborhood Association), the smallest level of governance in Indonesia. Association of several RT forms the RW (Rukun Warga – Community Association)

Fig. 1. Part of client profiling by YSA.

The ongoing studio in 2017

While the ADS course of 2017 batch is still running at the moment, the author sees it is pertinent to include the studio observation notes in this paper to enhance the analysis. The class shows a typical character of informal settlement chosen as the design target, i.e., the settlement is located on the riverbank, while the formal settlement is located in a mix of residential and commercial area. Students have decided to form 3 teams of 5 members. The composition of the team is depicted in Table 5.

Table 5. Composition of team and ongoing NOC.

Team Name	NOC	Formal kampung	Informal kampung
Jogoyudan	3		√
Ngampilan	3		√
Kaliwaru	2	√	

Note: NOC = Number of Clients.

Fig. 2. Hydrant installation – an existing example of government intervention to disaster mitigation at *kampung* Ngampilan.

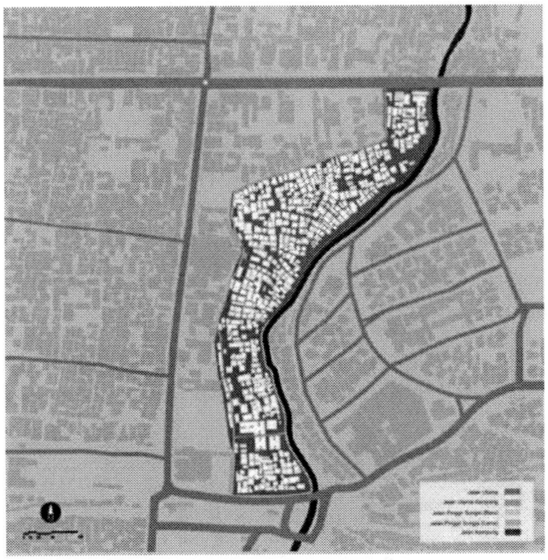

Fig. 3. A morphological map of informal *kampung* made by Jogoyudan team.

Table 5 shows the composition of the team and targeted *kampung*. The number of informal *kampungs* chosen, which exceeds the formal one, indicate that that informal *kampung* has been the preferred design target since the first delivery of Co-Design teaching in 2015. The average number of NOC gained in the case of informal *kampung* is higher than the formal one, despite its complex social and physical characters, presumably because the informal *kampung* tends to get more attention from the government and aid agencies through various improvement programs, which makes the community familiar with the process of action planning.

Conclusion

In response to the research questions stated earlier, the author argues that ideal Co-Design process is rather unrealistic to be applied in architectural design teaching. This is partly due to the limitation of teaching period, which in average last for 14 weeks, whereas time consumed for finding clients and engaging in the Co-Design process is unpredictable. For example, Table 5 shows that one of the design teams, i.e. Kaliwaru Team, could only gain 2 NOC within 7 weeks. NOC is an indication of the robustness of the Co-Design, the higher NOC involved in the design process will give more assurance in getting the right design decision.

Through discussions during design coaching, the author noted that introduction of Co-Design provides opportunities for students to directly observe the community needs. Students also learn how to select the reliable clients, who are not always formal leaders. The findings also suggest that previous experience in community engagement contribute to the gaining of shorter LOIC, which may give more room for the discussion on design process between students (designer) and clients. In other words, the shorter LOIC may lead to the longer LOCE.

Furthermore, the analysis also suggests that gaining the NOC in an informal settlement is relatively faster than that of formal settlement. The familiarity of public and private intervention in informal settlement seems to make access to the community easier than gaining NOC in the formal settlement.

This paper has proposed some new terminologies, i.e. LOIC (Length of Initial Contact), LOCE (Length of Community Engagement, and NOC (Number of Clients) to be used as indicators in determining the strength of Co-Design. Such terminologies may be further explored in the future research to see their relevance and reliability. The process of Co-Design is context-specific. This paper sees the term "Urban Co-Design" is apt to call the process of Co-Design within the field of Urban Design.

References

Albinsson, L, Lind, M. and Forsgren, O., (2007). Co-Design: an approach to border-crossing, network innovation, in Cunningham P. and Cunningham, M. (Eds) *Expanding the knowledge economy: issues, application, case study*. Amsterdam: IOS Press.

Barrie, J. and Edwards, M. (2013). *Co-Design -more than post-its and goodwill.* Retrieved from https://www.slideshare.net/DMA_Canberra/dma-presentation-13-1120.

Chisholm, J. (2017, 30 September). *What is Co-Design?* Retrieved from http://www.designforeurope.eu/what-co-design

Day, C. and Parnel, R. (2003). *Consensus Design.* Burlington, MA: Architectural Press.

Hagen, P. (2012). *Doing Co-Design.* Retrieved from https://www.slideshare.net/pennyhagen/hagen-rowlandcodesign-ux-australiaupload

Myers, T. (2012). *Co-Design for beginners.* Retrieved from https://www.slideshare.net/Optimal.Usability/co-design-for-beginners-13675913

Raharjo, W. (1993). *Towards an integrated model of architectural education in Indonesia.* Unpublished Master of Architecture dissertation. Halifax, NS: Technical University of Nova Scotia.

Rodriguez, M. and Ästrand, J. (1996). Organized small-scale self-help housing. *Building Issues, Volume* 8 Number 4, pp.-pp. 4-24

Schön, D.A. (1987). *Educating the reflective practitioner: toward a new design for teaching and learning in the professions.* San Francisco, CA: Jossey-Bass Publishers.

Using City Branding as the Part of the Intersection Point of Modernization and Tradition. Case Study: Surakarta City (Solo), Indonesia

Eko Nursanty[1,2,*], *Joesron* Alie Syahbana[2], and *Atik* Suprapti[2]

[1]University of 17 Agustus 1945 Semarang, Indonesia
[2]University of Dipenogoro, Indonesia

Abstract. To recognize a city is part of a city branding effort. By recognizing the dwelling, the city's inhabitants are able to create a spatial space in accordance with their easily recognizable cultural identity. In city branding, the memory of a city space is often the main ingredient in strengthening the identity. The efforts to use place making forces at this time are often used to solve the failures in multi-cultural spaces. Evolution in history is part of dynamic space planning as part of the identity of a city.The purpose of this research is to reveal the relationship between the formation of tradition spaces that are able to function in a modern way. The pattern of this relationship will produce spatial planning patterns with special characters as well. Humans as space users play a very big role as a space builder, where the decisions they make are often based on the tradition they have used for generations.The methods used in this study are qualitative, deductive using the theories that have been there before. Field research is used as empirical data in the form of interviews, photographs and field observations was conducted in various existing city spaces in Solo with the purpose to discover the existence of distinctive spatial patterns based on the typical behavior that has been passed down continuously as a result of the cultural process. The resulting discovery is a significant pattern of relationships between humans and their decision-making abilities in city space. This decision is based on self-awareness as part of the space they occupy. In a group, this awareness is part of the tradition's heritage that continues to develop in accordance with the human intellectual development.

Keywords: city branding, Surakarta, Solo, historical place, tradition place, urban heritage

Introduction

Major cities of the world are beginning to lose their ability to display their legal status at this time. Based on the existing laws, the cities no longer have special and separate powers. They follow the policies that have been determined by the government. Thus, the city managers attempt to translate the things that are still "unclear" from the delegated power (Frug, 1980).

The city becomes a productive region with its economic growth. It is not only a production center but also a distribution center and a collection center for local products. Furthermore, Frug's next research states that the history of city's growth illustrates how the city's freedom of opportunity develops itself through creative effort are the way to generate sustainable economic growth (Frug, 1999).

The world today has become a market system, where one city and another become a place to offer and sell its potential. Global growth means that every country, every city and every region must compete with one another and divide the world's consumers consisting of tourists, investors, students, entrepreneur, etc., with other countries that are also doing the same thing (Kotler, et al., 2010). The competitive ability of cities can be defined through several factors: the existing policies, the management institutions and the various strategies also processes that have been done. This will determine the level of sustainable productivity in a city, including economic, environmental and social issues (Eide, 2014).

The competitive ability of cities can be defined through several factors: the existing policies, the management institutions and the various strategies also processes that have been done. This will determine the level of sustainable productivity in a city, including economic, environmental and social issues (Anholt, 2007).

Because of the previous reasons, government and society need to find out what becomes the perception or picture of the world about their country, and then create a strategy to explore it. One of the keys to build a city's reputation is by describing itself openly, honestly, strongly, attractively, economically, socially and politically, also reflecting the spirit of local genius based on the aspirations of its people. The process and strategy to generate this positive reputation is by Branding effort in a city.

The quality of a brand in marketing is adding values to products and services in the invisible value form in order to win the existing market competition. It requires effort in extracting the values, which sometimes displays basic capabilities of the owner. It is easier to sell well-known products, rather than selling wholesale commodities or

* Corresponding author: santy@untagsmg.ac.id

general goods. Because people in general have understood that goods with certain quality are considered more valuable than those who do not or do not have a Brand (Anholt, 2005).

Still according to Anholt (2005), in the history of the world economy, many occurrences of a small country that looks unpromising, managed to become very rich. Meanwhile the country with a wide-scale and fertile land is not able to improve the ability to become a city that can improve its quality. Those are not able to improve the standards and qualifications of living for the inhabitants of the city. City space which owned in large quantities is unable to produce the quality of life that is dreamed by its citizens.

Various branding efforts today only become an element of the value of Brand made based on the wishes of the government. In some countries, city branding is only a universally applicable symbol. Brand should emphasize the invisible added value (intangible) on the product. In which this Brand is an effort of adding value so it will be higher or as an effort to ensure that the strength of Brand will last as long as the product exists.

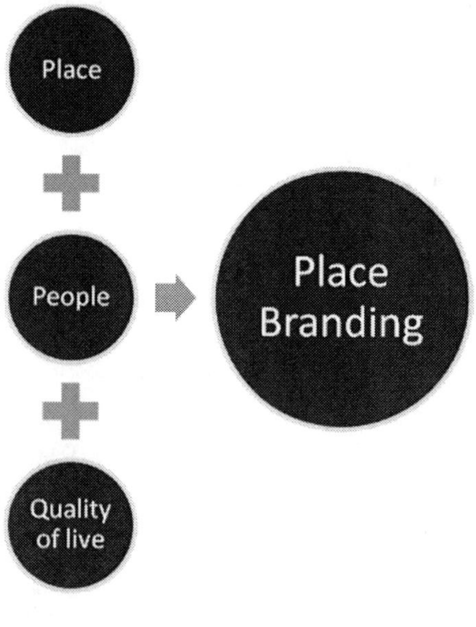

Fig. 1. Human and Qualified Space in Place Branding.

In Figure 1, it is shown that specifically in urban space; the most important element is the space user, in this case it is human who performs their activities in the urban spaces. The use of city space based on the concept of harmonization generates a good life quality; it can also give a good brand image in a city.

Meanwhile in the urban areas, which concerns the brand in the space and city is not only about communication but also about the policies and applications in it. The effort of this brand is done in relation that space or place also has the image as a product. These places depend heavily on the other things outside of itself as the power to create or improve those images for their progress and prosperity (Anholt, 2010).

Regional/spatial brands have several approaches in which enable increasing the speed, efficiency and effectiveness by using better imagery (Kotler & Haider, 1993).

City Competitive Advantage

In the general marketing world, there are 5 strategies that can be done to win the competition in the development effort of a company, namely: **First**, A low-cost provider strategy appeals, it is a strategy from the consumer's perspective to choose low prices products or services. On the company, this is a strategy to find the steps to run a business at a low cost. **Second**, A broad Differentiation, it is to make differentiation efforts of product against products offered by the competitor company which is a step from the buyer's point of view, thus the level of difficulty for the opposing company to do the duplicate process needs to be sought. **Third**, A best cost provider strategy, it is a strategy given to the consumers by providing more value for the amount of money they have paid. The goal is to have a cheaper price and compete with opponent companies, but offering products that have superiority over the others. **Fourth**, A focused strategy which is based on the low cost concentration as a target segment of buyers and come out from the existing competition by serving buyers with certain skills that are lower than buyers from competing companies. **Fifth**, A focused (or market niche) strategy that is based on differentiating concentration efforts on the target buyer and come out of the opponent's competition by offering the consumer to choose their own necessities (Drucker, 1999). It is believed to be able to meet the tastes and needs of consumers than similar products produced by the opposite party (Swain, 2010). In a diagram, the five elements to win the competition in the business are illustrated in figure 2 below.

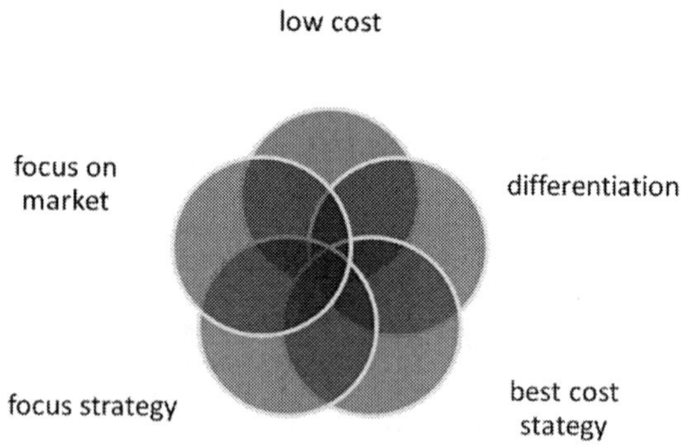

Fig. 2. Strategy of Winning the Competition in the Business. Source: Kotler, processed by researcher, 2017.

Some of the above elements can also be applied for the development of the city's ability to compete with other cities around it, namely the element of Brand ability and human excellence in order to create a unique urban space.

Essentially, the planning cycle is nothing more than a logic activity that directs marketing planning and applies it regularly on marketing objects, including the city (Braun, 2008).

City branding is a new concept introduced by some marketing figures as an attempt to win a competition between countries and cities in the current global era.

They argue that the struggle to get an attention for a brand is not limited to commercial goods and services, but applies equally to geo-political entities. Cities across the world are now competing to get income from various businesses, tourism and often tax sectors; sometimes this competition happens between one city and another. The supporters of City Branding's theory argue that a highly competitive environment becomes important for a space. It does not depend on its composition and size, but also to clarify the differences between them and convey its superiority. This is what they need to convey, which are the hidden values that are often called space branding (Anholt, 2005).

The implementation of the brand theory to the developing space on regional concepts such as country, district and city is starting to be considered important. It is now being realized that every city competes with other places to compete for markets, experts, investors, tourists, etc. The brand phenomenon in its initial space is an organic process of an organic process of a picture communication without strategy. City Branding is any activity aimed at transforming the city from an ordinary location into a destination. A successful Brand can transform the city into a place where people want to live, work and pay a visit to (Salman, 2008).

City Branding is very different from urban marketing. Urban marketing is a marketing theory that focuses on the consumer desires as the basic principles in conducting the operational process of an organization. Whereas in City branding, there are concerns about the chosen vision, mission and identity which play role in it. City Branding refers to the application of Branding techniques to geographic locations in a broad sense. It involves reputation, credibility, political power, security, comfort, etc. (Dinnie, 2008).

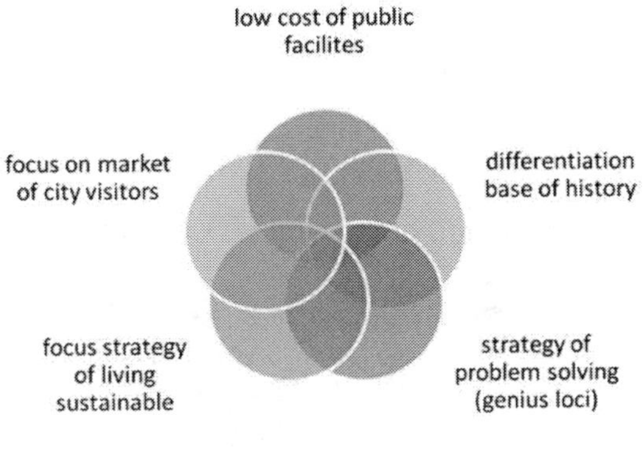

Fig. 3. Elements in Winning the City Competition.

Figure 3 illustrates the five elements to win the competition, which can also be done for a city. Although the city is not as production goods, but as

Thus, City Branding creates a picture of the city and more extensively on all that it offers and the interactions that occurs. From the consumer's point of view, a unique image is created in the city through all levels of interaction. City Branding is a valuable promise, a promise that keeps being maintained.

Methods

A brand, to be able to connect with human traits needs to attempt to build authentic DNA as a major marker that distinguishes it from others. This DNA will reflect the identity of a brand essential in the social network of its user community. Those brands that already have DNA will be able to build their sustainability character. The uniqueness that sets them apart from other brands has been understood by the consumers (Kotler, et al., 2010).

In the diagram, the research stages can be seen in the following Figure 4. Case Study in Solo City was observed starting from the human user space and the space itself. Through intense interaction for a long time, both of these elements produce a tradition space, modern and contemporary space.

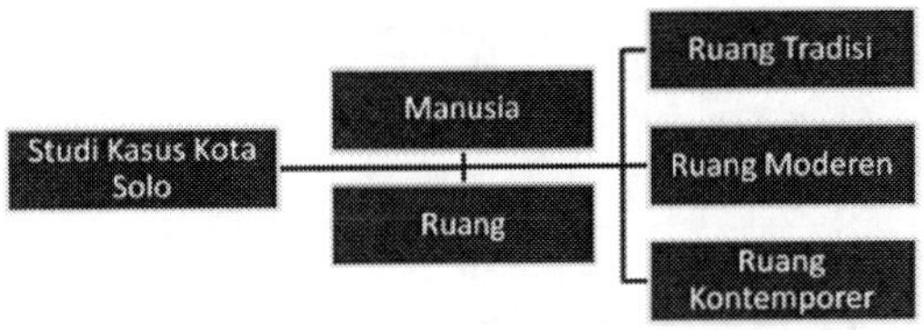

Fig. 4. Research Stages.

Furthermore, marketing researchers, especially for tourist destinations, need to understand the diversity of types and values contained in a place. This understanding will easily lead to an understanding of the essential character of each space, also known as space DNA. The authenticity of the brand space starts from the human. Learning the true identity can only be done through an understanding of the local people (Cuypers, 2016).

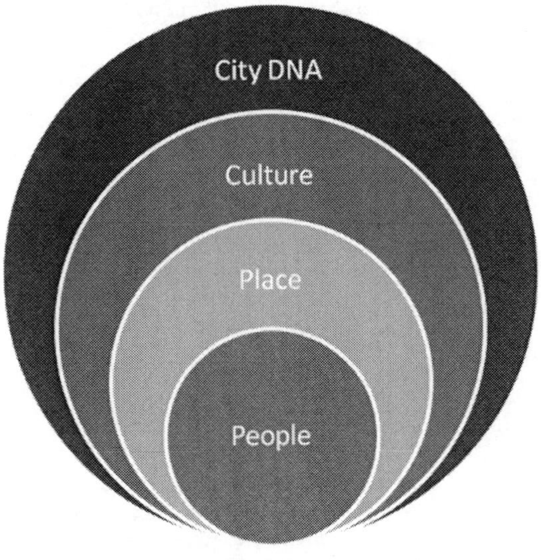

Fig. 5. Research Element Structure.

Figure 5 above describes the space DNA defined as the essential character. In the marketing of a city, it is often referred to as authenticity. A space is something tangible or real. It is a product which has consumptive meaning. Space

https://doi.org/10.1051/shsconf/20184104001

brings all the needs of space for space consumers, such as tourists, visitors, citizens, investors and scientists, both students and researchers. In addition of having the ability to meet the needs of tangible or real space, a space also has an intangible or abstract value. For example, Paris is known as: city of light; city of mode; city of love. This is a way of describing the essence of DNA through the physical environment of a city.

At this time, many government of the city are beginning to find the importance of creating a brand for the place where they live. Citizens make a serious effort to create local culture in their city by owning and operating their local businesses. It can describe the personality of the said city (Hayden, 1997).

Both of community groups and decision makers protect valuable aspects of their city, such as historic areas, activities that have taken place for generations, and make their city even more attractive as a place to stay. The city's wealth and uniqueness are sources of inspiration from brand strategy that can illustrate a challenge or complexity. The city is expected to be able to use the specific factors combined into a distinctive framework, personality, history, values and composition of settlements as well as community's wealth (Florek, Insch, & Gnoth, 2006).

City managers must understand the motives, expectations and needs both of their existing and potential residents; also ensuring their needs are fulfilled. A variety of personality and situational factors can influence the quality evaluation of the offered place and decrease the value of the place itself. The biggest challenge of City Branding is communicating different stakeholders into something united. Several cities in Indonesia have tried to capture the complexity that they have in a brand promise but failed. The effectiveness of city branding depends on the support and commitment of local elements such as communities, local businesses, government, and community groups. At the same time, City Branding must also be able to show more value to potential residents about the identity of the city.

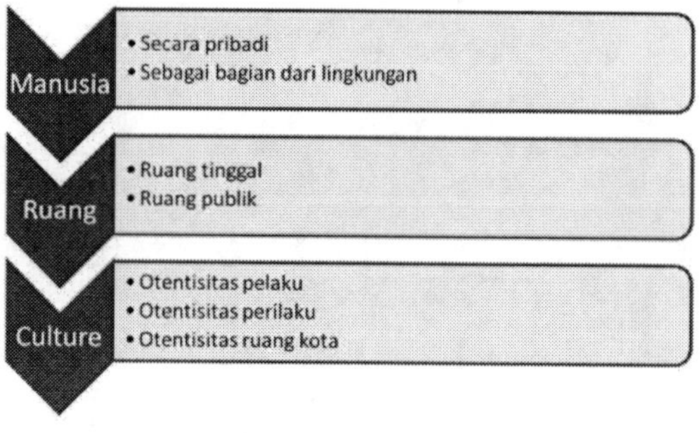

Fig. 6. Structure of Research Findings.

Figure 6 shows the structure of research finding. Human always behaves to maintain balance his daily life. Humans as individuals have the norms of developing themselves and their potential. Humans also structurally feel they are part of another human being around the community in their society and activity. On the outside of themselves and their society, humans also always keep harmony with the natural surroundings and believe they are part of the entire universe.

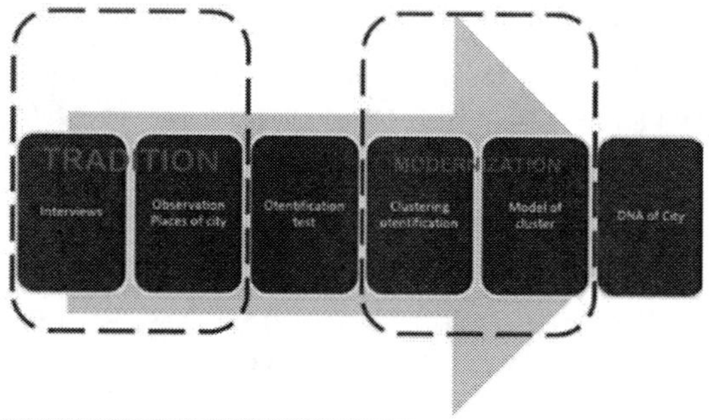

Fig. 7. Analysis Process of Research Findings.

In Figure 7 is the structure of research analysis which tradition process has existed since hundreds of years ago and developed into the modern era. The said process went through the existed process of authenticity. So that there are stages of change in accordance to the high and low level of authenticity owned.

Result

Surakarta (also called Solo or Sala) is a city located in the province of Central Java, Indonesia with 545,653 inhabitants and population of 12,390 / km^2.

This city has 44 km^2 areas which adjacent to Karanganyar and Boyolali regencies in the north, Karanganyar and Sukoharjo regencies in the east and west, and Sukoharjo regency in the south. The east side of the city is passed by Bengawan Solo river. Together with Yogyakarta, Surakarta is the heir of the Mataram Kingdom which was broken down in 1755.

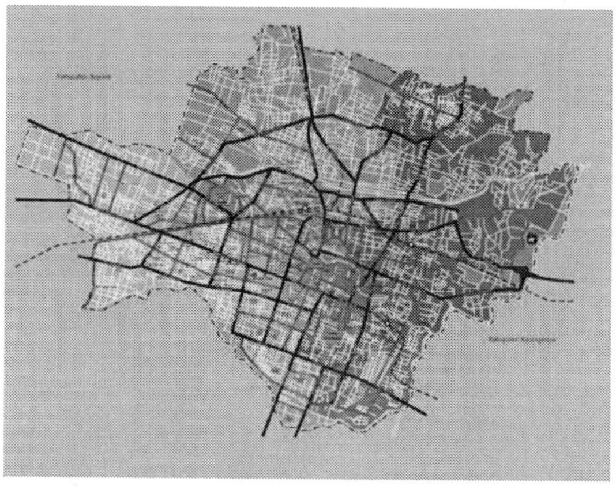

Fig. 8. Map of Surakarta City (Solo).
Source: Sumber: Solo Digital Map, 2010.

The current area of Surakarta city (Solo) can be described in the picture above, which consists of 5 districts and 51 sub-districts, as shown in Figure 8. Based on the following table in Figure 9, it is seen that the most crowded sub-district is Serengan Sub-district followed by Pasar Kliwon Sub-district, and Laweyan Sub-district as 3 of most crowded sub-districts among 5 districts in Kartasura. The location of this research was conducted in the 3 densely populated districts because they were the original regions of Surakarta (Solo).

Kecamatan	Luas Wilayah Km2	Jumlah Penduduk / Number of Population			Rasio Jenis Kelamin/	Tingkat Kepadatan Penduduk
District	Area Km2	Laki-laki / Male	Perempuan / Female	Jumlah / Total	Sex Ratio (%)	Population Density (Jiwa/Km2)
(1)	(2)	(3)	(4)	(5)	(6)	(7)
Laweyan	8,64	53 457	55 807	109 264	95,79	12 646
Serengan	3,19	29 981	31 198	61 179	96,13	19 178
Pasar Kliwon	4,82	44 646	46 576	91 222	95,86	18 926
Jebres	12,58	73 799	74 643	148 442	98,86	11 800
Banjarsari	14,81	86 395	88 984	175 379	97,09	11 842
Kota	44,04	288 278	297 208	585 486	97,00	13 294

Fig. 9. Statistics Table of Density of Solo City Area.
Source: Department of Population of Surakarta, 2014.

Surakarta and its satellite cities (Kartasura, Solo Baru, Palur, Colomadu, Baki, Ngemplak) are mutually integrated areas called Solo Raya. Solo Raya is unique because the area of Surakarta city is only 44 Km^2 and surrounded by its buffer cities, each of which is approximately half the size of Surakarta city and directly adjacent to form a unified central city area. The total area of Surakarta City and its buffer areas are currently about 150 km^2 with population about 1 million people.

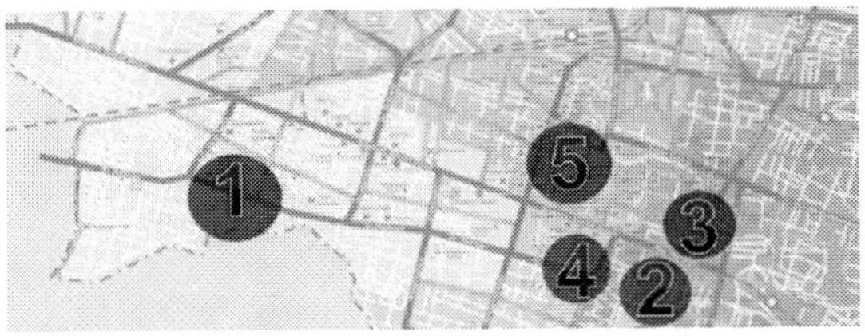

Fig. 10. Elements of Solo Heritage Room.

In figure 10 above, there are about the 5 elements of heritage space in Solo city found based on the preferences of tourists, namely: (1). Kampung Batik Laweyan; (2). Mangkunegaran Palace; (3). Kampung Batik Kauman; (4) Klewer Traditional Market and (5). Mangkunegaran Palace.

Based on the preferences of tourists, two groups of dominant space zones were also found as follows: Zone A is a heritage zone and zone B is a zone that is considered to represent modern symbols.

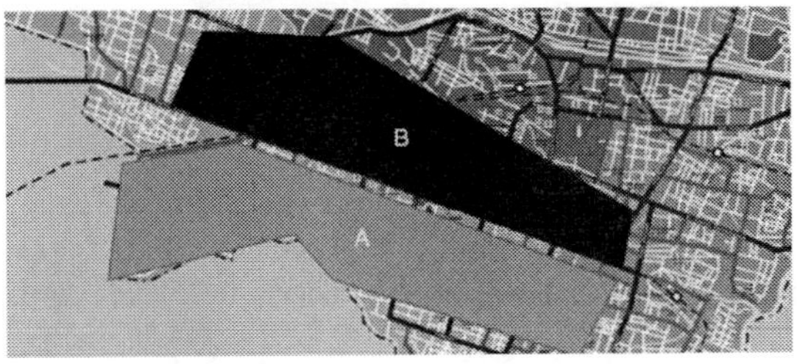

Fig. 11. Two magnetic symbols of the Space of Solo City.

Discussion

The centers of kingdoms in Java could only grow because they had favorable geographical factors. These geographical factors enable them to develop economic base of agrarian and trading, which in turn conquered other areas. The capital of the ancient kingdom of Java is always based on the concept of macrocosm and microcosm (Handinoto, 2015).

If cultural symbols get farther away from the center, those symbols will be weaker. However, there is often a counter-culture that emerges in the midst of subjects, which comes from their collective unconscious mentalities. Mangkunegara's position, which was closer to the axis, made using access to the Vastenberg Fortress, produced several recognizable symbols closer to the European symbol. These can be seen by the existence of Mangkunegara military tradition (Kuntowijoyo, 2016).

The following Figure 12 illustrates the structure of findings of space authenticity. In which there are elements of the efforts of the city to do city branding to its city as follows: **Modern spaces** consist of economic space. Contemporary spaces consist of public spaces. Contemporary space has a modern form but accommodate the activities of tradition which firmly rooted by the community. Meanwhile **settlement** is a part of space that is considered able to describe the integrity of the authenticity of space, be it through tangible or intangible.

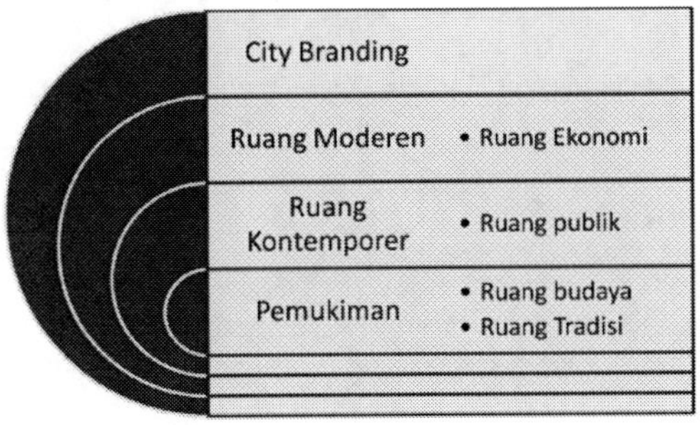

Fig. 12. Struktur Temuan Otentisitas Fungsi Ruang Heritage Kota Solo.

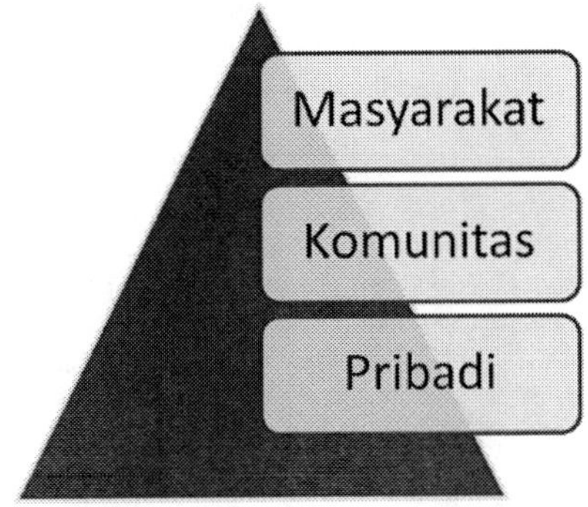

Fig. 13. Findings Structure of the Authenticity of Solo City Heritage Room Function.

Figure 13 illustrates the findings structure of the heritage city space users based on interviews with tourists. Tourists who travel freely and privately occupy the highest number in the pyramid. The second position of most tourists is those who come in groups of the community. They generally have an interest in similar travel types and city space. Meanwhile

 https://doi.org/10.1051/shsconf/20184104001

the public chose the tourist space on the amount and location of more specific types and amounts.

The following Figure 14 illustrates the zonation of the previous Figure 13. The findings structure of urban space users with self-contained tourism character occupies the largest urban space in zone C. Meanwhile the next urban space user is the community who occupies zone B in Figure 14. The last space user in which the user group is the local community, the preferred tourist space is the zone A.

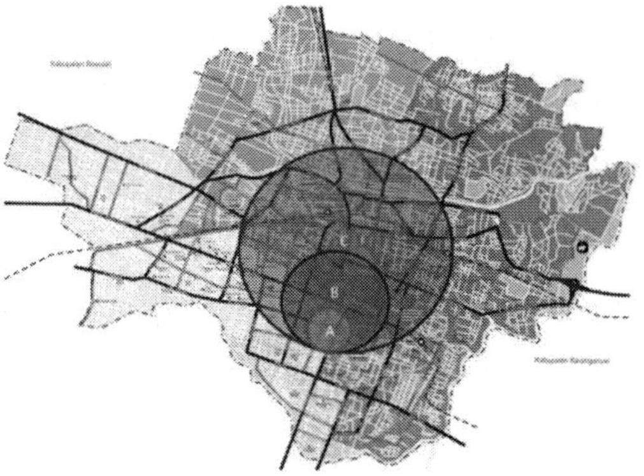

Fig. 14. Structure of Hierarchy Authenticity Public Space of Solo Heritage.

Conclusion

The encounter between heritage city space and modern urban space always generates a unique combination for a city. The uniqueness of the composition is analogous to the DNA of the city. Recognizing the city's DNA is able to provide a clearer understanding of the uniqueness of the city. In the end, the uniqueness of a city will give strengthening branding efforts to the city.

A city which has been identified by the DNA of its city, has grown and transformed into a modern city but has never lost its form and identity of both space users and the urban spaces in which people are located and interacting with each other.

The composition of all city elements in varying degrees of created authenticity creates a force for the city to be able to compete with the other cities around it. It is able to survive the stages of authenticity it has. Although the city changes according to the needs of the development of the era, the parts of the city still have the value of authenticity structurally that forms the ability to compete for the said city.

References

Anholt, S., 2005. *Brand New Justice How branding places and products can help the developing world.* Burlington: Elsevier Butterworth-Heinemann.

Anholt, S., 2007. *Competitive Identity. The New Brand Management for Nations, Cities and Regions.* New York: Palgrace Macmillan.

Anholt, S., 2010. *Places Identity, Image and Reputation.* London: Palgrace Macmillan.

Braun, E., 2008. *City Marketing Towards an Integrated Approach.* Rotterdam: Erasmus University.

Dinnie, K., 2008. *Nation Branding. Concepts, Issues and Practice.* Burlington: Elsevier.

Drucker, P., 1999. *Management Challange.* Burlington: Elsevier.

Eide, E. B., 2014. *A report of the Global Agenda Council on Competitiveness, The Competitiveness of Cities.* Swiss: World Economic Forum.

Frug, G. E., 1980. The City as a Legal Concept. *Harvard Law Review,* 93(6), pp. 1057-1154.

Frug, G. E., 1999. *City Making. Building communities without building walls.* Jersey, United Kingdom: Princeton University Press.

The Built Environment of Kampung Arab Sugihwaras Pekalongan

Astari Wulandari[1,*], *Loekman* Mohammadi[1], and *Anwar*[1]

[1]Architecture Study Program, University of 17 Agustus 1945 Semarang, Indonesia

Abstract. Since the arrival of Arabian ethnic to Sugihwaras, they settled and built their own neighborhood identity. The built environment is an example of a system on which human beings seek to express their own identity as a collective group. Every built environment consists of distinctive features that relate to social, cultural, economic and environmental parameters inherent in that particular built-environment (Al-Hokail, 1995). Based on that background, the research question of this study is "what is the distinctiveness of built environment in Kampung Arab Sugihwaras Pekalongan?". The purpose of this article is to identify and explain the built environment in Kampung Arab Sugihwaras by its social and physical parameters. Using both qualitative and quantitative approach, this article examines the social parameters by its four social scales: neighborhood, family and individual, the physicalparameters by its urban fabrics. The result shows that community who lives in Kampung Arab Sugihwaras especially the Arabian descendant preserving Islamic value in their social lives. But, modern urban culture faded some social rules away. While the Kampung Arab Sugihwaras urban fabric shows that neighborhood had grid pattern with two inner main streets and buildings withIndische Architecture style in that corridors. The further result shows the street hierarchy by overlapping the social and physical parameters analysis.

Keywords: Kampung Arab Sugihwaras, built environment, social, and physical

Introduction

The trade route between Hadhramaut and Nusantara has been established since the 7th century shortly after the entry of Islam (Arnold, 1935). Within the next period, the Hadhramaut's came gradually until reached its peak in 18 century. Based on statistic record in 1985, there was six big colony of Arabian ethnic in Java. They settled in Batavia, Tegal, Pekalongan, Semarang, Surabaya, and Madura (Van den Berg, 1988).

Pekalongan became a destination since it well known as a paddy and spices producer (Yudhohusodo dkk, 1991). There were ancient port in Pekalongan, Bandar and Doro ports which more reputable than Pekalongan itself. The existence of that port in the past is corroborated by a note in 18 century.

It stated that a Chinese group (which call *Ma Huan*) who stayed over in Kampung Sampangan which located in near Kupang River springhead. That location became a place where the loading boat tethered from the port (Asa,, 2009).

Islam came to Pekalongan through three phase. First phase happened in 14 to 15 century which related to the Chinese and Arabian populations growth. Second phase happened in 15 to16 century when Islam brought by *Wali Songo*. The third phase, Islam was come along with Hadhramaut *Ulama'* and local *Ulama'* who known as *Alawiyyin Ulama'* (Asa, 2009).

Sugihwaras is a district where Arabian ethnic inhabitant mostly found in Pekalongan. The Arabian ethnic who brought their own culture, tradition, and religion influences the housing layout and built environment. Every built-environment consists of distinctive features that relates to social, cultural, economic and environmental parameters inherent in that particular built-environment (Al-Hokail, 1995).

The Problem

The buildings with Indisch Empire architecture style still found in Kampung Arab Sugihwaras, especially in Surabaya street. Nowadays, the city development and economic growth influence the built environment in Kampung Arab Sugihwaras directly. Most of inhabitant are enterpreneur who run family bussines. Therefore, there was a change in their houses. Both on the layout and building façade. According to the Buchler (1979), the built environment reflect who we are and what our stand should be towards the natural world. Built environment is the focal point of a network of commitments and opportunities—a way of life—that leave the present and future generations indebted. Based on its importance and the background of *Kampung Arab Sugihwaras*, analyzing built environment by its social and physical parameters is necessary.

* Corresponding author: ast_wulandari@yahoo.com

Social Principles in Traditional Islamic Built Environment

For Muslim, Islamic *sha'ria's* should practice in their lives as a commitments and responsibilities. Mortada (2003) analyzing the traditional Islamic built environment by its social framework and classified into four scales: society, neighborhood, family, and individual.

In Islam, social solidarity (*'ummah)* is the foundation to build the strong community. As proposed in Mortada's social framework, it evolved to the social interaction and social justice. Muslims as a big community should notice these principles in their daily life to represent their commitment and faith in Islam. They also suggest avoiding any communal action which unties the social solidarity between them.

Neighborliness have important rule in community as stated by Mortada (2003),

"Islam exceptionally looks at neighborliness as the backbone of 'ummah. This is exhibited in the provision of a set of ethics which themselves are obligations that enhance and control neighbourliness. These ethics are found in two main principles: strong neighborly relationships and the preservation of neighbors' rights."

The importance of neighborliness narrated in Al-Qur'an as well. It stated to do good to parents, kinsfolk, orphans, those in need, neighbors who are near, neighbors who are far, the companion by your side, the wayfarer (ye meet), and that your right hands possess' (Qur'an, Su. 4:36). Based on that rule, Muslims are required to maintain the good relationship with the neighbors.

At the family scale, the social principles could be identified by the strength of family ties and the existence of extended family. Family became an important foundation in Islam society especially its role to build entire socio-cultural structure. Moreover, a family is a self-sustaining institution which ensures ideological and cultural stability over the entire spectrum of society on the one hand, and in time, past, present and future on the other (Mortada, 2003).

An Individual represents the Islam society where every Muslims have to show the finest quality of behavior and against individualistic lifestyle (Levy, 1938). Muslim individual should sustain an amicable relationship with his fellow Muslims and be humble in his behavior and lifestyle and moderate in spending (Mortada, 2003).

Physical Principles in Traditional Islamic Built Environment

The application of Islamic value in built environment and Islam cities reflected on particular principles which heading to elevate the Islamic social framework. Both, social and physical principles build upon Al-Qur'an and *as-sunnah* while another value and social norms growth in Muslims society. The influence of social principles Islamic to the built environment is clearly state by Mortada (2003),

"From the birth of Islam in the seventh century until the end of the nineteenth century Islamic law was a major force shaping that environment, providing stability and continuity in its urban planning and design"

Based on those phenomena, Mortada (2003) analyzing the physical principles by its scale: natural environment, the urban environment and the house.

There is a limitation in this article, particularly in analyzing scale of physical princ0iples. In this article, to analyze the physical principles of built environment limited on the neighborhood and the house scale.

Built Environment

The variables which related to the Muslim built environment are the Muslim house form that adjacent and clustered (Al-Shishtawe, 1982), street hierarchical function that accommodate the needs (Al-Bukhari, 1960), spatial hierarchy of open space that changes its size character and function (Mortada, 2003), and the mosque existence based on its hierarchy and function (Mortada, 2003).

The House

By Al-Quran and *As-Sunnah,* Islam delivers a systematic review that indicates the principles to build or to develop a house. The Muslim houses represent the family member's faith of Islam and its value. It comprise the privacy setting, the straight barrier between personal and social live inside the house (Patherbridge, 1987), the inner open space and room organizing (Mortada, 2003).

Hablumminallah, Hablumminannas, and Hablumminal'alamien Principles

The application of Islam principles in architecture also mention by Munichy (2010). As a Muslim architect, the application of Islam principles in architecture should be ensure the relation between God, Human, and Environment. *Hablumminallah* describe the relation between human and God. *Hablumminannas* is the relation between human and another human (the social life). *Hablumminal'alamien* describes the relation between human and the environmental. That concept will produce the Islamic architecture concept which stands on Al-Quran and *Hadith.* This concept deliver five important variables, function, form, technique, safety, and comfortably which all of them have consider the contextually and efficiency.

Methodology

This article use both quantitative and qualitative approach. The quantitative method is used to analyze the social principles in Kampung Arab Sugihwaras by its variable. The qualitative approach is used to describe and analyze physical principles which related to the built environment in Kampung Arab Sugihwaras.

The random sampling technique is used to analyze the social characteristic by using the above theory as a framework. The informants are the inhabitant of Kampung Arab Sugihwaras. While to analyze the physical principles, observation method is used to identify the physical character.

The Beginning of Sugihwaras

The arrival of Arab community from Hadhramaut in Pekalongan dominated by Sayyid, a called for the highest society who capable in religious knowledge and had a settle economic condition Van den Berg (1989) and Steenbrink (1984). They married to the headman daughter and became the core of big colony in Sugihwaras.

Habib Hussein Al-Attas was the first one who came to Sugihwaras in year 1800. At that time Sugihwaras was a jungle. He built a mosque as a place to pray and a place to introduce Islam as well. That act represented what *Rasullullah* did when *hijjrah*. The mosque became an embrio to the social interaction in *Sugihwaras* (Astuti, 2002). Nowadays, that mosque is well known as Wakaf Mosque which located in Surabaya Street.

The name *Sugihwaras* developed in colonial era. *Sugihwaras* is consists of two words in Javanese, *sugih* and *waras*. History of *Sugihwaras* related to the existence of pawnshop and Doctor Agus Mulyadi house, at the past. The pawnshop had important role, supporting habitant financial fund. Most of habitant was a batik and sarong trader and they avoid the interest-bearing loans from the bank to expand their business as obedience to their faith in Islam. Furthermore, the colonial bank would be confiscating the habitant asset if they could not afford to pay the loan and interest. Meanwhile, dr. Agus Mulyadi was well known as a generous and his humanity. The sick habitants prefer went to the doctor's house than to the hospital which run by Dutch Government (Astuti, 2002).

Based on that background, the history of Sugihwaras describes the convenience of funding so that the habitants became wealthy (*sugih*). While the generosity and humanity from the medical doctor support the people's health (*waras*).

Social Life of Kampung Arab Sugihwaras

To analyze the social condition in Kampung Arab Sugihwaras, in this article is classified in society, neighborhood and family.

Society

The dominance of Islam in Arab Sugihwaras makes Islamic culture deeply rooted in society. All of respondent was agree that charity (*sedekah*) as a habit, hospitality to guest, and good deeds to others reflected the strong social interaction. It state on Al-Qur'an.

And serve Allah. Ascribe no thing as partner unto Him. (Show) kindness unto parents, and unto near kindred, and orphans, and the needy, and unto the neighbor who is of kin (unto you) and the neighbor who is not of kin, and the fellow-traveller and the wayfarer and (the slaves) whom your right hands possess. Lo! Allah loveth not such as are proud and boastful (QS 4:36)

One of the community act that show the strong social interaction in Sugihwaras is a charity to the poor which organized by the locals organization and the toddler as well by distributing the formula milk and foods. Related to charity, all of respondent understand that charity is required to every Muslim which not limited to the money and materials. Abu Hurairah (ra) reported that the Messenger of Allah (sas) said

"Every joint of a person must perform a charity each day that the sun rises: to judge justly between two people is a charity. To help a man with his mount, lifting him onto it or hoisting up his belongings onto it, is a charity. And the good word is a charity. And every step that you take towards the prayer is a charity, and removing a harmful object from the road is a charity." (Al-Bukhari and Muslim)

Neighborhood

Islam obligates to enhance and control neighborliness. Prophet Muhammad said that,

"The best of companions with Allah is the one who is best to his companion, and the best of neighbors with Him is the one who is best to his neighbor. He also said, "The one who harms his neighbor will not enter Paradise." (Al-Bukhari and Muslim)

Neighborhood life reflected the people's social condition by its norms and another intangible value which indirectly take control. In Kampung Arab Sugihwaras 81% respondent had relatives who lives around Sugihwaras district. That phenomenon shows likely to strengthen neighborliness ties. Occasionally they visit the neighbor around to greet or deliver the food. They believe the good deeds not only limited to the next door neighbor, but also extended to those further away (Mortada, 2003).

Related to the preservation of neighbors' rights, the habitants of Kampung Arab Sugihwaras showed when build a building. An individual who plan to build usually ask permission to the neighbor next door and the neighbors whose wall coincide with the building to be built. The pictures below show the building height condition that most of buildings in Kampung Arab Sugihwaras have similar height.

Fig. 1. The Similar Building Height.

The similar building heights in Kampung Arab Sugih Waras represent the equality. There is no building which height is block the air circulation and the sunlight to another building. Like The Prophet also said (Karim, 1938 in Mortada, 2003),

'Do not block his (the neighbor's) air by raising your building higher without permission. Harass him not. Give him a share when you buy fruit or at least do not throw the peelings outside the door so offending your neighbor.'

Family

Family had an important role as a foundation of social-cultural structure in Islam to the past, present, and the future.

https://doi.org/10.1051/shsconf/20184104002

The family bond in Sugihwaras identified by its strong family ties and the big family concept.

In the tradition of Arab descendants, in determining the spouse is preferred to choose a person who is a relative or a distant relative to avoid marriage with someone who is not of Arab descent. This picture bellow showed the marriage scheme in the highest social class (*Sayyid*) in Pekalongan.

But now, there is a social transformation in spouses determining in Arab community in Sugihwaras. The analysis found 17 % respondent who agreed to married to their relatives.

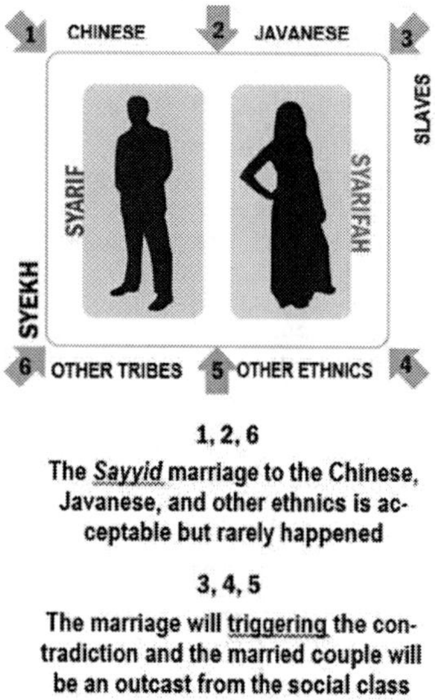

Fig. 2. The Sayyid Marriage Scheme (Redraw from Astuti, 2002).

In Islam city, mostly found a house where three or four generation lives. Ahmed (1987) stated that extended family consists of relations emerging from consanguinity, affinity and foster-nursing. On the contrary, now in Sugihwaras, 61% respondent thought the extended family is not a common thing.

Sugihwaras Built Environment

Sugihwaras built environment characteristic will be explained by two scales; neighborhood scale by its pattern, circulation, and open space; house scale by its privacy and spatial order. In addition, in this chapter will explain the role of Wakaf Mosque to Sugihwaras as well.

Neighborhood

The Pattern

The neighborhood physical pattern would illustrate the conformity between physical and social space. Based on the figure ground, Sugihwaras tend to be homogeneus.

Kampung Arab Sugihwaras have a gridiron pattern where two main roads, Surabaya Street and Semarang Street, split this district. Most of the building oriented to the street. In both corridors, the atmosphere of the area is pretty quiet, not many motor vehicles passing by. Most of the users of the vehicle are the people living in the vicinity. In the area of life can be captured the impression that the area is less "alive" because the activity of most people of Arab descent reside in the house.

Both corridors are formed from ranks of road-oriented building. Most of the buildings on both corridors have an *Indische Empire* architectural style. Even there are found some buildings that change on its facade. On the next page shows the corridors of Surabaya and Semarang which are dominated the homes of Arab descendants.

Circulation

Based on classification there is local roads, neighborhood roads and alleyways. The road network classified to the local road is Sultan Agung Street, Bandung Street, Patiunus Street and Agus Salim Street. The main network is a road with asphalt pavement with a width of about 4 - 5 meters. While the road network that classified to neighborhood roads is the corridor of Semarang Street and Surabaya Street in the form of asphalt road with width 3.5 meters. For alley and dead end roads are scattered throughout the region. In general, the alley in the area has a width of less than 2.5 meters and most can only be passed by bikers, cyclists, and pedestrians. Based on observation, the entire street in Sugihwaras had two functions, supporting the accessibility and becoming the social space, even though the social interaction intensity mostly found in alley around Sugihwaras district.

Open Space

There is no open public space found in Sugihwaras. Therefore to celebrate the important event (celebrating the Independence Day, or religious event) the street is used. Surabaya Street is the place where the important event (neighborhood event, local event, or regional event) was held. While the alley limited to the neighborhood event.

The House

The similarities of architectural forms of the Arab descendants of Sugihwaras begin by the three rich Arab families, the Argubi, Yahya, and bin Shihhab families. At that time home rental businesses was a promising business and deliver high profits. The three rich families built many houses in Kampung Arab Sugihwaras, in addition to rent also for the habitable. The similarities in architecture style, colonial style, indicated that house developing at past duplicated the architectural recent styles (Astuti, 2002). Van den Berg (1989) stated the same idea. The Arab community who migrated to Pekalongan was a rich tradesman. The implication of their wealth, their houses built by adapted the recent architecture style.

Fig. 3. The Pattern of Kampung Arab Sugihwaras.

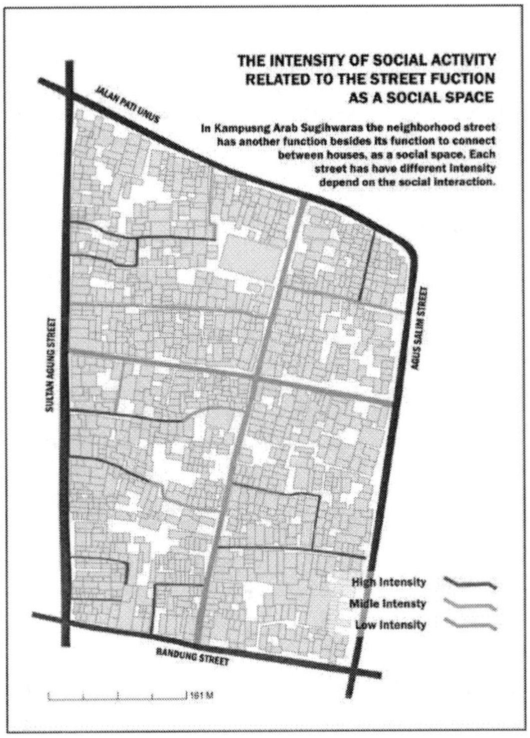

Fig. 4. The Intensity of Social Activity Related to the Street Fuction as a Social Space.

Fig. 5. Influence of Indische Empire Style.

Spatial Organization

The spatial organization in the homes of Arab descendants is based more on social and personal interests, rather than on activities that take place in the home. For that the spatial order inside the house is based on the concept of hijab. The concept of hijab is not only related to social and personal life, but the concept also relates to the personal lives of each resident (Astuti, 2002).

That concept manifestation seen on the spatial zoning based on its character, private or public. Every house in Kampung Arab Sugihwaras had separated guest room and family room where usually devided by another door or partition. Bedroom as private room is well planned also. Parents' bedroom and children bedroom are separated as well.

Privacy

In the past, the existence of an alternative door in the home of the Sugihwaras Arab Village community was used by women to go into the house or referred to as a woman's path (Astuti, 2002). Generally behind the door - the alternative is a small road like a hallway. There is a rule for every Arab woman not to enter the house through the front door, if the non-mahram guest is visiting. It is based on the presence of male guests who are welcomed by the house. The existence of an alternative door can be found in every home of Arab descendants in Sugihwaras that have not changed. Nevertheless there are some homeowners who do not use the alternative door. The picture below is a few alternative doors located in Sugihwaras Arab Village and a small road (alley) to go directly to the private room inside the house.

Fig. 6. The Woman's Path.

Another arrangement to maintain the privacy aspect inside the house is the high wall in each house that ensures to block the sight from outside. The existence of the second door between the guest room and family rooms became another manifest of privacy aspect.

The Role of Sugihwaras Wakaf Mosque

In the development of the mosque is a sign and symbol, belief, ritual and religious embodied in religious socio-cultural institutions that are contained in the physical formation and the people in it. The mosque community is a product produced and brings traditional and modern elements. On the other hand, through the architecture can be explained the concepts that exist and interpret broadly. Thus the mosque can also be interpreted as a symbol of community life that is embodied in architecture as a religious expression (Imammudin, 1985). The Wakaf Sugihwaras Mosque was built around 1852. in the 1800s the city of Pekalongan was under Dutch Colonial rule. Indirectly the building of Wakaf Mosque Sugihwaras also has a building architecture style that is developing at that time that is colonial style of building or commonly referred to as Indische Landhuizen. It is seen from the symmetrical features of the building, has a shield roof, open impression, there are pillars on the front porch and back (Astuti, 2002).

Fig. 7. Wakaf Mosque Sugihwaras.

Wakaf mosque Sugihwaras, as a place of worship, became a catalyst for community development. Some of these activities include regular recitation and discussion. Certain spaces within the mosque, the open space, the veranda, the praying hall, are used for religious activities and social activities. The activities which held in the mosque area show different character depend on the space that cover it. The closer to the *mihrab* the more intense and more important the activity could be.

Conclusions

The result shows that community who lives in Kampung Arab Sugihwaras especially the Arabian descendant preserving Islamic value in their social lives. But, modern urban culture faded some social rules away. While the physical principles based on Islamic value still applicable in their house especially for the privacy aspect and spatial order inside the house. Another result shows from the role of the Sugihwaras Wakaf Mosque. The activities which held in the mosque area show different character depend on the space that cover it. The closer to the *mihrab* the more intense and more important the activity could be.

References

Ahmad, Shaikh Mahmud dan M. Akram Khan. 1984. "Some Economic Issues: Reflections and Answers" dalam M. Tariq Quraishi (ed), *Islam: A Way of Life and a Movement*. Indianapolis: American Trust Publication

Abulughod, Janet. 1980. "Contemporary Relevance of Islamic Urban Principles". Dalam Ekistics Volume 47 Nomor 280 Januari/Febuari 1980

Al-Bukhari, Muslim. 1960. "Saheeh Al-Bukhari" dalam Abdulbaqi, M (ed) *Al-Maktabah Al-Falsafiyyah*. Beirut

Asa, Kusnin, dkk. 2009. *Mozaik Sejarah Pekalongan*. Pekalongan: Pemerintah Daerah Kabupaten Pekalongan

Astuti, Sri Puji. 2002. "Rumah Tinggal Etnis Keturunan Arab di Pekalongan". Tesis tidak Dipublikasikan. Semarang: Magister Teknik Arsitektur Universitas Diponegoro.

Habraken, N Jhon. 1988. "Types of Social Agreement". The Collection of ACA-3 Conference Paper. Asian Congress of Architect. Korea

Imammudin, Abu H, Shamim Ara Hasun dan Debrashir Sarkar. 1985. Community Mosque: A Symbol of Society. Exploring Architecture in Islamic Culture. Regional Seminar in the Series Exploring Architecture. Bangladesh:17-22 Dec

Karim, Fazlul.1938. Al-Hadis: An English Translation and Commentary of 'Mishkatul Masabih' (compiled by Weli ¯ al-Di ¯n Muhammad Ibn 'Abd-Allah al-Khatibal-Tabrizi (d. 737/1337), al-Mektab al-Islami li al-"ibå'ah wa al-Nashr, 1961),Vol. I. Calcutta

Karim, Fazlul.1938. Al-Hadis: An English Translation and Commentary of 'Mishkatul Masabih' (compiled by Weli al-Dinn Muhammad Ibn 'Abd-Allah al-Khatibal-Tabrizi (d. 737/1337), al-Mektab al-Islami li al-"ibå'ah wa al-Nashr, 1961),Vol. I. Calcutta

Mortada, Hisham. 2003. *Traditional Islamic Principles of Built Environment*. London: Routledge Curzon

Petherbridge, Guy. 1987. *Vernacular Architecture: The House and Society in Architecture of The Islamic Wolrd*. London: Thames and Hudson

Rapoport, Amos. 1969. *Home, Form and Culture*. London: Prentice Hall Inc

Steenbribk, Karel A. 1984. *Beberapa Aspek Tentang Islam di Indonesia*I. Jakarta: Bulan Bintang

Van Den Berg. LCW. 1989. *Hadramaut dan Koloni Arab di Nusantara*. Terjemahan: Rahayu Hidayat. Seri INIS Jilid 3. Jakarta: Perpustakaan Nasional INIS

Yusuf Al-Qaradawi. 1960. *The Lawful and the Prohibited in Islam (AlOalalWal aaram Fil Islam)*. Indianapolis: American Trust Publication

https://doi.org/10.1051/shsconf/20184104003

Characteristic Ideal Form to Architecture Traditional Makassar with to Humanist Shelter Development in Settlement Bricks Processor at Gowa Resident, South Sulawesi

Imriyanti[1,*], *Shirly* Wunas[1], *Mimi* Arifin[1], and *Idawarni* J. Asmal[1]

[1]Hasanuddin University, Indonesia

Abstract. The ideal form architecture traditional Makassar there is to ideology form, function and architecture mean. Form, function and mean can is known to utilization macro space and micro space at house traditional Makassar. Architecture traditional being culture identity tribes because in there is life society. The society traditional Makassar it is that society still holding long habit by true in daily life. Humanist shelter referring to human needs especially to the occupants, there are human needs that connected with that environment, is sense of security, comfortable, hygienic and economic. The settlement bricks processor at Gowa resident identic with shelter traditional but with growth development in various ways sector than change shelter bricks processor too experiencing good change from at form, function and mean. The settlement bricks processor to Iowa resident at being exactly in Dusun Giring-Giring Kalase Rena village, Bontolangkasa village and Maccini baji village. The change form, function and mean to house/shelter bricks processor than the problem and that found how is ideal form architecture traditional Makassar there is development humanist shelter in settlement risks processor at Gowa resident. This research purpose is element analysis ideal form architecture traditional Makassar that is can to shelter bricks processor at Gowa resident. This research in descriptive explore. The tory character that is done by direct observation to house/shelter bricks processor at Gowa resident. The results of this study that shows ideal form architecture traditional Makassar there is to macro space a house traditional Makassar in settlement bricks processor at Gowa resident. Ideal form to macro space architecturetraditional Makassar there is to house orientation, patterns of home footprint with to utilization as bricks processor at Gowa resident. With utilization macro space at house traditional Makassar can then to form humanist shelter for bricks processor at Gowa resident, South Sulawesi.

Keywords: ideal form, architecture traditional Makassar, humanist shelter, settlement bricks processor

Introduction

Culture in Indonesia is very diverse and has its own characteristics and character in each region. Culture consists of everything that is learned from the needs, the normative patterns of my manners, that is to include all the ways or patterns of thinking, feeling and human acting. The needs of society are largely fulfilled by the human ability in cultures which is the result of his creations to meet all his needs. (Soerjono and Sulistyawaty, 2014).

Culture owned by each region has one characteristic through its traditional architecture in an integrated manner. Traditional architecture is one element of culture that grows and develops along with the growth of a tribe. Traditional architecture is the cultural identity of a tribe, because it contains all the life of society.

The existence of traditional architecture is closely related to the daily life of the people who still adhere to the collective life order, that is to have harmony and harmony between the macrocosm (universe) and the microcosm (building) (Soeroto, 2002). In the traditional architecture is contained in an integrated form of ideal, social form and material form of a culture (Mardanas, 1985). The form of culture is a system of ideas, concepts and results of human activities.

The development of cultural values of a region is able to strengthen the identity of the community or community groups and become one of the preconditions of creating a space of meaning in the dwelling which has a function in the utilization by the residents of the house. In the residential area of the bricklaying management of Gowa regency dominant inhabitants of Makassar tribe which has the ideal form in its traditional architecture is the shape of the stage house inhabited macro changes, so this raises the problem that is how the ideal form of traditional architecture of Makassar to the development of humanis housing in settlement settlements brick district of Gowa, this is connected with the profession of society as pengelolah brick that can form a modern

* Corresponding author: imrianti@gmail.com

dwelling. The settlement of brick management in Gowa Regency is located in Dusun Giring-Giring, Bontolangkasa Village, Bontonompo Sub-District and Maccinibaji Village, Bajeng Sub-District. These three locations are the largest brick producing areas in the province of South Sulawesi.

Culture Concept In Traditional Architecture

Culture is a commonality and unity for the people of a region and at the same time become a human character in shaping individual behavior and personality. Culture has a big function for people and society. (Soekanto and Sulistyowati, 2014). Culture is a very broad of the life of society so that culture can be a character that is required by every individual and make a stage of life development of each community group in the form of activities.

To manifest individual human characters can be in form through activities or human activities themselves so that formed in an ideology embodied in the form of his work. Human works are reflected through the social culture of society as a whole.

Rapoport in Altman *et al.*, (1980), dividing the culture into three main views. First, culture as a way of life is the same from a group. Second, as a system of symbols or symbols transmitting through symbolic codes. Third, as an adaptive strategy unit to survive, relate to ecology and resources. Man's work embodied in a socio-cultural form is reflected in the ideology expressed in his daily life in a place or space. The statement of space and time of all life of a cultured society is revealed in architecture. Traditional architecture is the embodiment of the culture and thought patterns of reflection in the relationship between the universe and the creator.

Traditional Architecture of Makassar Tribe

Traditional architecture is part of the policy and wisdom of living community development. Its existence is closely related to the daily life of the traditional society that still adheres to the collective life order. There is harmony and harmony between the macrocosm (universe) and the microcosm (building) that must always be nurtured (Soeroto, 2003). Traditional society in question is a society that still holds the old habits firmly in everyday life. Traditional societies are often regarded as a society that only lives in an atmosphere of solely ancestral beliefs that are influenced by exclusive local "cultural ethos" and have special characteristics.

Makassar tribe, is the name of a tribe that has a large population in Sulawesi. According to Wahid (2008), that the society of Makassar tribe as a unity of people who live together for a long time, hereditary by itself will show elements that are traditional. The tribal community of Makassar holds that the house is part of the traditional architecture that has the belief and think that life is only achieved when between the macrocosm and microcosm is always established a harmonious relationship. This view is reflected in Makassar traditional house in the form of a stage with a rectangular base shape.

The Ideal Form

The ideal form is part of the form of culture. The Ideal form have in common with ideology, where ideology is a collection of ideas and ideas derived systematically in a form. According to Destutt de Tracy, ideology can be regarded as a comprehensive vision as a way of looking at things in general in everyday life and in some directions philosophically (1796). The value of ideology in culture is indicated through cultural phenomena. According to J.J. Honingmann, there are three cultural phenomena, namely ideas, activities and artifacts (in to Koentjaraningrat, 1976). Koentjaraningrat offers four forms of culture, namely: culture as an ideological value; culture as a system of ideas; culture as a system of behavior and patterned actions; culture as a physical object (artifact) (Koentjaraningrat, 1976).

Humanity Occupation

Humane dwelling has a meaning as a place of settlement that leads to humanity and can be united with the earth or the environment. The influence of humanism in architecture, present strongly in the era of modern architecture. Architecture when it looks very trying to humanize the architecture, by looking at human needs in the world. Humanisme bringing a state that points to all human needs must be obeyed regardless of anything else. According to Hali (2008), humanism is a plot of discourse that constantly reminds and defends human beings as the main agent of choice for one's own life.

Methodology

This research is a descriptive explorative and interpretative study that study theoretically by observing directly the ideal form of brick processing settlement in Gowa regency of South Sulawesi. The ideal form observed is the macro space of brick management settlements in Gowa district. For the macro space observed is the orientation direction of the dwelling, the pattern of the home foot brick management

Population is the whole object that is the source of information / data (Yusuf, M; 2015) argued that the population is a generalization region consisting of objects / subjects that have certain qualities and characteristics set by the researcher to be learned and then in conclusion. According to Lexy. M.J (2001) argue that, with regard to the size of the population that a researcher is concerned, it is necessary to distinguish between two different populations: experimental or observationally accessible populations and target populations. The population of this research is in Dusun

Giring-Giring of Kalase'rena Village and Bontolangkasa I Village of Bontonompo Subdistrict and Maccainibaji Village of Bajeng Sub-District of Gowa Regency.

Sample were taken from all three locations or populations. According to Prasetyo (2005), sample quantities can be determined based on the Slovin formula. The sample taken purposively were carried out on sample of pinggawa houses and pajama houses, as these two samples were the main sample of research related to brick-making work.

Pinggawa and pajama house consist of traditional houses (house on stilts), mixed house (house on stilts together with stone house) and modern house (house made of stone structure). For Dusun Giring-Giring the number of sample is 64 house, Bontolangkasa Village as many as 88 house and Maccinibaji Village 78 house.

Rumah pinggawa dan pajama terdiri dari rumah tradisional (rumah panggung), rumah campuran (rumah panggung yang menyatu dengan rumah batu) dan rumah modern (rumah yang strukturnya terbuat dari batu). Untuk Dusun Giring-Giring jumlah sampelnya 64 rumah, Desa Bontolangkasa sebanyak 88 rumah dan Desa Maccinibaji 78 rumah.

Discussion

Location of Bricklaying in Gowa Regency of South Sulawesi

Brick processing settlements in Gowa regency, the Giring-Giring village of Kalaserena village, Bontonompo sub-district, Bontolangkasa village, Bontonompo district and Maccinibaji village, Bajeng sub-district. The location of this study is an area that has brick-making materials from the natural resources of the environment such as in Dusun Giring-Giring and Desa Maccinibaji while Bontolangkasa Village purchases natural resources / stone-making material from Maccinibaji Village.

These two sub-districts are sub-districts that have a clay excavation area based on the License, Business Placements issued by the Gowa Head of Regional Administration in 1978. Gowa regency is a lowland area that has natural resources for development is very much including sand, clay, and gravel, all of which are the results of the mine class C.

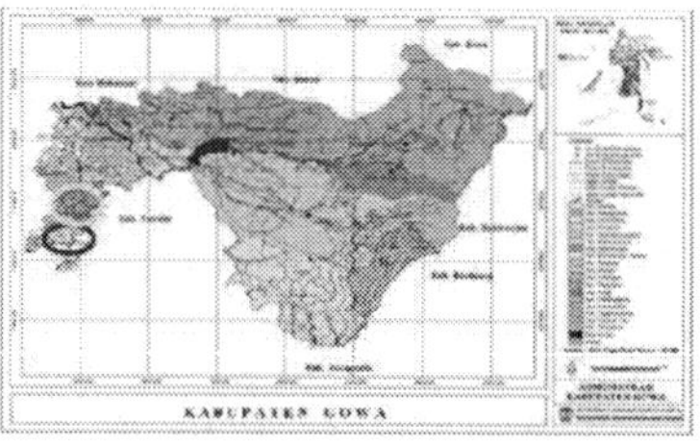

Fig. 1. Map of Gowa Regency Administration.
Source: LP2B Study of Gowa District of South Sulawesi, 2016.

Gowa Regency is located at 119.37730 West Longitude and 120.03170 East Longitude and 5,08293428620 North Latitude and 5.5773054370 South Latitude; where the area is located in the southern part of South Sulawesi Province with an area of 1,883.33 km², or equivalent to 3.01% of the total area of South Sulawesi Province.

Dusun Giring-Giring Village Kalase'rena District Bontonompo Gowa Regency.

Dusun Giring-Giring Kalase'rena Village has a rice field area of 120 ha, 60 ha of land there are 2 RW and 4 RT, the distance Giring-Giring Hamlet to the capital District 2 km and the distance to the capital District 21 km. Population density is 1317 people, and the average number of household members is about 6 people.

Kalase'rena village is a plain area with boundaries:

North side : Bajeng Sub-district
East side : Takalar District
South side : Tamallayang urban village
West side : Bontolangkasa Village I

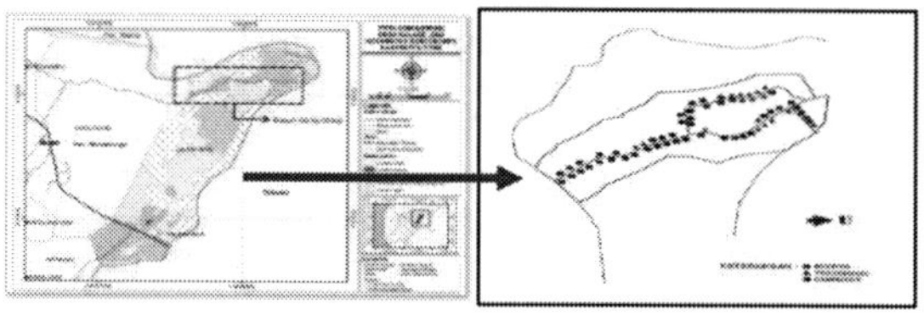

Fig. 2. Map of Kalase'rena Village and Map of Giring-Giring Hamlet (Source: LP2B Gowa District and Survey Result, 2017).

Natural Resources (SDA) brick material in the hamlet Giring-Giring Kalas'rena village can be obtained in the neighborhood of community settlements because this hamlet is in the highlands.

Village Bontolangkasa I District Bontonompo Gowa Regency.

The total area of Bontolangkasa village ± 2.45 km2 with the population density 1.127 per km2, this village has a tropical climate with irrigated rice fields. The village of Bontolangkasa is at an altitude above sea level of 19.00% with an area of 2.45 km2. The distance from Bontolangkasa village to the district capital ± 1 km and distance to the district capital ± 18 km.

Bontolangkasa village with area the following limits:

North side : Kelurahan Tamallayang
East side : Takalar District
South side : Barembeng Village
West side : South Bontolangkasa Village

 https://doi.org/10.1051/shsconf/20184104003

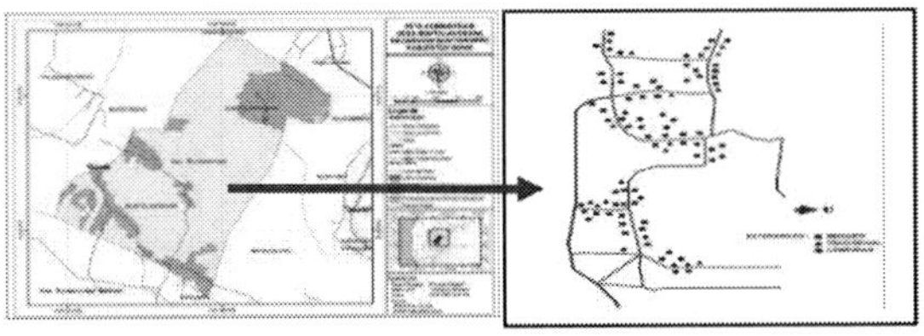

Fig. 3. Map of Village Bontolangkasa I District Bontonompo Gowa Regency
Source: LP2B Gowa District and Survey Results, 2017.

Percentage to sub-district by administrative of Bontolangkasa village is 8,06%. Bontolangkasa village has 4 hamlets, 9 RW / RK and 16 RT, with a population of 2,683 people. Natural Resources (SDA) in the village of Bontolanagkasa I was obtained by buying from Maccinibaji village, Bajeng sub-district.

Village Maccinibaji Bajeng District Gowa District

Maccinibaji village is located in Bajeng District of Gowa Regency. The area of Maccini-baji village is ± 4.32 km2 with a population density of 1.045 per km2 of the total population of 4,513 people, this village has a tropical climate with irrigated and non irrigated rice field conditions. Maccinibaji village is at an altitude above sea level of 12.00% with an area of 7.19%.

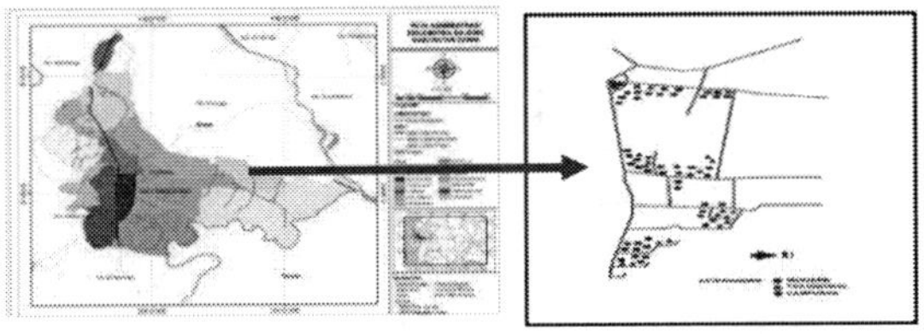

Fig. 4. Map of village administration Maccinibaji Bajeng District Gowa Regency
Source: LP2B Gowa District and Survey Results, 2017.

Maccinibaji Village Bajeng District is a highland area with the following limits:
North side : Pallangga District
East side : Paraikatte Village
South side : Panyangkallang Village
West side : Kalebajeng Village

The distance from Minasabaji village to the district capital ± 3 km and distance to the district capital ± 15 km. For brick material in Maccini Baji village, Bajeng sub-district is found in its residential neighborhood and sold to another village, one of them is Bontolangkasa village in Bontonompo district. Natural Resources (SDA) in the village of Maccinibaji Bajeng subdistrict is located in the community settlement area so that bricklayers can produce bricks and also sell brick material to Bontolangkasa village.

Analysis of Macro Room in Ideal Orientation

Table 1. Ideal Behavior through Home Orientation in Brick Plantation Areas of Gowa Regency.

Ideal Form		Location								
Home orentation		Giring-Giring			Bontolangkasa			Maccinibaji		
		Traditional	Mixed	Modern	Traditional	Mixed	Modern	Traditional	Mixed	Modern
SDA (Natural Resources)	Materials Bricks	8	1	5	10	19	36	8	11	1
	Garden	1	8	4	3	5	5	0	3	15
	Rice fields	1	0	2	0	0	0	0	6	0
	Etc	3	9	21	1	0	8	7	4	23
Amount		14	18	32	14	24	49	15	24	39
Street	Trail (2 m)	8	4	8	0	2	1	0	0	0
	Secondary Local I (3 m)	2	0	3	5	10	8	11	7	4
	Secondary Local II (5 m)	4	14	19	9	8	38	3	4	8
	Collector (7 m)	0	0	0	0	0	0	0	9	24
	Artery (8 m)	0	0	0	0	0	0	0	0	0
	Etc	0	0	0	0	4	2	1	4	13
Amount		14	18	32	14	24	49	15	24	39
Points of the compass	East	8	5	10	4	6	19	2	7	18
	West	5	11	14	2	4	14	4	7	10
	South	1	2	4	6	6	6	3	6	9
	North	0	0	4	2	8	10	6	4	4
	Etc	0	0	0	0	0	0	0	0	0
Amount		14	18	32	14	24	49	15	24	39
Holy place	Mousque	0	1	0	0	1	2	0	0	0
	Mushollah	0	0	0	0	0	0	0	0	0
	Etc	14	17	32	14	24	47	15	24	39
Amount		14	18	32	14	25	49	15	24	39

Natural Resources

Orientation of house with natural resource (Natural Resources) variable and indicator of material brick, garden, rice field and others.

In Dusun Gamping-Gamping Traditional house dominant leads to the source of brick material. Houses with mixed shapes and many modern leads others like the street. Brick material contained in the community settlement in Giring-Giring Hamlet so that brick making becomes a supporting livelihood besides farming and gardening in the hamlet.

Fig. 5. Form of traditional, mixed and modern house in Giring-Giring Hamlet.

Fig. 6. Natural Resources (SDA) adjacent to brick processing in Giring-Giring Hamlet.

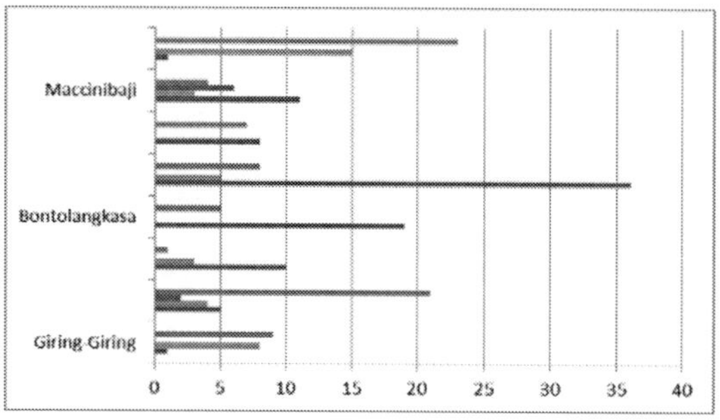

Fig. 7. Graph of home orientation towards Natural Resources.

Orientation of the house to the natural resources in the village of Bontolangkasa, as many as 36 houses in the modern form of natural resources leads to this matter is influenced by the placement of brick material in place in front of the house so the residents of the house can observe and simultaneously make a brick in the yard home.

Fig. 8. Orientation of houses in Bontolangkasa village that leads to brick processing site.

In the village of Maccinibaji Bajeng subdistrict, the house with the modern form is not oriented to the SDA because in the village brick manufacturing is far from the brick processing house. While 11 houses mixed shapes and 8 traditional houses oriented to the natural resources.

Street

The orientation of the house that leads to the road in the three brick processing locations of Gowa Regency is dominant in the direction of the local, secondary, I and II roads but in Maccinibaji village there is a mixed and modern house that leads to the collector road as well as in Bontolangkasa and Maccinibaji villages. orientation to the neighbor's house because the pattern of settlements in the village is a centralized pattern.

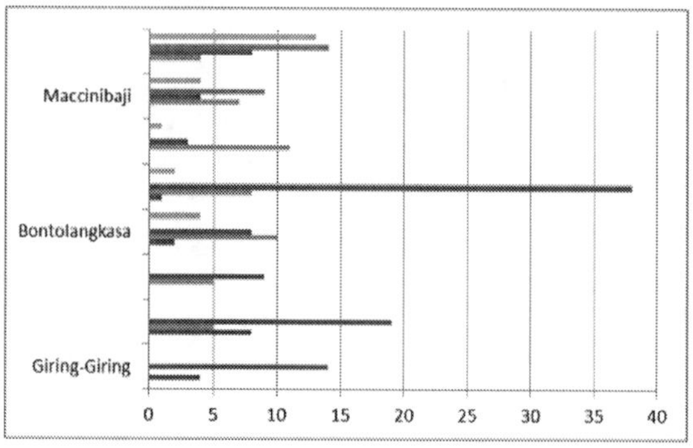

Fig. 9. Graph of house orientation to road in brick processing settlement of Gowa Regency.

Points of the compass

The orientation of the house to the winds in the three research sites tends to lead east, west and south but there are also traditional, mixed and modern houses heading north in the villages of Bontolangkasa and Mac-cinibaji villages but for brick-processing plants not facing north

.

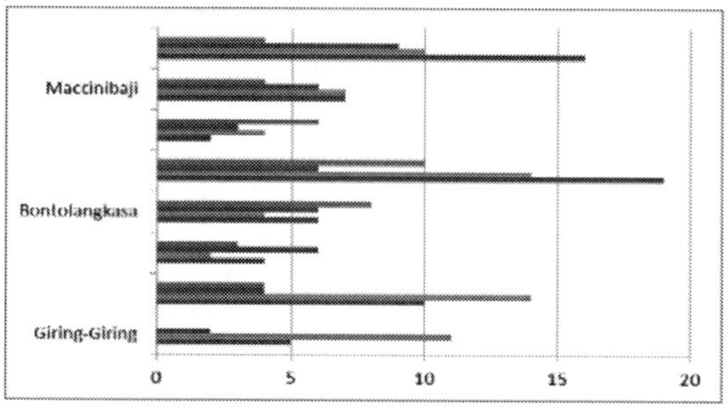

Fig. 10. Graph of home orientation toward the wind in the brick processing settlement of Gowa regency.

The orientation of the house does not lead to the north in the three brick processing locations influenced by the meaning of living owned by the tribe of Makassar in the three brick processing locations because the north gives the meaning of sadness.

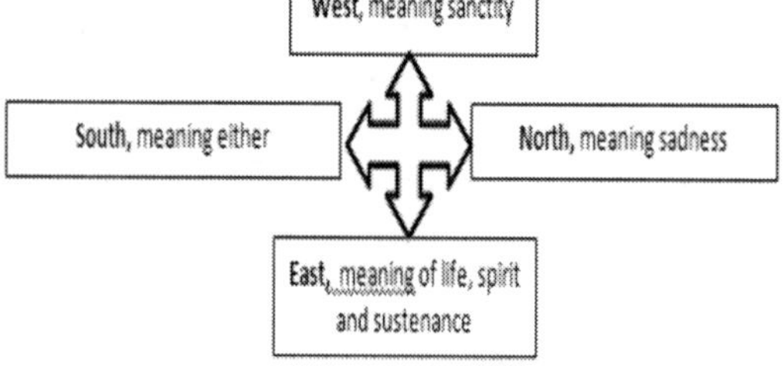

Fig. 11. Meaning of the direction of the wind that is believed Makassar tribe society.

This may also be adapted to the naming of the divided Bontolangkasa village, where the village of Bontolangkasa North and South should be but the villagers of Bontolangkasa North do not use the name but the name of Bontolangkasa I village.

Holy Place

For the orientation of the house to the holy places such as mosques and mushallah only a few houses that lead to the shrine, this is influenced by the pattern of settlements in the hamlet Giring-Giring, village Bontolangkasa and Maccinibaji village pattern centered, spread and linear in the direction.

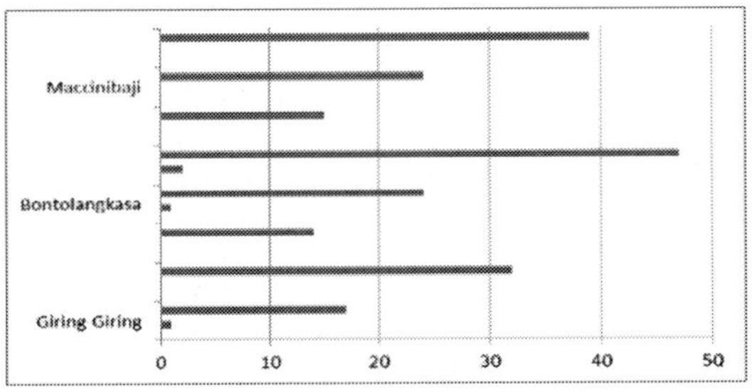

Fig. 12. Graph of home orientation to the shrine in the brick processing settlements of Gowa Regency.
Source: Analysis Result, 2017.

Home Tread Pattern

The ideal form for home tread patterns is observed through the variables of outer space with kinship indicators, wells / sinks, trees, casual / bale-bale and others. Work location variable of brick processing have indicator of house distance with brick processing location worker while work group variable brick processing with indicator individual, family and group.

Table 2. Ideal Being through Household Pattern in the Brick Plantation Area of Gowa Regency.

Ideal Form		Location								
Patterns of home footprint		Giring-Giring			Bontolangkasa			Maccinibaji		
		Traditional	Mixed	Modern	Traditional	Mixed	Modern	Traditional	Mixed	Modern
Outdoor Room	Kinship	3	7	7	8	13	19	3	12	21
	Well / Wash	2	3	12	0	2	3	4	0	2
	Tree	0	2	5	4	5	11	4	3	2
	Place of Relax / Bale-Bale	8	4	8	2	4	11	4	9	3
	Etc	1	2	0	0	0	5	0	0	12
	Amount	14	18	32	14	24	49	15	24	39
Work Location Brick Processing	< 500 m	10	6	17	10	24	49	8	3	6
	> 500 m - 1000 m	4	12	15	4	0	0	7	8	12
	> 1000 m - 2000 m	0	0	0	0	0	0	0	5	10
	> 2000 m - 3000 m	0	0	0	0	0	0	0	8	8
	> 3000 m - 5000 m	0	0	0	0	0	0	0	0	3
	Etc	0	0	0	0	0	0	0	0	0
	Amount	14	18	32	14	24	49	15	24	39
Working Group Processing Brick	Individual	2	2	8	8	16	20	7	7	3
	Family	8	6	4	2	8	28	4	8	27
	Group	4	10	20	4	0	1	4	9	9
	Etc	0	0	0	0	0	0	0	0	0
	Amount	14	18	32	14	24	49	15	24	39

Outdoor Room

The outdoors analyze the ideal form of the aspect of the place where the social relationship occurs so that the indicator of the outdoor space variables are kinship, well / wash, tree, relaxation and others.

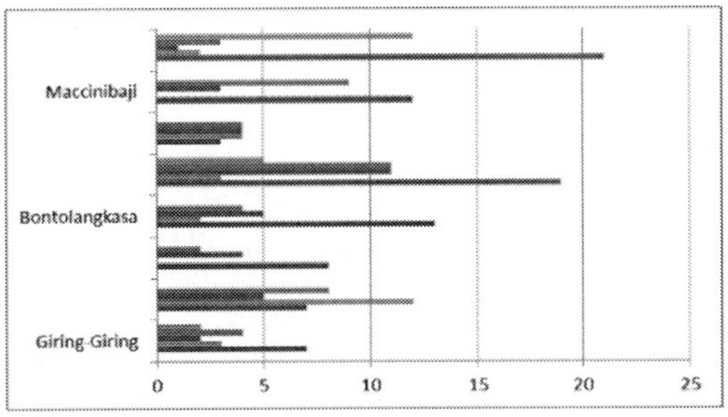

Fig. 13. Graph of home tread pattern to outer space of house in brick processing settlement of Gowa Regency.

Social relation of bricklayer society is happening at kinship level so brick making process can also be done through cooperation between relatives like neighbors, family who live close to brick processing house.

Work Location of Brick Processing

The location of brick processing work can also affect the pattern of home tread in the ideal shape because the pattern of brick processing behavior adjusted to the ideology of brick processing activity can occur at the location of the work. Location work <500 m dominant is in the area of modern house in Bontolangkasa village, it is influenced by the level of brick management activity is very influential with the daily activities where brick-making activity can be done anytime by penglolahnya.

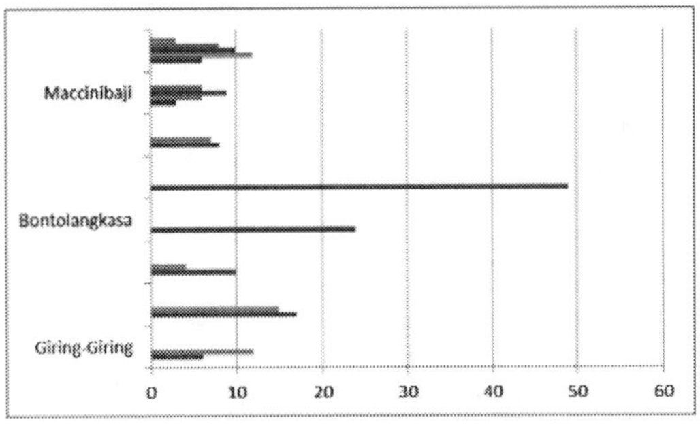

Fig. 14. Graph of home tread pattern to work location in brick processing settlement of Gowa regency.

Fig. 15. Processing location and brick material away from the schoolhouse located in Maccibaji village.

The location of the farthest bricklaying workshop is located in Maccinibaji village for a modern house with a distance of> 3000 m - 5000 m, this is influenced by the distance of the house and the bricklaying place is likely to merge with the place of SDA (Natural Resources) or

 https://doi.org/10.1051/shsconf/20184104003

brick material which usually occurs the process of selling the material to other villages.

Working Group of Bricks

The working group of bricklayers in Gowa Regency is divided into individuals, families, groups and others. In the hamlet Giring-Giring ideal form through working groups are done in groups because in this hamlet the source of brick material is very easy to get because it is in the area of settlement pengelolah so done in groups for the collection of materials and manufacture. This is done by residents who live not far apart or there is a kinship.

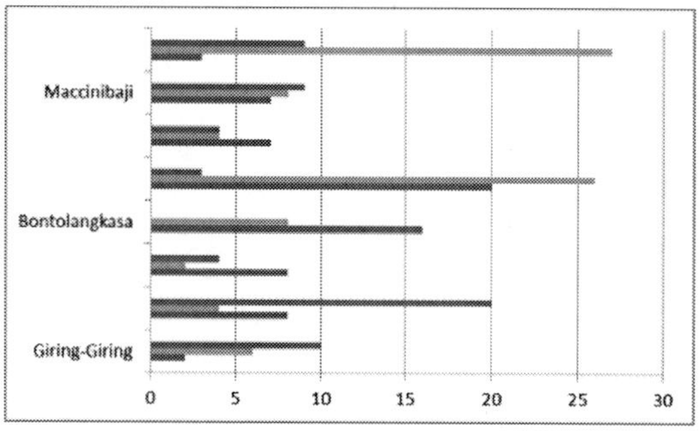

Fig. 16. Graph of home tread pattern to working group in brick processing settlement of Gowa Regency.

For Bontolangkasa and Maccinibaji villages, the brick processing process is mostly done in families because the brick material is in the yard so that the brick management comes from the inhabitants of the house, it is very much happening in the modern house which also do the brick making business in supporting the home economy stairs.

Characteristic of Ideal Behavior of Bricklaying House in Shaping Humanist Shelter

Residential bricklayers in Gowa Regency, especially in Giring-Giring, Bontolangkasa and Maccinibaji villages, have generally undergone changes both in appearance and form of the house where originally shaped traditional houses ie stage or not the stone has been transformed into a mixed and modern form that made of stone. So that the formation of humanist housing in the settlement is obtained from the ideal form of occupancy (orientation and tread pattern). These two ideal beings are the first part of the process of forming humanist dwelling, because humanis occupancy can occur when the activity and the daily behavior of the brick management can conform with the nature for the human.

Through the analysis of the ideal form of macro space (orientation and tread pattern) residential brick management in shaping humanis shelter, ie in terms of orientation:

1. Natural Resources, affecting the change of bricklaying house because home orientation tends to lead to the place of brick material, there is also the orientation of the house leads to the garden and rice fields so that the ideal form in shaping humanis still obtained by pngelolah stone brick because in economical terms very close to source of supporting livelihood.
2. The road, as the shaper of humanis is obtained through the orientation of the house to the road. The dispersed pattern of settlements directed by the road shows the limits of sensitivity to space and territory that are still intimate to humans, the structure of the village road network based on the circulation paths formed within a traditional village sphere, leading to the limitation of land / soil to be reached by the activity community bricklayers in Gowa District.
3. The wind, which influences the orientation of the bricklaying house still holds on the customs of the Makassar tribe people in the west, east and south. The west, east and south directions give a good meaning to the inhabitants of the house while for the north it gives the meaning of sadness so that many of the brick management elements do not lead north. When connected with the pattern of occupancy and wind direction that comes the wind direction from west to east while the brick processing place in the south so dust and wind from the brick processing direction does not lead directly towards the dwelling. This is in accordance with one of the characteristics of humanist housing that is hygienic.
4. The holy place, located in a brick-processing settlement area, is always placed in a section not in the direction of the brick management because holy places such as mosques and mosques are always expected to be in a clean state, this is very united with the characteristics of humane dwelling.
5. Outer space, consisting of kinship, wells / sinks, trees and lounges / bale-bale is part of the outer space of the bricklaying house that is part of humanist.
6. The location of brick processing work in Gowa Regency is dominant to be close to its residence due to the close location of brick making is very supportive to humanist residential aspects in terms of economic.
7. Working groups of brick processing, mostly done in families so that the prospect of making is done for generations and sustainable to improve the household economy.

Characteristics of the ideal form in shaping humanis in the residential area of bricklaying management can be formed despite changes in shape / appearance of the building. From the macro aspect of the humanist macro space many provide economic and social sustainability value.

Conclusion

Humanist shelter can be known through the behavior and habits of the community, this can occur in the area of bricklaying settlements although the form of his house has undergone changes and development.

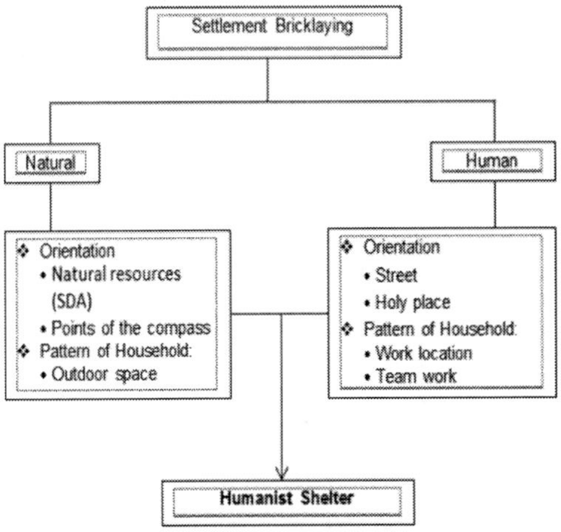

Fig. 14. Humanist Shelter Schemes in the Brick Plantation Settlement of Gowa Regency.

From the ideal form of bricklaying house in Gowa regency which form humanis occupancy characteristic can be known through macro space that is orientation and pattern of home tread which still maintain environmental, social and economic aspects in brick processing settlement area.

In shaping humanist residence in the residential areas of brick processing begins from the behavior and activities of society or human in everyday life.

References

Altman, Irwan and Chemers, Martin. 1980. *Culture and Environment*, First Published by Canbridge University Press, California.

Hali, Damianus J. (2008). Humanism and Global Civilization, in Humanities and Its Relevance to Education. Sugiharto, Bambang, ed, Jalasutra , print I. Yogyakarta

Koentjaraningrat., 1976. Human and Culture in Indonesia, Publisher Djambatan, Jakarta.

LP2B Study Kab. Gowa South Sulawesi, 2014

Mardanas, Izarwisman and friends. 1985, Traditional Architecture of South Sulawesi Region, Dep. P and K, Jakarta.

Moleong, J., Lexy. 2001, Qualitative Research Methodology, Rosdakarya, Bandung.

Muri Yusuf. 2015. Quantitative Research Methods, Qualitative and Joint Research.

Prasetyo, Bambang and Jannah LM (2005), Quantitative Research Methods, PT. Raja Grafindo Persada, Jakarta.

Soerjono S and Sulistyawaty. 2014. Sociology an Introduction. PT. Raja Grafindo Persada. Jakarta.

Soeroto. 2003. Jambi Malay Settlement Pattern, Thesis, Diponegoro University, Semarang.

Wahid, Sugira. 2007. Makassar Man, Reflection, Makassar.

Architecture Studio Learning: Strategy to Achieve Architects Competence

Ahmad Saifudin Mutaqi[1,*]

[1]Associate Professor in Department of Architecture, Universitas Islam Indonesia, Indonesia

Abstract. In most Schools of Architecture, Architecture Studio is at the core of the architectural learning process. In the process, students are trained to have the skills of architectonic spaces design based on the study of the site, its function, and its aesthetics. Students are also trained to have awareness and understanding about the impact of their design on the surrounding environment, both physically and socially. Also, students are trained to present their designs in various forms such as visual graphics, verbal narratives, and three dimensional model animations. Indonesian Association of School of Architecture (AP-TARI Asosiasi Perguruan Tinggi Arsitektur Indonesia) and Indonesian Institute of Architects (IAI - Ikatan Arsitek Indonesia) has formulated an education Standards, Curriculum, and Achievements of Architect Professional Program to be referred by Ministry of Research, Technology, and Higher Education (KEMENRISTEKDIKTI – Kementerian Riset, Teknologi, danPerguruanTinggi) as the guidance for the implementation of Architect Professional Program (PPA - Pendidikan Profesi Arsitek) in Indonesia. One of the eight recommendations is the PPA Content Standard which contains the learning for the achievement of IAI Architect Competencies through the recommended study materials. However, the recommended study materials did not indicate the activity of the Architecture Studio learning model (Final Report of APTARI Part II and IAI). Will architect's competence be achieved if the learning process withoutarchitectural studio learning model? The formulation of the curriculum that is developed independently by the IAI recommends the learning of Architectural Studio as Professional Studio. The size of the SKS is large enough to enable someone who follows the lesson to intensively gain experience in designing the building as a real architectural work. This Architecture Studio learning model is interpreted by PPAr organizer universities with various forms, among others: (1) apprenticeship of architects; (2) supervised studios; And (3) project simulation studio. From various models of Architecture learning model mentioned above, all aim to achieve 13 Architect Competence as formulated by IAI. Which model is effective in learning the Architectural Studio mentioned above? This exploratory study would like to compare the three forms of Architecture Studio learning model to see how much the achievement of the targeted competencies by measuring the success of Competency Test activities still use the standards implemented by LPJK by involving assessors from IAI. The results of the comparison will show the compatibility of the implementation of the Architecture Studio learningmodel what is considered effective. These findings will certainly benefit the development of future PPAr implementation, especially if the Architecture Studio model can be commensurate with the studios developed by the School of Architecture in various countries, at least in the region of 21 member countries ARCASIA.

Keywords: Studio Architecture, Apprenticeship, Supervision, Simulation. Blended Learning in Heritage Conservation Course: Cultural Mapping and Google My Maps Platform

Introduction

What is the Architectural Design Studio?

The architectural design studio as a learning environment, including its pedagogy, history, and interaction, focusing on what makes the architectural design studio unique. Dutton (1984) stated that "compared to typical classroom scenarios, studios are active sites where students are engaged intellectually and socially, shifting between analytic,synthetic, and evaluative models of thinking in different sets of activities (drawing, conversing, model-making)" (p. 16). Although scholars have long praised the design studio as being a unique environment, there have also been criticisms (Ledewitz, 1985). For instance, Ledewitz said, "The lack of clarity over the purpose of the design studio reflects its complexity as a teaching learning setting" (p. 2). Another case with Salama is different withLedewitz's opinion. Salama (2005) claimed that "the design process is intended to function based on intuition, logical

* Corresponding author: ahmadsaifudin@uii.ac.id

treatment, and rigorous reasoning" (p. 1). Nevertheless, the intent may not align with reality because instructors tend to teach how they were taught when they were in school (Salama, 2005). So, contrary to the literature on problem-based learning (PBL) and constructivism, even though design studio educators focus on a project-based technique of teaching (Salama, 1998) and it is obvious that design students should be responding with PBL actions, students' actions end up being constrained by the instructor's teaching format (Salama, 2005).Other scholars have viewed the design studio as a reflexive learning environment (Agryis, 1981; Anthony, 1991; Austerlitz, Aravot& Ben-Ze'ev, 2002; Schön, 1985).

This reflexive quality is what makes the design studio stand out in higher education as a different educational environment (Schön, 1985) and is why, in architectural education, Kolb's learning style inventory (Kolb, 1984) and the idea of experiential learning is most commonly used by scholars to study learning in the architectural design studio. An example of this type of research is a study based in China that found biased correlation between the academicsuccess of students with different learning styles. The investigators concluded that students who were convergers (are students who are more apt to be successful in conventional systems, such as tests, and are more geared toward the physical sciences) were less successful in the architectural design studios being studied than assimilators (students whose strength lies in their ability to create theoretical models) (Jia&Kvan, 2004). Another study by Kvan and Yunyan (2005) expanded on the aforementioned study, attempting to relate the design process to the process of experiential learning. Demirbas and Demirkan (2003) also embarked on the same journey as Kvan and Yunyan (although Demibras and Demirkan's focus was the design process through Kolb's learning styles and not experiential learning). Additionally, in design education, cognitive styles, described by Messick (1984) as "characteristic self-consistencies in information processing that develop in congenial ways around underlying personality trends" (p. 61), are studied in relation to student progression in learning in architecture (Roberts, 2006, 2007); student understanding of architectural concepts in architecture (Saalman, 1990); artifact production in architecture, planning, and engineering (Akin & Akin, 1996; Purcell &Gero, 1998); and design process and cognition (Chan, 1995; Lui, 1996).[i]

Furthermore, Lueth explains in his book (p. 60) on the history of architectural design studio learning.The origins of the design studio are attributed to two art movements*: the École des Beaux-Arts* (School of Fine Arts) and *the Bauhaus*. Before the establishment of the design studio environment, design was learned through an apprenticeship model or a pupilage model, and the design professions were organized in guilds (Fisher, 2000; Kostoff, 1977). This type of education ensured that the master had control over the student (Kostoff). To provide a more structured education, deeming artists more credible through the development of art exhibitions, different académies (academies) were created. The most renowned of these, the French Académie Royale de Peinture et de Sculpture (The French Academy of Painting and Sculpture) (Draper, 1977; Kostoff; Van Zanten, 1980), was established in 1648 (Lagasse, Goldman, Hobson, & Norton, 2001). Around this time (the 17th century), organized group education was established. Regardless of this group sensibility, the dominance of the master as the sole provider of knowledge remained.

Students later rose up against the École system because of the emergence of the modern movement. This student uprising in conjunction with the rise of the modern movement in architecture became a catalyst for the abandonment of the Beaux-Arts system (Littmann, 2000). The architect Walter Gropius was a prominent German leader of the modern movement (an architecture movement in the 20th century) that started in 1925 (Fitch, 1960; Nerdinger, 1985), and within the modern movement, the Bauhaus, established by Henri Van de Velde, arose (Frampton, 1985). The modern era's educational principles were geared toward training students as craftsmen, combining art with the modern technology of the time (Nerdinger), and the modern era itself was characterized by the use of steel as a building material and the use of art and architecture to serve people's special needs (Frampton). The modern-era teaching methods ranged from lecture courses to workshops where students would learn how to build from the day they stepped into the workshop (Gropius, 1937, 1968), while the Bauhaus workshop pedagogy, established for hands-on, interactive, and integrative learning, reinforced the design studio model as a place for all student activities to occur (Nerdinger).

The history of design education illustrates some important characteristics of the current design studio mode from its inception; however, Austerlitz, Aravot, and Ben-Ze'ev(2002) outlined four characteristics that make the modern day design studio a different learning environment than it was in the past: (a) the reflective learning component; (b) the personalized design process, which implies creativity; (c) the instructor's influence on the product of the project; and (d) the fact that a student's actions, personality, and feelings are laid out in the open. In contrast to this list, the traditional classroom has characteristics such as (a) the student as a blank slate mentality, (b) a non-requirement of creativity, (c) the instructor not having a direct influence on the process students use to produce work, and (d) an attitude that students' personalities are unimportant, mainly due to large classroom sizes. Indeed, other characteristics suggested by Lueth (2003) also make the design studio a unique educational environment unlike other environments: (a) the influence that students have on each other; (b) the influence that students could have on the instructor (in terms of creating an environment that may or may not be conducive to teaching); (c) the influence of the physical environment; and (d) the influence that the products created during class time and outside of class time might have on the students' learning. The educational environment in the design studio, therefore, is defined as the components of the physical, pedagogical, and virtual (the classroom through the internet) space that have an effect on the education of the participants (students and instructors).

In his book, Leuth concludes that there are a number of important statements related to what effects arise and what impacts.Participants spent a great deal of time talking about themes that fell under the category of interdependent learning. In the descriptions of their learning experiences, interdependency can be likened to the idea of constructionism, where knowledge and meaning are individually created, yet at the same time, there is a dependency on the relationships in the learning environment. The dependency on relationships allowed for a sharing of individual constructions and the generation of other meanings due to these interactions. These meanings can end up becoming shared meanings.

In the above study can be understood what is meant by Architecture Design Studio. The teaching and learning process that occurs in it involves a teacher and a number of students. Teachers can function as facilitators and / or instructors who have the authority to assess the work of students who follow the lectures. The process of interaction occurs naturally, taking place between them: the interaction of teachers and students and the interaction of students and students. The studio participants can start freely, be able to learn about understanding a problem, learn about finding solutions to a problem they've got. While the teacher as a facilitator can help students show important sources to understand the problem or to get the solution.

Architectural Design Studio Model

Researchers observed the teaching and learning process of Architectural Design Studio which was held in Architecture Profession Education Study Program, Universitas Islam Indonesia. There are three studio learning models: (1) Studio Internship Architecture; (2) Architectural Studio Supervised and (3) Studio Architecture Simulation Design.

Studio Internship Architecture

The Studio Internship Architectureare involving senior architect and intern architect. An intern architect or architectural intern is a person who is working professionally in the field of architecture in preparation for registration or licensure as an architect. An intern need not have attained a professional degree in architecture to begin accruing experience hours, but said degree is a prerequisite for licensure.In the United States, Canada, and other countries, an intern architect is enrolled in a regulated program, such as the Intern Development Program (IDP) in the United States or the Intern Architect Program (IAP) in Canada, while working under the supervision of a licensed architect and preparing for professional registration exams. The use of the title "architect" (or any derivation thereof) is legally protected in the United States, Canada, and other countries. Most U.S. states and all Canadian provinces, however, allow the use of the terms "intern architect" or "architectural intern" for a person enrolled in an architectural internship program."The term intern refers to any individual in the process of satisfying a registration board's experience requirements. This includes anyone not registered to practice architecture in a U.S. or Canadian jurisdiction, graduates from NAAB-accredited programs, architecture students who acquire acceptable experience prior to graduation, and other qualified individuals identified by a registration board." National Council of Architectural Registration Boards (p. 2).

At the Chabot School of Architecture, the learning of an Architecture Internship is arranged with specific requirements.Architecture Internship in Architecture department setting approved by Architecture faculty as related to student's architecture major or classes at Chabot. Cooperative effort between student and architecture firm supervisor to accomplish agreed upon work objectives and broaden experiences. Student provides verification of service experience hours during the term. Students will get an architecture firm approved by architecture faculty and make arrangements for hours and duties directly with architecture firm supervisor. Students will meet with architecture instructor one hour per week on campus for input and hands-on experience discussion focused on architecture firm structures, project procedures, design developments and construction documents.Before entering the course, the student should be able to[ii]:

1. Discuss construction using masonry, steel, concrete, and wood frame construction;
2. Demonstrate ability to use the Uniform Building Code for non-residential projects;
3. Prepare complete working drawings for small masonry, steel, and wood frame buildings;
4. Identify and execute critical details for construction documents;
5. Discuss the roles working drawings and specifications have in construction contract administration;
6. Use computer-aided design & drafting, CD-ROM based information, and Internet resources for compiling construction documents.
7. Define in words and diagrams the central concept or part of a design;
8. Use both traditional and digital graphic and model building techniques to study design strategies;
9. Apply the role of human activity as determinant of and constraining element in design decisions;
10. Use functional objectives and constraints to affect the form and organization of architectural elements;
11. Manipulate architectural elements (floors, walls, roofs, openings) to define form and space that support the communication of intended meanings and concepts;
12. Adapt architectural design in response to a variety of site factors, including climatic, daylighting, and acoustic conditions;
13. Describe how cultural values and beliefs inform the perception and meaning of built environments;
14. Devise architectural designs in response to the cultural and anthropometric needs of occupants;

15. Apply knowledge of structural potential and limitations as a design factor.
16. CAD organization and system components
17. Open, retrieve, and save drawing files in CAD;
18. Draw lines and geometric shapes using various coordinate entry systems in CAD;
19. Select and modify drawing entities in CAD;
20. Control lines, layers, and colors using AIA (American Institute of Architects) standards;
21. Place text with appropriate fonts and sizes in drawings by using CAD;
22. Dimension drawings using architectural standards in CAD;
23. Use external reference systems and commands in CAD;
24. Plot hardcopies with appropriate scales in CAD.

Taking into account the practice of apprenticeship of architecture above it seems that learning activities should be done together with the bureau of architects who work to serve the needs of architectural services. The requirement to follow the learning is very strict and skill issues should not be an obstacle for those who are dutiful to do architectural apprenticeship.

Architectural Studio Supervised

The supervised Architecture Studio is a studio workplace by apprentice architects managed and supervised by lecturers on campuses. The apprentice architects study and work with senior architects to resolve the real issues facing the demands of their contextual needs and contextual sites. the communication process undertaken by senior architects and interns is monitored directly by the supervisors.Learning and work activities undertaken in architectural design studios located on campuses are influenced by the order of design process undertaken by senior architects in managing their clients. The characteristics of each building planning project have different situations but the process stages can still be grouped into categories: programmatic design (feasibility studies and design concepts); schematic design (and preliminary design); design development and construction document (detailed engineering design)

Table 1. Subjects and Activity Topics[iii]

Subjects	Activity Topics
Programmatic Design 1. The method of tracking and preparing the design program for a design project.	• Conferences with the user +owner • Surveyandcollecting structuring data • Analysis of Project Requirement (program analysis and concepts, site analysis, space analysis, • Building code information • Diagram studies of Space Requirement • Assembly of utility and Survey data
Schematic Design 2. The relationship between people and buildings and between buildings and the environment. 3. Architectural design that meets the aesthetic and technical requirements.	• Schematic design studies and recommended Solution • Schematic design plan • Sketches and study model • General Project description • Engineering system concept • Preliminary cost estimate • Presentation 0f SD documents to Owner
Design Development 4. Knowledge of the construction industry, organization, rules and procedures for the implementation of design concepts into integrated and comprehensive building planning.	• Conferences with the user +owner • Refinement of project requirement • Formulation of civil engineering system • Formulation of Structural system • Formulation of Electrical& mechanical system • Selection of major building material • Preparation of DD document: plan, elevations, building • profile section, outline specification, • Perspective, Sketches, or model • Preliminary cost estimate • Equipment schedule • Reviewing plans with applicable agencies • Presentation of DD documents to Owner
Construction Document 5. Skills required to meet requirements within the range of development and building regulatory constraints.	• Conferences with the user +owner • Development of major detail condition • Diagram study of major mechanical and electrical system • Diagram study of major civil and structural system • Architectural working drawings, specifications • Civil working drawings, specifications • Structural working drawings, specifications • Mechanical working drawings, specifications • Electrical working drawing, specifications • Built-in equipment working drawings, specifications • Update construction cost estimate • Submission of construction documents to applicableagencies • Presentation of CD documents to Owner

If observed the study of architecture studios supervised above can be seen the existence of more controlled process guarantees by higher education institutions as education providers. The working atmosphere in an apprentice studio would not be as ideal apprentice studio would not be as ideal as studios owned by architectural consulting firms. But the control of the learning process step by step more can be guaranteed to be achieved. The process of education between lecturers and students and communication between senior architects and apprentice's architects can be more secure.

Architectural Studio Simulation Project

Architect work[iii]: designing, not drawing. Images are simply communications mediums between architects and clients, contractors, or others involved in design and construction. To be an architect, students who are studying at the School of Architecture,Institut Teknologi Bandung (ITB), followed a series of designing training activities on Architecture Design Studio course, from semester 3 to semester 8. Designing exercises, starting from small buildings with simple problems. The semester is getting up, the building is getting bigger, the problems are solved in the more complex design, and the drafting exercise is not just designing the building but also the tread / environment around the building. In the studio, students practice designing while drawing, making a model or making 3D simulations. Explanation of activities undertaken by the School of Architecture ITB describes the working atmosphere in the project simulation studio.
Characteristics of architectural studios that use project simulations include:

1. The working order of the studio is set within the term of the reference of the project.
2. The technical requirements of the building are set according to the assumption of difficulty level that will be charged to the students.
3. Supervisor is positioned as the project owner.
4. Building codes are sometimes manipulated to get a quasi-ideal state.

Learning activities through the Project Simulation Studio have no real risk, but architectural problem solving can be done ideally, especially overriding the budget factor and building regulations formally.

Architect Competency

A competency is the capability to apply or use a set of related knowledge, skills, andabilities required to successfully perform "critical work functions" or tasks in a definedwork setting. Competencies often serve as the basis for skill standards that specify thelevel of knowledge, skills, and abilities required for success in the workplace as well aspotential measurement criteria for assessing competency attainment.Competencies define the applied skills and knowledge that enable people tosuccessfully perform their work while learning objectives are specific to a course ofinstruction. Competencies are relevant to an individual's job responsibilities, roles andcapabilities. They are a way to verify that a learner has in fact learned what wasintended in the learning objectives

Architect Competence marks an architect having certain skills in the field of architecture achieved through long learning process at higher education and professional education. Higher Education and Professional Education are held jointly between the University and the Association of Architects Professions within 5 (five) years. Equality recognition of the achievement of competence and trust in the implementation of education is done through accreditation process by independent accreditation institution of architectural field (LAM Architecture).

Union Internationale Architect (UIA)

Members of the architectural profession are dedicated to standards of professionalism, integrity, and competence, and thereby bring to society unique skills and aptitudes essential to the sustainable development of the built environment and the welfare of their societies and cultures. Principles of professionalism are established in legislation, as well as in codes of ethics and regulations defining professional conduct:

1. **Expertise**: Architects possess a systematic body of knowledge, skills, and theory developed through education, graduate and post-graduate training, and experience. The process of architectural education, training, and examination is structured to assure the public that when an architect is engaged to perform professional services, that architect has met acceptable standards enabling proper performance of those services. Furthermore, members of most professional societies of architects and indeed, the UIA, are charged to maintain and advance their knowledge of the art and science of architecture, to respect the body of architectural accomplishment, and to contribute to its growth.
2. **Autonomy**: Architects provide objective expert advice to the client and/or the users. Architects are charged to uphold the ideal that learned and uncompromised professional judgment should take precedence over any other motive in the pursuit of the art and science of architecture. Architects are also charged to embrace the spirit and letter of the laws governing their professional affairs and to thoughtfully consider the social and environmental impact of their professional activities.
3. **Commitment**: Architects bring a high level of selfless dedication to the work done on behalf of their clients and society. Members of the profession are charged to serve their clients in a competent and professional manner and to exercise unprejudiced and unbiased judgment on their behalf.
4. **Accountability**: Architects are aware of their responsibility for the independent and, if necessary, critical advice provided to their clients and for the effects of their work on society and the environment. Architects undertake to perform professional services only when they, together with those whom they may

engage as consultants, are qualified by education, training, and/or experience in the specific technical areas involved.

The fundamental requirements for registration/licensing/certification as an architect as defined above, are the knowledge, skills, and abilities listed below that must be mastered through recognized education and training, and demonstrable knowledge, capability, and experience in order to be considered professionally qualified to practice architecture[iv].

1. An ability to create architectural designs that satisfy both aesthetic and technical requirements
2. An adequate knowledge of the history and theories of architecture and the related arts, technologies and human sciences
3. A knowledge of the fine arts as an influence on the quality of architectural design
4. An adequate knowledge of urban design, planning and the skills involved in the planning process
5. An understanding of the relationship between people and buildings, and between buildings and their environment, and of the need to relate buildings and the spaces between them to human needs and scale
6. An understanding of the profession of architecture and the role of the architect in society, in particular in preparing briefs that take account of social factors
7. An understanding of the methods of investigation and preparation of the brief for a design project
8. An understanding of the structural design, construction and engineering problems associated with building design
9. An adequate knowledge of physical problems and technologies and of the function of buildings so as to provide them with internal conditions of comfort and protection against the climate
10. The design skills necessary to meet building users' requirements within the constraints imposed by cost factors and building regulations.
11. An adequate knowledge of the industries, organizations, regulations and procedures involved in translating design concepts into buildings and integrating plans into overall Planning
12. Awareness of responsibilities toward human, social, cultural, urban, architectural, and environmental values, as well as architectural heritage
13. Adequate knowledge of the means of achieving ecologically sustainable design and environmental conservation and rehabilitation
14. Development of a creative competence in building techniques, founded on a comprehensive understanding of the disciplines and construction methods related to architecture
15. Adequate knowledge of project financing, project management, cost control and methods of project delivery
16. Training in research techniques as an inherent part of architectural learning, for both students and teachers

Ikatan Arsitek Indonesia (IAI)

Association of Indonesian Architects (IkatanArsitek Indonesia) as the only professional association of architects recognized by law has formulated 13 Architects' competence.Through the 13 points of competence that become the standard of certification qualification certification fulfillment of architects. Any architect applying for a new certificate is required to answer questions about these standards, as one of the evidence of deepening and involvement in any proposed project as a benchmark[v].

1. **Architectural Design**
   The ability to produce architectural design that meets the aesthetic and technical requirements, and which aims to preserve the environment (Ability to create architectural designs that satisfy both aesthetic and technical requirements, and which aim to be environmentally sustainable)
2. **Architectural Knowledge**
   Adequate knowledge of the history and theory of architecture including art, technology and human sciences (Adequate knowledge of the history and theories of architecture and related arts, technologies, and human sciences)
3. **Art Knowledge**
   Knowledge of art and its influence on the quality of architectural design (Knowledge of the fine arts as an influence on the quality of architectural design)
4. **Urban Planning and Design**
   Adequate knowledge of urban planning and design as well as the skills needed in the planning process (Adequate knowledge on urban design, planning, and the skills involved in the planning process)
5. **Relationship between Human, Building and Environment**
   Understanding the relationship between people and buildings and between buildings and the environment also understands the importance of linking the spaces that are formed between people, buildings and the environment for human and human needs (Understanding of the relationship between people and buildings and the buildings and their environments, and the need to relate spaces between them to human needs and scale.)
6. **Knowledge of Environmental Supporting Power**
   Mastering adequate knowledge on how to produce the appropriate design of environmental carrying capacity (An adequate knowledge of the means of achieving environmentally sustainable design.)
7. **The Role of Architects in Society**
   Understand the professional aspect in the field of Architecture and realize the role of architects in the community, especially in the preparation of terms of reference which take into account social factors (Understanding of the profession of architecture and the role of architects in society)
8. **Preparation Work Design**
   Understand the search method and the preparation of a design program for a design project (Understanding of the methods of investigation and preparation of the brief for a design project.)

9. **Understanding Inter-Disciplinary Problems**
   Understanding structural, construction and engineering problems related to the design of buildings (Understanding of the structural design, construction, and engineering problems associated with building design.)
10. **Physical Knowledge and Building Physics**
   Mastering adequate knowledge of physical and physical issues, technology and building functionality so that it can equip them with internal conditions that provide comfort and protection of the local climate (Adequate knowledge of physical problems and technologies as well as internal conditions of comfort and protection against climate.)
11. **Implementation of Budget Limits and Building Regulations**
   Master the skills required to meet the requirements of building users in the construction and building construction cost constraints (Necessary design skills to meet building factors)
12. **Knowledge of Construction Industry in Planning**
   Mastering adequate knowledge of industry, organization, regulations and procedures relating to the process of translating the concept of design into building structures and the process of integrating the layout of the plans into a comprehensive planning (Adequate knowledge of the industries, organizations, regulations, and in translating design concepts into buildings and integrating plans into overall planning.)
13. **Project Management Knowledge**
   Mastering adequate knowledge of project funding, project management and development cost control (Adequate knowledge of project financing, project management and cost control).

How to Achieve Architect Competency?

Architectural competence is achieved in three levels, namely: **awareness; understanding** and **ability**.

Table 2. Relationship of Studio Model with IAI Competency [iv].

IAI Competency	Internship Studio	Supervised Studio	Simulation Project
	Off campus	**In campus**	**In campus**
1. Architecture Design	*ability*	*ability*	*understanding*
2. Architecture Knowledge	*understanding*	*understanding*	*awareness*
3. Art Knowledge	*awareness*	*understanding*	*awareness*
4. Urban Planning and Design	*understanding*	*awareness*	*awareness*
5. Relationship between Human, Building and Environment	*understanding*	*understanding*	*awareness*
6. Knowledge of Environmental Supporting Power	*ability*	*ability*	*awareness*
7. The Role of Architects in Society	*ability*	*understanding*	*awareness*
8. Preparation WorkDesign	*ability*	*ability*	*understanding*
9. Understanding Inter-disciplinary Problem	*understanding*	*understanding*	*awareness*
10. Building Physic and Physical Knowledge	*ability*	*ability*	*understanding*
11. Implementation of Budget Limits and Building Regulation	*ability*	*ability*	*awareness*
12. Knowledge of Construction Industry in Planning	*understanding*	*understanding*	*awareness*
13. Knowledge of Project Management	*understanding*	*understanding*	*awareness*

For each architecture studio model has a different level of ability than any required competence. Working atmosphere in an intensive internship studio has an effect on every individual working in it. Daily meetings with senior architects and working relationships with colleagues put pressure on them. Nevertheless, working in a studio on campus is no less intensive.

The supervising lecturer controls each day on the work instructions provided by senior architects each time periodical meetings. Working atmosphere is not obtained by students who study in the studio in campus, but the atmosphere of learning to work more intensively can be felt. Students realize that they are not solving the same problem, but each participant is busy with their respective affairs according to the project.The learning outcomes they get that illustrate the relationship of achievement of competence with the studio model in which they work are described in the table below [vi].

Awareness is the ability to directly know and perceive, to feel, or to be cognizant of events. More broadly, it is the state of being conscious of something.

Understanding is often, though not always, related to learning concepts, and sometimes also the theory or theories associated with those concepts. However, a person may have a good ability to predict the behavior of an object, animal or system—and therefore may, in some sense, understand it—without necessarily being familiar with the concepts or theories associated with that object, animal or system in their culture. They may, indeed, have developed their own distinct concepts and theories, which may be equivalent, better or worse than the recognized standard concepts and theories of their culture.

Ability include of Intelligence, logic, abstract thought, understanding, self-awareness, communication, learning, having emotional knowledge, retaining, planning, and problem solving. Ability also about knowledge, a familiarity with someone or something, which can include facts, information, descriptions, or skills. Skill, the learned ability to carry out a task with pre-determined results.

Conclusion

Learning process activities conducted in off-campus internship studios have a positive impact on students who develop themselves to work as architects. Internship activities to senior architects give many new experiences in dealing with real problems. The working atmosphere in the studio encourages students to work even though each one is doing different things. Students are 'forced' to work to resolve unscheduled issues caused by clients' requests or other reasons that make it necessary to work overtime. Officers off-campus apprentices may occasionally acquire undisciplined students, but the situation there forces students to follow the rules of the game. Achievement of learning competence still depends on the learning process that occurs between Supervisor, Senior Architect and Student. These three actors are very influential on the achievement of competence due to the delivery process, communication, problem complexity and sharing mechanisms in learning to work.

Studio internships and supervised studios are recommended for learning to achieve the architect's competence. Discipline of all actors (Students; Lectures; Architects) in the studio design and order as well as the commitment of teaching and learning process from them in the learning process become the key of success to achieve the competence of architect.

References

---

[i] Lueth, Patience LamunuOpiyo, (2008), The architectural design studio as a learning environment: a qualitative exploration of architecture design student learning experiences in design studios from first- through fourth-year, Iowa State University.

[ii] Downloading at September 29, 2017: http://www.chabotcollege.edu/searchresults.asp?q=internship+architecture&imageField.

[iii] Downloading at September 30, 2017: https://ar.itb.ac.id/2010/12/13/studio-perancangan-arsitektur/

[iv] UIA Accord On Recommended International Standards Of Professionalism In Architectural Practice Ammended August 2014 At The Xxvi General Assembly (Durban, South Africa 2014)

[v] Downloading at September 30, 2017: http://www.iai.or.id/sertifikasi/13kompetensi

[vi] Result of evaluation of learning process at Archival Profession Program of Universitas Islam Indonesia for 3 years (2013, 2014, 2015).

Sulapa Eppa As The Basic or Fundamental Philosophy of Traditional Architecture Buginese

Syarif [1,*], *Ananta* Yudono[2], *Afifah* Harisah[1], and *Moh* Mochsen Sir[1]

[1]The Architecture Department Faculty of Engineering Unhas, Makassar, Indonesia
[2]PWK Department Faculty of Engineering Unhas, Makassar, Indonesia

Abstract. In general, traditional buildings in South Sulawesi Province are built based on the hopes for better future. Therefore, the house is considered as the manifestation of the natural universe (macrocosms). House and settlements are ones of the highest cultural heritage works, in Buginese communities. The value of sacredness and robustness raises the meaning of sulapa eppa philosophy, which is considered as the philosophy to realize Buginese traditional house. The purpose of this study is to discover the philosophy of sulapa eppa that underlines the formation of Buginese traditional architecture. The elements of the house present a philosophical meaning. Panrita bola serves as the informants of sulapa eppa philosophy. This research is qualitative, descriptive and explorative methods as well as naturalistic paradigm approach. The population is traditional Buginese buildings, and the samples arepost habitation, or temporarily built Buginese traditional buildings. The sampling techniques are sampling proposed and sampling incidentally. Research location is in Soppeng regency, South Sulawesi Province. The results of this study are sulapa eppa as the basic of Buginese formation traditional architecture, either physical or nonphysical. Physically, the formations of architecture are in the form, size of proportion, and building elements. And non-physically, it also can be obtained from Panrita bola, and Lontara Bugis manuscripts.

Keywords: Philosophy, sulapa eppa, Buginese traditional architecture

Introduction

Forms and models of traditional buildings that are still preserved in Indonesia, in general have strong characteristics of culture and customs that are inherited from their ancestors. Not only they have the forms and models as mentioned previously, traditional buildings also have adaptive brand on their surrounding environment; for example, using building materials taking from the nature, or forest that they protect. processing and production manners of the materials are based on the spiritual elements or norms, ancestor religions, tradition as well as local geographic conditions.

Traditional architecture can be categorized as *tangible* cultural work; as manifestation of honesty and sincere statement, having ability to express past cultural impressions. Architecture means culturing because traditional architecture grows and develops based on ancestor heritage from concerned nation. Existence of traditional house is a product of an evolution process, based on human past empirical experiences expressed its interaction with the nature. So, it is considered as manifestation of traditional community mind set and lifestyle in the past; definitely containing knowledge about wisdom in the interaction with natural environment. At least, it is intellectual description about the initial time of a culture and its architecture (Shima, 2006).

Architecture culture is formed by a trust at ancient mythology expressing that traditional community concept, spiritual aspect, symbols and values provide significant effects on building forms and motives. This is always taken as the basic for every life in which they live.

The significant cultural effects on traditional society will also influence on environmental pattern, house spatial pattern and its civilization.

Village pattern by Buginese in the past was grouping in dense and spreading manners. Dense grouping pattern could be mainly found at lowland, which its location was close to paddy field areas or sea side or lake. Meanwhile, the spreading pattern could be easily found at mountainous areas or plantation areas. So, manifestation of the village in the past was mainly related to their place of works so that it could be called as *pallaon ruma* (farmer village) and *pakkaja* (fishermen village or fish catcher village). In general, a house can face to one of the wind directions (may face to the West, East, South or North). This is related to Buginese cosmologist view assessing this world as a rectangular. Therefore, these four wind directions have the same position (Mardanas, 1986).

* Corresponding author: syarif.beddu@gmail.com

Buginese life view on their house can be seen from their cultural behaviour and pattern; in a period of time, human was so close to their natural environment, so that their culture is cosmos manifestation (cosmography). In this issue, human living in this world considers himself as a part of the universe (macro cosmos); human life in the level of macro cosmos must 'obey' to macro cosmos control and rules. Thus, human will always try to make harmonious interaction with their nature so that they can obtain better, saver and peace life (Saing, 2010).

In Buginese traditional belief, house is considered as the natural manifestation as the center of human life cycle. It is as a place for human to be born, brought up, got married, and died. Therefore, it is considered to be sacred and greatly respected. So, it can provide peace, health, safety, welfare, and respect for its dwellers (Shima, 2006). Therefore, an existence of a house (settlement) for Buginese society is unseparated part with itself; then it is very important for home presence or ownership because the house form and manifestation symbols will be manifested on its dwellers.

Based on the discussion results above, it shows that the existence of house (settlement) of Buginese society cannot be separated from philosophical elements ranging from selection of material types to be used; implementation of field work is by *panre bola* (carpenter of Bugis stage house), *mappatettong bola* (establishing the stage house), until new *menre bola* (climbing new house) and *maccera bola* (house anniversary party). All of these events are based on the philosophical elements of *sulapa eppa* (perfect, pleasure, health, welfare and safety elements).

Looking at Bugis society house (settlement), with stage architecture and wooden construction are called as *knock down* system; so that it creates an assumption that behind the strength and sturdy are based on intangible rules or advices. This is considered as philosophical basic by *panrita bola/sanro bola* (Bugis traditional house architects), in implementing its tasks as an architect (Bugis traditional architect).

Then, the problem is how the philosophy of *sulapa eppa* as the basic of Bugis traditional architect study is and how its application on wooden stage buildings that are inhabited and owned by Bugis tribe society is.

This research aims to find out philosophical elements that have been longer as study basic of Bugis traditional architecture formation. This philosophy ability is understood and mastered by *panrita bola/sanro bola* (Bugis traditional building architecture).

Study Review

Bugis Language and Literature

Bugis literature written in *lontara* script has been started in the 16th century, in general by people of Southeast Sulawesi. The *lontara'* scripts as well as Bugis literature works (classic) have close relation to life problems manifested in *panngaderreng* (form of Bugis culture). There is such a belief taken by Buginese that creation of script and literature were taken from one philosophical background of *panngaderreng* as one unity (Mattulada, 1993).

So, at the beginning, Buginese literatures were written on leaves of lontar/siwalan *(Borassus sp.);* results of these writing collections were then called as *lontara'*. The *Lontara'* presents sacred literature, in the forms of mythologist beliefs and spells. Then, there was also a development of literature works having characteristics of world life, this was based on the development of *lontara'* and society life behavior as well as its culture.

Related to the creation of sound signs called as *lontara'* script, there is an assumption that this was derived from Bugis people mythologist belief and view, considering this universe as *sulapa' eppa' wolasuji* (rhombus). Definition of *wolasuji/walasuji* is a partititor having function as an accessories in tribal ceremony or Bugis wedding party; it is made of bamboo skin woven diagonally forming a shape of rhombus (figure 1).

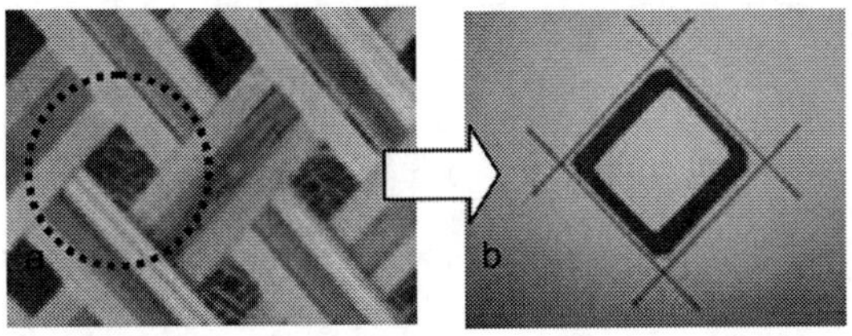

Fig. 1. Production Process of Letter *"sa"*, (a) *wolasuji* (woven bamboo), (b) letter *"sa"*.

Figure 1 shows the production process of letter *"sa"* (Bugis lontara script), taking the motif from *"wolasuji"* (woven bamboo that is made specifically for tribal ceremony). Motif of letter *"sa"* (shape of rhombus) is the pioneer philosophy of *sulapa eppa*, with its four angles shine scared spiritual values.

Fig. 2. application of wolasuji, *(a)wolasuji* (ladder fence), (b) production process of *wolasuji*..

Figure 2 shows application of *wolasuji* as *"railing"* or fence, in the ladder of Bugis stage house. It indicates that the house dwellers have had party.

Philosophy of *sulapa eppa* is also manifested in human, often called as *"sulapa eppa na taue"* (philosophy of human rectangular), this is in line with an opinion by Pelras (2005) stating that portrait of Bugis people mentality is dominated by four characteristics. In the *lontara'*, it is mentioned four characteristics namely *sulapa' eppa'* (rectangular) that should be owned by each good leader (figure 3). A prospective one to be a good leader should not only be taken from proper

 https://doi.org/10.1051/shsconf/20184104005

descendant, he or she should also be *warani* (brave), *macca* (intellect), *sugi* (rice), and *panrita* (kind).

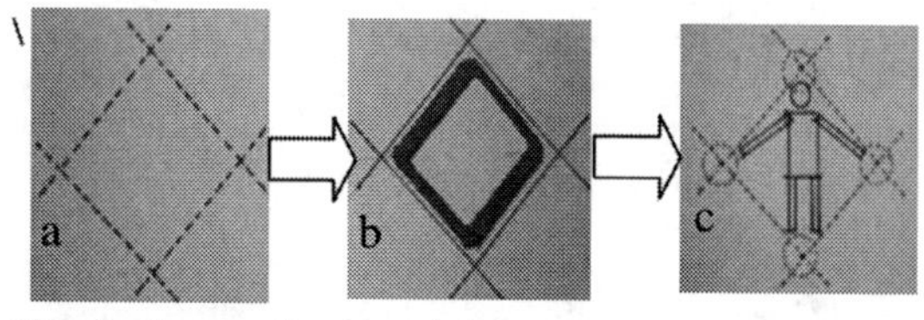

Fig. 3. Creation Process of *Sulapa Eppa'na Taue* (human rectangular) (a)Schematic of letter *"sa"*, (b) script letter of lontara *"sa"* (c) *Sulapa eppa'na taue.*

It can also be seen totality in daily expression indicating that a man can be perfect or have *sulapa' eppa'*, if he can bind his spouse in a marriage, and in Bugis belief that 'only if you are assigned to be a leader, then you must have *sulapa' eppa'*.

Balance of Four Side Areas

According to Pelras (2005), in-depth diachronic study about mythology and literature creates patterns of ethos and beliefs formed in early South Sulawesi society. This can be seen from symbolic images of cloth motifs having strong links with the concepts of real and abstract spaces in Bugis society. The order in Bugis society is shown through the philosophy of *sulapa 'eppa'*, which literally means "four sides", a four-sided world view showing totality.

According to Morrell (2005) that similar philosophy can also be found in many Austronesian societies, and in South Sulawesi, historical manuscripts indicates that this philosophy structures various types of beliefs and practices both in spiritual and material forms, in terms of cosmology, politics, and social life. Furthermore, there are four classifications of the natural environment elements, human life and efforts, for example division of world base into wind, water, fire and soil, leading to totality concept giving mutual correlation which its significance can still survive in contemporary communities (Robinson, 2005).

Throughout the world, the four sides governing the universe, usually following the four wind directions, and often shining from the center of spiritual, can commonly be found in ancient societies, although the universal nature of the concept is modified based on different interpretations, adopted based on specific worldview (Abidin , 1969).

According to Mattulada (1993) that there is an ancient saying that supposedly derived from Arung Palakka proposing a question about the four elements that actually are required as a condition for him to be sworn in as king. The saying goes: "do not count on these four things, namely: courage, proficiency, descent, wealth. Because all of which are corrupted wood; unsuitable in three places; put on the ground, these will be surrounded by powder; soaked in the water, these will be quickly infused by water; put in a charred, the fire will be soon got lost; soaked in the water, these will be quickly infused by water.

Concept of Cosmos Shape

According to Bugis assumption, this universe is in the shape of "rectangular". The concept of a rectangular is so dominant that it influences on the lives of Bugis, so it is their philosophy and life view so called as *sulapa eppa* (four sides), and has the principle of life balance measurement. The philosophy of *sulapa eppa* shows that the Bugis view of life indicates the effort for self-perfect that has been taught by hereditary in an oral manner (Saing, 2010, Abidin 1969).

Based on the source of *lontara* (Bugis manuscripts), one related to the philosophy of *sulapa eppa* states village, *appabolang* yard (plot) and building plan of Bugis stage house; then, all of which will be considered perfect if these are rectangular. So, all of the floor plans and home yards of Bugis stage houses have always pattern of rectangular shape.

In accordance with the philosophy of *sulapa eppa*, so in the past, Buginese, would be called to be perfect if they meet four kinds of natures, namely; wind, fire, water and soil (table 1).

Table 1. Meanings of *Sulapa Eppa* Elements.

Sulapa Eppa Elements	Colors	Meanings (symbols)
Wind	Yellow	Greatness
Fire	Red	Courage
Water	White	Holiness
Land	Black	Magical power

Based on table 1 above, the wind is personified as a yellow color that symbolizes "greatness", fire is the personification of red that symbolizes "courage", while water is the personification of water that symbolizes "holiness", and black is the personification of land that symbolizes darkness and is considered to have 'magical power' (figure 4). The properties that are personified through certain colors also reflect certain meanings as well. This plays an important role in the life of Buginese, as in certain ceremonies, there are the uses of white, yellow royal umbrellas; great color symbolizes the kingdom of heaven as the origin of *Batara Guru* that goes down to earth, then breed as the earth dweller (Saing, 2010).

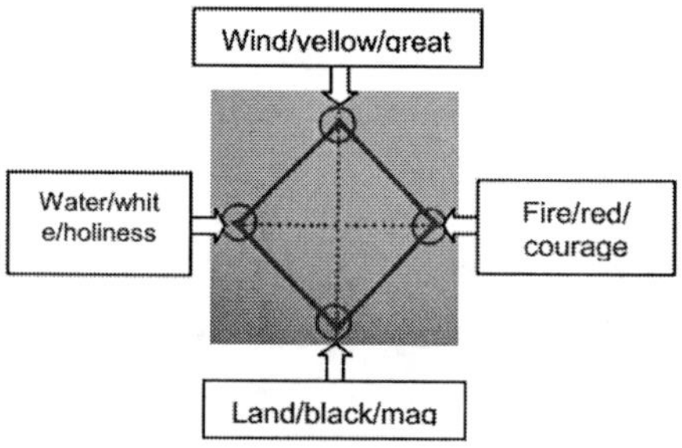

Fig. 4. Personification of *sulapa eppa'* (philosophy of perfectness).

The personification of the *sulapa eppa* colors above can also manifested from the unique way of serving food by Buginese, for example *songkolo* (steamed glutinous rice), usually intentionally colored based on its assignment. Yellow *Songkolo* is presented for "master of wind", red *songkolo* is presented for "master of fire, white *songkolo* is presented for "master of water", and black *songkolo* is presented for "master of land" (Pelras, 2005).

Table 2. Offerings of Sulapa Eppa Elements.

Foods *(songkolo)*	Offerings
Yellow	Master of wind
Red	Master of fire
White	Master of water
Black	Master of land

In table 2 above, the colors of food are also inspired from philosophy of *sulapa eppa*, symbols of the *songkolo* food colors have long been used as part of tribal ceremony procession in Buginese society, and still can be found nowadays.

Stage House as Reflection of Human Manifestation

The house built by the Bugis tribe aims to fulfill their needs, both spiritual and pragmatic, so the establishment of the house is constantly inspired by the values of life, in the forms of character and trends, desires and ideals. The house speaks out human being, it is symbolizes all humans, beauty and greatness of its owner. House is a reflection of a dignified human language (Saing, 2010).

The symbol of the human manifestation (*rupa-tau*), is reflected in the shapes and structures of the house, namely the pole / foot of the house (*aje-ball*) as the poles, the body of the house (*ale-ball*) is the living space, while the head of the house (*ulu-bola*) is the roof (figure 5), and the center of the house (*posi-bola*) is the mast navel of the house. The dimensions of the house (height, length and width of the house) are based on the size of the body parts of the house owners (husband and wife). As the height of the house column is equal to the height of the husband plus the length of the cubit (cycles), the height of the house wall is equal to the height of the wife plus the length of the cubit. The length of the house multiples of the length of the fathom (*reppa*) of the husband, the width of the house is a multiple of fathom (*reppa*) of the wife. Thus, the proportion of the shape of the house is the reflection of the physical sizes (dimensions) of the husband-wife as the owner of the house (Shima, 2006).

The manifestation of the human symbol on Bugis stage house form vertically illustrates that there is greatly important relationships between people and their homes. Their home is not just a residential building but also a reflection of their own identity. In terms of macrocosm and human form, Bugis ancestors combined in form; size, proportion, and structure of their own homes. So that, it can be personified interrelationship between human forms, home forms and macrocosm shapes into a single, binding and integrated trinity. This trinity manifestation is closely related to the philosophy of *sulapa eppa*, which has long been used as the architectural principle for the Bugis community. The philosophy of *sulapa eppa* in Bugis society can be a reference in various fields; such as leadership, social culture, community, housing, settlement and so forth.

Fig. 5. House as the reflection of human manifestation (foot, body, head).
Source: Adaptation by Shima, 2006.

In figure 5, it shows reflection of human manifestation on their houses, as if the houses establish so strong as a human (dweller).

Trinity relation of macrocosms to microcosms and house dweller presents meaning and hope that the house and its dwellers will create harmonious relation with their environment and prevented from any natural disasters, such as earthquake, flood, and tornado. The expectation of dwellers on their house have been realized since the very beginning, started from the implementation of construction, until climbing on the new house); the expectations are delivered by *panrita bola/sanro bola* through any manners and rules that have been applied, in the building in concerned. The advices by *panrita bola/sanro bola* in general are the *intangible* ones, having philosophy of *sulapa appa'* aspects.

Research Methodology

This research is done by qualitative, descriptive, as well as holistic manners (complete, cannot be separated), so the qualitative researcher will not determine the research only based on the research variables, but on the whole social situations under the study including the place, actor and activity aspects that interact synergistically (Sugiyono, 2014). The "*folklore*" analysis method is part of collective culture, which is spread and passed down from one generation to another generation, collectively, traditionally in different versions, both in oral and exemplary forms that are completed with gestures or reminders. Part of a culture called as the *folklore* can be in the forms of people architecture, arts, folk painting, folk music and so forth (Bungin, 2011). The population in this study are traditional Bugis houses, and the samples are new or built houses (place). While, the informants are *Panrita Bola* / architect (actor), and *Panre Bola* / carpenter (activity).

 https://doi.org/10.1051/shsconf/20184104005

Results and Discussion

Research Location

The research location is focused on Bugis community concentration in Southeast Sulawesi Province, namely Soppeng Regency (figure 6). Bugis tribe is quite dominant inhibiting Soppeng Regency, stage houses have been their settlement option.

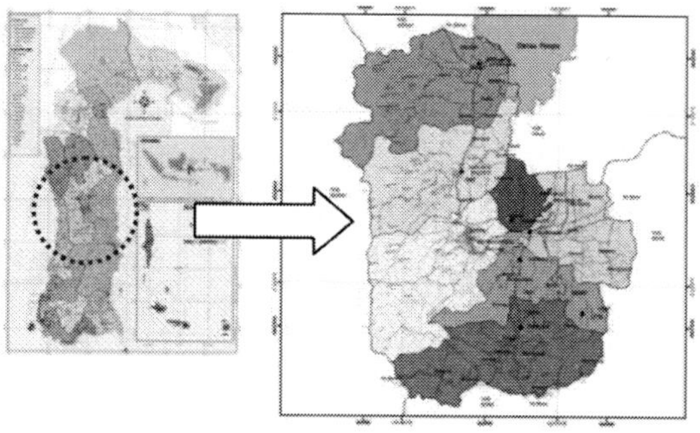

Fig. 6. Map of Soppeng Regency.
Source: Spatial Office Soppeng Regency, 2017.

Based on BPS Soppeng 2016, around 80% (table 3) of Soppeng community inhabits or has wooden stage houses, distributed until to the village. And all of the districts in Soppeng regency can be reached by motorcycles.

Table 3. Types of Settlement and its Materials.

No	Types of settlements /materials	Percentage
1	Wooden stage house	80%
2	Roofing	97,08%
3	Wooden wall	83%
4	Bamboo wall	4%
5	Board floor	80%
6	Bamboo floor	3,83%

Source: BPS, 2017.

Philosophy of *Sulapa Eppa* in Bugis Traditional House

Selection of Plot *(Appabolang)*

Selection of plot or *appabolang* in Bugis Tribe is always presented in a rectangular-shaped field (geometric pattern). There is a concrete shape between the shape of the plan and the shape of the plot, the layout position will be "placed" on the ground in such a way that it forms a positive space outside the layout of the floor plan. There are no negative spaces because the floor plan is always positioned in the center of the plot.

This happens because the water running from the roof is expected to always fall on their own land, and it is strongly prohibited that the water running from the roof fall on any neighboring lands. This condition results in building coverage (BC) by 30: 70 (30% of built area and 70% of yard area), the yard is wider so there are enough rooms to grow fruit and vegetables.

The form of square plot is the most preferred shape, with the shortest side is parallel to the roadside in front of it (figure 7). The rectangular land can be philosophized as *sulapa eppa*, which its elements (wind, water, fire and earth) present the values of life for mankind.

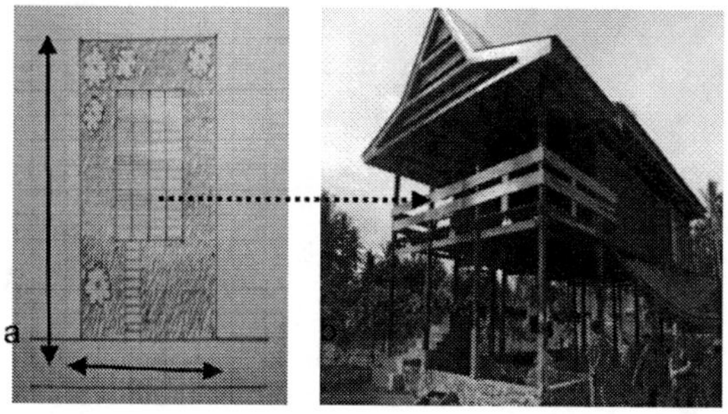

Fig. 7. Bugis Traditional house plot (a) Rectangular shape land , (b) Bugis traditional house.

Plan Shape Pattern

Bugis traditional building plan has always rectangular pattern, namely with wide side and long side. The plan is formed by row of poles (*aliri*) in longitudes and transversely connected to each other by pegs (*pattolo*) between a pole and other pole (figure 8).

The layout of the rectangular embodies the philosophy of *sulapa eppa*, with its elements (wind, water, fire and earth) presenting the values of life for its inhabitants.

As a form of perfection of a new settlement, at the time of climbing on a new house, it is required that every corner of the house is "*diazani*" simultaneously by four people, at sunset.

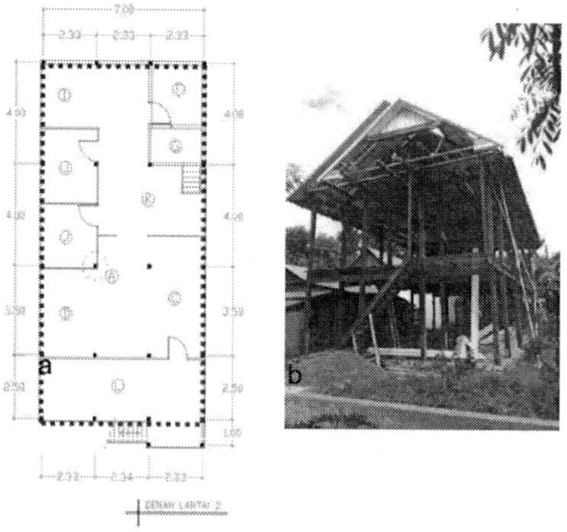

Fig. 8. Bugis Traditional House Plan (a) Rectangular Shape Plan ,(b) Bugis Traditional House.

Structural Shape

The structure and construction of Bugis traditional buildings are arranged to form a construction with *knock down* system. According to Pelras (2006) stating that the form of Bugis stage building construction presents a series of structures in the form of letter "H" (figure 9). The series of the pole (aliri) and the flat peg (pattolo) form a solid stiffness and is the place of load force distribution to the foundation. These structural spaces can manifest the shapes of rectangular, which present the values of sulapa eppa philosophy.

Fig. 9. Series of Rectangular Construction *(sulapa eppa).*

Figure 9 shows rectangular patterns that always dominate the structure and construction of Bugis traditional buildings. The phyilosophy of *sulapa eppa* is the principle of Bugis society.

Spatial Shape Pattern

The spatial patterns that can be found in traditional Bugis buildings generally do not have any chambers but it uses the term of room divider or *"pallawa tengnga"*. The Pallawa tengnga is a divider space, between the front room and the living room, and between the living room and the back room. It is made of thick fabrics with dark color, and flexible because at any time, it can be dismantled or moved, if there is a ceremony or wedding ceremony.

The pattern of spatial shape is always rectangular, following the pole column module position *(aliri)* (figure 10).

Modular-shape space is based on the rows of poles that are transverse and longitudinal. There are three divisions of space abstract functions namely; the front space serves as guest living / room for taking a rest; the medium part is for family room, and the back room serves as girls room and kitchen / dining room.

Fig 10. Bugis Traditional house rooms (a) rectangular shape rooms ,(b) Bugis Traditional house.

Material Dimension Shape

The profile shapes (dimensions) of the Bugis traditional building materials are always rectangular and square. The shape of the material dimension is determined by the function or its loads. For the vertical load distribution, the material dimension is square (pole), while for horizontal load distribution, the material dimension is rectangular (figure 11).

Fig. 11. Material dimension shape.

Figure 11 shows dimensions of material profile having rectangular and square shapes *(philosophy of sulapa eppa).*

Pattern of Door and Window Elements

The pattern of door and window elements in traditional Bugis buildings is always in rectangular shape following the philosophy of sulapa eppa; which contains the meaning of "perfectness" and "life" (figure 12). Windows and doors serve as holes for light and air going into the room.

Fig. 12. Pattern of Door and Window Elements.

Conclusion

The philosophy of *sulapa eppa* is a sacred "term of reference" (TOR) for Bugis community, so that this philosophy can manifest itself as a basic or fundamental in Bugis traditional architecture.

The philosophy of *sulapa eppa* is used as the basis for determination of selecting plot (land), pattern of the floor plan, structural shape, sphere pattern, material dimension, and shape of window / door; which aims to achieve the perfection of life.

The philosophy of *sulapa eppa* is also used as the guidance in the field of leadership (government), social, socio-cultural, welfare and prosperity, housing and settlement, life values and self-perfectness.

References

Abidin, Mr. A. Zainal. (1969). *Falsafah Hidup Sulapa Appaka Orang-orang Bugis-Makassar (Pandangan Hidup Segi Empat).* Bingkisan Budaya Sulawesi Selatan Nomor 12 tahun II, Agustus Ujung Pandang : Yayasan Kebudayaan Sulawesi Selatan dan Tenggara.

Adimihardja, Kusnaka dan Parnama Salura, (2004) *Arsitektur dalam Bingkai Kebudayaan*. Jakarta : Architecture and Communication.

Bungin, Burhan. (2011). *Metodologi Penelitian Kualitatif Aktualisasi Metodologis ke Arah Ragam Varian Kontemporer.* Jakarta : PT. Raja Grafindo Persada.

Dian, Mas. (1999). *Pengamatan terhadap Arsitektur Tradisional di Asia.* Simposium Nasional Naskah Arsitektur Nusantara : Jelajah Penalaran Reflektif Arsitektural. Surabaya : ITS.

Hamid, Abu. (1978). *Bingkisan Budaya Sulawesi Selatan.*Ujung Pandang : Yaysan Kebudayaan Sulawesi Selatan.

Hamzah, Ny.Aminah P dkk.(1984). *Monografi Kebudayaan Bugis di Sulawesi Selatan.* Ujung Pandang : Kanwil P dan K Provensi Sulawesi Selatan.

Mardanas, Izarwisma. (1986). *Arsitektur Tradisional Daerah Sulawesi Selatan.* Ujung Pandang : Proyek Inventarisasi dan Dokumentasi Kebudayaan Daerah Sulawesi Selatan.

Mattulada. (1995).*Latoa satu lukisan analitis terhadap antropologi politik orang Bugis.* Ujung Pandang : HasanudinUniversity Press.

Morrell, Elizabeth. (2005).*Simbolisme, Ruang, dan Tatanan Sosial* (Tapak-tapak Waktu Kebudayaan, Sejarah, dan Kehidupan Sosial di Sulawesi Selatan, penyunting Kathryn Robinson dan Mukhlis Paeni). Makassar : Ininnawa.

Moleong, Lexy J.(2000). *Metodologi Peneltian Kualitatif.* Bandung : PT. Remaja Rosdakarya.

Pelras, Christian.*Manusia Bugis.* (2005). Jakarta : Nalar bekerja sama dengan Forum Jakarta-Paris.

Robinson, Kathryn. (1993). *The Platform house : Expressions of a regional identity in the modern Indonesian nation. In Virginia Matheson Hooker (ed) Culture and Society in New Order Indonesia.* Kuala Lumpur : Oxford University Press.

Robinson, Kathryn. (2005) Tradisi Membangun Rumah di Sulawesi Selatan (Tapak-tapak Waktu Kebudayaan, Sejarat, dan Kehidupan Sosial di Sulawesi Selatan. Makassar : Ininnawa.

Saing, Ama, (2010). *Arsitektur Tradisional Rumah Adat Bugis-Makassar (suatu tinjauan arkeologis).* Makassar : Indhira Art.

Shima, Nadji Palemmui, (2006). *Arsitektur Rumah Tradisional Bugis.*Makassar : Badan Penerbit UMN.

Sugiyono. (2014). *Metode Penelitian Kuantitatif Kualitatif dan R & D.* Bandung : Alfabeta.

Waterson, Roxana. (1993).*Houses and the Built Environment in Island South-East Asia* : Tracing some shared themes in the in the uses of space.

 https://doi.org/10.1051/shsconf/20184104006

Social Spaces Territory of Bajo Tribe's Settlement

Syahriana Syam[1,*], *Ananto* Yudono[1], *Ria* Wikantari[1], *Afifah* Harisah[1]

[1]Hasanuddin University, Indonesia

Abstract. Kampung Bajo along the bay of Bone South Sulawesi, inhabited by Bajo tribe and Bugis tribes coexist so that there is social-cultural interaction. Since 1905 the life of the Bajo tribe has changed,they began to open themselves to relationships, initially exchange the staple for its survival. As a result of these relationships, dynamics arise in the use of space both micro space, meso and macro space in the scale of settlements, making it very difficult to see the limits of territory in the use of space. Territory deals with spaces with a certain extent where individuals or groups use and defend their exclusive territory. Seeing this, the research is important and interesting to do, focused on the concept of the territory of social space in the context of the interaction of two ethnic Bajo tribe and Bugis tribe in the form of settlement, using phenomenology method.

Keywords: Territory, Settlements, Social Space, Bajo tribe Bugis tribe

Introduction

An understanding of the cultural heritage treasures of archipelago architects is needed for practitioners of architecture for the development of creative quality. The increasing exclusion of local wisdom of archipelago architecture, encouraging the effort to save him with documentation and in-depth research. Especially local wisdom in the field of residential architecture that has a unique high value for environmental science, for the development of new theories, because it has a local space. (Sudaryono, 2003)[1]. History proves the existence of the Bajo tribe, beginning with its arrival at the coastal location, then settled and developed for generations to have the uniqueness as a marine man.

The changing environmental conditions and lifestyles of the Bajo Tribe, according to Rapoport (1969)[2] that change is the result of the process of adjustment activity carried out by man himself so that his needs can be fulfilled. . The change of the sea into the settlement of Kampung Bajo has an impact on the change of building form which produces social space as a container of interaction. Lefebvre (1991: 26)[3] that the social space of a social product, which is a specific representation of the social interaction between production and reproduction. The perception of space between them can be clarified with territory to know the limits of mastery of space clearly.

Altaman and Chemers (1984: 121)[4] territory is the control over the use of a place or object, the control of an unauthorized person or the use of others who are permitted to a limited extent. Territory concept defined as a spatial and physical boundary or marker is primarily the possibility of other party intervention, so that each group has a different expression in the phenomenon of territory and its privacy. The concept of territory of social space in the context of the interaction of the two ethnic Bajo tribes and the Bugis tribe in the form of settlement, is, a comprehensive study of the social aspects of settlements that will study the formation of territory, which will enrich the concept of territory in the treasury of architectural theory.

This research uses the phenomenology paradigm, which an empirical study that investigates social phenomena in the context of settlements in Bajo with locus in Kampung Bajo, Bone regency of South Sulawesi. The result of this research is to find out the concept of social space territory that is formed and show the development of architectural theory about the concept of territory in Bajo tribe settlement formed by social aspect (interaction, behavior, social structure of society) and culture.

Characteristics of Bajo Tribal Settlements

Rossi (1982: 59) explains that the components of settlements include roads, open spaces and types of buildings that involve physical elements of nature and non-physical (human) both individually and community groups along with all life activities, then Rossi explained that the neighborhood is a combination between footprint, event and sign (sign). Therefore, the settlement environment indicates the existence of physical and non-physical impact on the site which as a whole is a sign of a certain event, so that it can be characteristic of a settlement environment.

The elements of nature, environment, culture and residence are a unity, in which the natural environment shapes the culture of society, and culture is manifested through the shelter formed by the condition of the natural environment. The element of the natural

* Corresponding author: sary_archi@yahoo.com

https://doi.org/10.1051/shsconf/20184104006

environment is the most important element and can affect other elements, where the natural environment has territorial boundaries in nature. So that characteristics can be obtained from the physical condition of the environment, culture and social life related to the history of society through space and time.

Fig. 1. Research Sites of Bajo Settlements of Tanete Riattang Timur Subdistrict of Bone District of South Sulawesi.

Figure 1. Provides an overview of Kampung Bajo neighborhood in Teneteriattang Timur District of Bone District. Characteristics and uniqueness of the Bajo Tribe as a study area, geographically, located along the coast of Bone Bay, and the sea as its main orientation. The history of the existence of fishermen settlements began from the arrival of a group of ethnic Bajo tribes in the waters of Bone Bay, living in groups on the boat, then develop from generation to generation coexist with Bugis tribe.

The settlement areas on water from boats to houses on stilts, dominated by fishing settlements, are built with their local knowledge that is influenced by geographical conditions, sea orientation. Furthermore, the building developed into three zones, namely; buildings on the water, between water-land, and on land.

Social Space of Bajo Tribal Settlements

Some traditional Indonesian settlements originating from locality, meaning macro-micro cosmos, functional and socially symbolic meaning can be seen from beliefs, customs, traditions. rituals to form relations between residence and environment according to the cultural characteristics of the community. They will depend on the condition of the natural environment, as does the Bajo Tribe with the sea.

Rapoport (2005: 24)[5], that the environment can be understood as organization of space, organization of time, meaning, communication or interaction, Rapoport (2005: 32)[5], there are three elements forming a setting that is, fixed element in the form of infrastucture, remains surrounded by delimiters that are relatively fixed and not easily shifted. Semi fixed element, in the form of furniture and all the completeness of the environment either inside or outside the environment, can move. Non-fixed elements of human form and existing activities, behavior, clothes, hair shape, animal, or animal. Existing properties emphasize the functionality and users of the space components. The hierarchy of space formed in the settlements of the Bajo tribe, oriented towards the direction of the growth of settlements from the sea to the mainland. Sequentially, increasingly toward the sea (water zone) inhabited by Bajo tribe, and towards the mainland inhabited by Bugis tribe, but facilities and infrastructure are shared, peaceful coexistence. In table 1, explains some of the interaction spaces used by Bajo villagers

Table 1. Elements of the Social Space of Bajo Tribal Settlements.

Elements of Social Interaction Space			
Public Space Area	**Favorite Places**	**Territory Symbol**	**Activities**
On The Land	Environment Street	Non-fix	Gathering, Chat, Playing, and Parking Place
	Footpath	Non-fix Flower pot, clothesline	Sitting and Chat
	House Front Yard	Non-fix	Sitting, Gathering, and Chat.
	Stairs	Non-fix	Sitting, Chat, look for fleas on each other.
	Lego-lego	Non-fix Table chairs	Sitting, Gathering, and Chat.
	Stall	Non-fix	Gathering, Buying and Selling Activity.
	Workshop	Non-fix Cupboard	Sitting, Motor and Bike Service.
	Empty Land	Semi fix a pile of stones, sand, animals	Drying Seas Product, Sea weed, and Play ground for kid.
	Under The House	Fix Column house, wall	Food Stall, Cooking, Resting, Eating, and Gathering.
Transition Water - Land	Tetean	Fix bamboo, wood	Acces for a residential above the water, sitting and playing for kid.
	Levee	Fix streets, stone walls	Sitting, Rilexing, Playing, and Eating.
	Bridge	Fix	Gathering and Playing.
	Dock	Fix	Boot Mooring, Gathering

			and Playing.
	a Mound of rocks above the sea	Fix	Sitting and Fishing.
	Lopi-lopi	Semi-fix	Sitting, Gathering, and Resting.

The house is located on the water (based on the tide of sea water, that the position between one row (saf) with another row in the house there are boat streets, and akse directly into the open sea through the culvert under the bridge, so that each boat can be moored next to their homes.

To connect one house with another house on the water, made tetean from bamboo or wooden planks. While the existing house on land use paving block as a hardening for access path. The layout of the buildings lined up along the coast, each with access to water. With the agreement of the people and on the basis of mutual assistance, the pile of rocks is made as breakwaters in the event of high tide. In addition, the stone tumbukn also serves as the boundary of their land plot, so that the land is also used as an area where tether their lopi-boat (small boat) at the time of berthing (not down to the sea).

Table 2. Periodization of Bajo Tribal Settlement Environment.

Dev. Period	Up to 1800s	1850s	1900s	1920s	1950s	1970s	1980s	1990s-at present
Location	Cellu	Bajoe	Lassareng	Lassareng	Lassareng	Bajoe	Bajoe	Bajoe
Way of living	Nomadic	Nomadic	Nomadic	Semi-settled	Semi-Settled	Settled	Settled	Settled
	On-sea water	On-sea water	On-sea water	On-sea water	On-sea water	On-sea water	On-sea water	Coastal water (intertidal)
				Coastal water (intertidal)	Coastal water (intertidal)	Coastal water (intertidal)	Coastal water on-land	On-land On-sea water
	Cluster	Cluster	Cluster	Linear rows	Linear rows	Linear rows	Linear rows	Linear rows Cluster
	Seafaring	Seafaring	Seafaring	Seafaring	Seafaring	Seafaring	Seafaring	Seafaring Others
Type of dwelling	Boat (bidok)	Boat (bidok)	Boat (bidok)	Temporary shelter (Babaroh)	Temporarys helter (Babaroh)	Semi-permanen shelter (*papondok*)	Semi-permanen shelter (*papondok*)	Permanent house (rumak)
					Semi-permanent shelter (*papondok*)	Permanent house (rumak)	Permanent house (rumak)	
Territory symbol	Taguk Pulih Connection 				Taguk Pulih dan Tetean Connection  			Connection tetean / road, bridge tetean / road, bridge /tetean / road, bridge

Bajo tribal settlements begin when they feel safe to settle in groups living on a boat (boat) consisting of 15-20 families in the bone bay. Each house is declared as one household although there are more than one family in it. The composition of his house was built in such a way, in accordance with their philosophy patepik-tepik dipadijer (gathering, lined and blessed).

According to Table 2, the history of the changes that took place in the Bajo tribal settlements began with the nomadic life of Lassareng Island, a coral island far from the mainland, and eventually settled in Kampung Bajo and lived side by side with the Bugis. As they live in the sea, spatial batters are used as markers of territory, their kampung habits. Around 1970, the Bajo tribe began to settle and built a stage house (papondok), then in 1980 became a permanent home, in 1990 until today settlement land on water and on land based on tides.

Although in general the orientation of the tribe of Bajo houses faces the sea (east), but in fact found a house on the land there is facing the road. By reason of respect for the road. In their efforts to keep the belief that their home should face the sea as a tribute to the sea that is considered their only source of life, they also create a special door that opens directly toward the sea, so the back of the sea is not firmly apparent.

The seven zoning spaces in the picture above in figure 2, can be found at the house of the tribe of Bajo. The first zone, an outer space consisting of awa ball, addeneng (take home, and ladder elements), which functioned as access to territory on the outside of the house without waiting for permission from homeowners. The agreement was understood by the whole Bajo tribal community, and formed temporal territory. But in this study, is focusing on the space of social interaction, resulting in social territory.

Users in the Bajo home tribe's settings are five household heads, tagga housewives, family members, relatives, neighbors / guests. This can illustrate how the relationship and complexity of activities that take place within the house. Activities that are in the house and related to the sea merupan adaptation of zonation results that form social territory on Bajo tribal settlements.

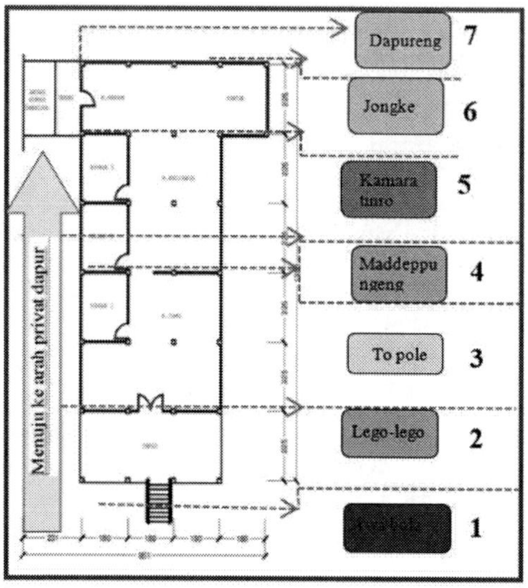

1. Information:
   1. The bottom of the house (yard, stairs)
   2. Terrace
   3. Living room
   4. Family gathering room
   5. Bedroom
   6. Service area, dining room, bath / Toilet
   7. Kitchen

Fig. 2. Hierarchy home space Suku Bajo.

Of the seven zonations, the lego-lego section is an area that is considered to have an important role in the social interaction activities of the community. The territory of the lego-lego room is accessible to all family members and guests. Various activities that exist in this place, such as a place to gather, sit, rest, a place to store the seafood such as fish, drying places, stalls or selling. Someone who comes to visit or visit, homeowners can give access to enter the territory topole (living room).

Overall, social space of the Bajo tribal settlements can be categorized into two zones ie on land and between water-land, favorite locations such as (sea, roads, bridges, tetes, lego-lego or the front of the house) . In this space the activities of trade, social, gathering, rest and have created a strong emotional bond between citizens, thus obscuring the territory of space.

Conclusion

Territory formed in Bajo settlement is inseparable from historical values, natural environment condition, sea as main orientation and privacy demand in the form of motivation of economic interest. Territory categories that have been unique to the uniqueness, social aspects and similarities of marine and economic cultural backgrounds can foster a strong sense of kinship, thus eliminating the boundaries of space and time in their daily interactions and activities. Bajo tribal people enjoy doing private activities together, with awareness or unintentionally intervene in the privacy space by passing the territorial boundaries.

References

Altman.I, Chemens, M 1984. Culture and Environment, Brooks/Cole Publishing Company Monterey, California.

Lefebvre, Henry (1991), *The Production of Space*, Blackwell Oxford, UK & Cambridge, USA

Lowe, 2003, The Magic of Place: Same at Sea and on Land in Sulawesi, Indonesia, in Bijdragen tot de Taal, Land-en Volkenkunde, 2003, Leiden, http: www.kitly-journls.nl

Mulyati, Ahda, Sama in Lao as the Foundation for the Establishment of Bajo Tribal Architecture in Central Sulawesi's Kabalutan Island, Dissertation, PPS of Gadah Mada University Yogyakarta.

Purbadi, YD, 2010, Tribal and Spatial Arrangements on Dawan Tribal Settlement Architecture in Kaenbaun Village on Timor Island, Dissertation, PPS Universitas Gadah Mada Yogyakarta.

Rapoport,A, 2005, Culture Architecture and Design, Locke Science publishing Company, Inc, Chicago, Illinois,USA.

Said, Ratriana, 2016, Diversity and Utilization of Public Space in RusunawaMariso, Makassra-Indonesia Jurnal JAABE.

Syam, Syahriana, 2016, Community Attachment to Public Space Vernaculer on Ethnic Settlement, The 10th International Symposium on City Planning, Makassar Indonesia.

The Main Elements of Historical Objects as "Spirit" of Cultural Heritage in Yogyakarta City

Pawitra Sari[1,*], *Aris* Munandar[2], *Indung* Sitti Fatimah[2]

[1] Magister Student of Faculty of Agriculture, Bogor Agricultural University, Indonesia
[2] Lecturer of Faculty of Agriculture, Bogor Agricultural University, Indonesia

Abstract. The history of Yogyakarta City still holds many "millions" of tangible and intangible meaning. The history has brought Yogyakarta City known as a City of Philosophy that is realized in the spatial layout of the city, one of them is the philosophy axis. In this research, the historical objects have been studied the main elements that become the icon and to be grouped into four objects. There are TuguArea, Malioboro Area, Zero Kilometer Area, and South Square Area located on the philosophy axis.Another consideration, these four objects are supported by other space forming elements become a place that has a "spirit" of cultural heritage to gather and interact with local communities and tourists in the public space. The purpose of this research is enriching knowledge about the main elements of historical objects. Besides that, the purpose of this research is to support Yogyakarta City to realize world Heritage City by UNESCO about cultural heritage. This research uses mix methods. Qualitative methods by conducted the survey and collected the literature. The observation and interviews towards informers and involves the perception of local communities and tourists with 327 respondents in Yogyakarta City. While the quantitative method for a non-parametric test with different test Mann Whitney U using SPSS 17 program. This research uses spatial and descriptive analysis to explain the main elements of historical objects and examine their relationship with local communities and tourists perception. The result of this research that Malioboro Area has the important corridor in historical value that associated with historical events and historical figures, style building, and culture. As well as Malioboro public space that has the meaning of the spirit of life for urban residents and the heart of the city, which means the inheritance of all things. The results of this research are expected to provide of place attachment towards all mainly historical objects "spirits" of cultural heritage for people to gather and interact. The importance of the psychological dimension of the environment to help the psychological well being in urban as a concern for architects, planners in shaping, developing a city in the future, especially public space. Also, the results of this research are expected Yogyakarta City to keep become a magnet for the traveler as Tourism Destination Object.

Keywords: cultural heritage, historical objects, public space, spirit of place

Introduction

Yogyakarta is one of the cities that has a wealth of history and culture that are still found until today. History of the founding of Yogyakarta began from the Mataram Kingdom in the 17th century until the entry of Dutch colonial. The collapse of the Mataram Kingdom was marked by the signing of the Giyanti Agreement between Pangeran Mangkubumi with VOC (Vereenigde Oost-indische Compagnie) (Dutch) on 13 February 1755. In this agreement has explained that Mataram was split into two regions, there are Surakarta and Yogyakarta. The concept of spa-tial planning of Yogyakarta City cannot be separated from Pangeran Mangkubumi who later titled Sri Sultan Hamengku Buwana I.

Yogyakarta City as a City of Philosophy, in spatial layout, was applying the philosophical value of culture as the image of the city. Lynch (1960) was mentioning that in the development of a city, the image of the city has a role as a former of the city identity, and as an adder to the attraction of the city. Therefore, the image of the city has a clear and strong will affirm the identity and face of the city so as to make the city has an attraction and interesting. In this research, the main elements of historical objects are also the mental map elements of the city (cognitive map) which important components of physical (tangible) have special meaning in history and culture and other important values, and the formers of the image of the city.

The historical objects have been studied in this research are the main elements that have been grouped into four objects. The four objects have a special (high value) that are on the philosophy axis. Besides to the primacy of the value of historical objects, each of the four objects is an icon or symbol of Yogyakarta City

* Corresponding author: rrpawitrasari@gmail.com

which has a spirit of place. Tugu, Malioboro, South Square, and Dutch colonial buildings at Zero Kilometer located in the heart of the city that has a spirit of place for gather people (public). A spirit of place of the four objects now has provided a public space for gathering and interacting, especially for local communities (natives and newcomers). Also, refers to the main elements of historical objects sur-rounded by other space forming elements (support/secondary elements) physically.

Public space in this research is being in a revitalization effort by the Yogyakarta Government, such as Zero Kilometer Area. While Tugu and Malioboro Area have been completed the revitalization phase. Currently, the four objects have a public space that becomes a place for the communities to gather and interact as well as a cultural tourism for the tourists. In this research, the purpose is enriching knowledge about historical objects. In this paper, the researcher describes how substantial the local communities (public/society) and the tourist's assessment towards the historical object, mental map elements of the city (cognitive map), and a spirit of place based on the perception of the four objects. Certainly also sharpened by involves informers[1,2,3] (resource persons) who had competence in the disciplines of urban spatial, culture, and architecture.

Literature Review

In 1972 the UNESCO General Conference adopted the Convention Concerning the Protection of the World's Cultural and Natural Heritage in article 1 is explained that cultural heritage by the following classifications: monuments, groups of buildings and sites. In operational guidelines for the implementation of the World Heritage Convention in article 4 is explained that "the cultural and natural heritage is among the priceless and irreplaceable assets, not only of each nation but of humanity as a whole. The loss, through deterioration or disappearance, of any of these most prized assets constitutes an impoverishment of the heritage of all the peoples of the world. Parts of that heritage, because of their exceptional qualities, can be considered to be of 'Outstanding Universal Value' and as such worthy of special protection against the dangers which increasingly threaten them" (UNESCO, 2005).

"OUV means cultural and/or natural significance which is so exceptional as to transcend national boundaries and to be of common importance for present and future generations of all humanity. As such, the permanent protection of this heritage is of the highest importance to the international community as a whole. The Committee defines the criteria for the inscription of properties on the World Heritage List. (UNESCO, 2005)". The process of world heritage assessment consists of three main pillars, nominated inheritance must meet at least one criterion established by UNESCO, must meet the criteria of integrity and authenticity, and must meet the criteria for protection and management.

[1] Mr. Yuwono (Scope of Spatial Planning)
[2] Mr. Teguh (Scope of Culture)
[3] Mr. Ruli (Scope of Culture)

Department for Culture of Yogyakarta Government (2015) that Yogyakarta City has fulfilled three criteria contained in the OUV. There are criteria I, represent a masterpiece of human creative genius; criteria II; exhibit an important interchange of human values, over a span of time or within a cultural area of the world, on developments in architecture or technology, monumental arts, town-planning or landscape design, and criteria VI; be directly or tangibly associated with events or living traditions, with ideas, or with beliefs with artistic and literary works of outstanding universal significance (Yogyakarta City of Philosophy, 2015). The formation of Yogyakarta City is essentially an implementation of the concept of Pangeran Mangkubumi (Sri Sultan Hamengku Buwana I) based on the understanding of the philosophy of human life in the form of setting the architectural of the building, vegetation arrangement, cultural values order (philosophical-ritual), and government regulation (Department of Culture, 2015).

In a broad sense, cultural heritage is a result of human processes and activities rather than a biophysical product (Aplin, 2002). Heritage can grow individuals who are associated with the main elements of the city with regard to cultural heritage. Inheritance can be an important life as part of the hereditary history and tradition in preserves cultural capital (Bourdieu, 1984). It can be said that the object of heritage value is a former of the city identity that re-quires a long time process and sequence of occurrence in shaping it. Lynch (1960) has the concept of legibility or ease of a place under-stood cognitively and 'read' by a person so that the concerned can be oriented in the environment, related to paths, nodes, edges, districts, and landmarks has become the basic infrastructure for various advanced studies, and an emphasis on the physical order of space. The physical identity that is easily captured by the observer is an object that is used as a reference (navigator) of the region. Environments identity is not only based on the buildings but also can be components such as street elements and others.

The main elements of historical objects in this research have an appeal in the historical aspect, the importance of historical objects, uniqueness and authenticity element, as well as other 'magical influences'. The 2008 ICOMOS Quebec Declaration on the Preservation of the Spirit of Place (ICOMOS, 2008) sought to connect ideas of intangible heritage with non-tangible concepts of material place, such as genius loci, under the concept of the spirit of place. Norberg-Schulz (1980) describes the genius loci, reflecting the uniqueness of a place, distinguishing one place from another. Without the presence of genius loci, a place has no meaning and will arise events that are not personal but general. It can be fully revealed that the genius loci is the soul/energy of place. These spirits inhabit all places and give meaning, keep and inspire the place with feelings. The genius loci symbolize individual strengths that complement a place with personality and character in the form of a quality that is more than a fact. The task of the genius loci is to maintain a feeling of an animating presence.

A spirit of place can affect the quality of life better for the public welfare. This matter enabling mutual interaction of residents, and thereby improving their quality of life (New-man,1980). Related with social value, culture may provide a key contribution to the city's social assets and cohesion and to their social sustainability (see, e.g., Tavano Blessi, Tremblay, Sandri, & Pilati, 2012), as well as to the development of a sense of place (e.g., Sandler, 2007). Duany, et al. (2000) says the presence, for example, green open space or waterfronts, not only provides people with a place to engage in various recreational activities but also offers to escape from the routine of daily life, making more happy, more healthy, and more relaxed, and ultimately improving the quality of life. Like the square, a type of civic places that allow people to socialize and inter-act with each other in the public (Carmona, Heath, Oc, & Tiesdell, 2003; Childs, 2004; Cooper-Marcus, C. & Francis, 1998; Lynch, 1981; Tibbalds, 1992).

There is a deterministic belief among architects and urban planners that well-planned and developed public spaces generate socially desirable behavior and attitudes (Alexander, 1977; Gehl, 1987; Project for Public Spaces, 2000; Whyte, 1980). Truly public spaces encourage social interaction among individuals with diverse interests, opinions, and perspectives (Young, 1990; Németh, 2006). Most of the research in this area focuses on 'flagship' urban public spaces, and the changes in their level of publicness following a redevelopment or improvement process (e.g. Akkar, 2005a, 2005b; Madanipour, 1995, 2003; Van Melik, Van Aalst, & Van Weesep, 2007). The UK government also recognizes the links between the built environment – particularly built heritage – and quality of life (Tweed & Sutherland, 2007). With considering how built heritage contributes to the satisfaction of human needs.

Methods

This research was conducted in Yogyakarta City which is along the philosophy axis. The philosophy axis is a straight line that longitudinal north-south from Tugu Yogyakarta to Panggung Krapyak which is right in the center or core of Yogyakarta City. The philosophy axis is used as the basis for taking the object of this research, there are four objects selected in this research, that is;

a) Tugu Area,
b) Malioboro Area,
c) Zero Kilometer Area, and
d) South Square Area.

The four objects located in the cultural heritage of Yogyakarta City which becomes 'the silent witness' the establishment of Yogyakarta City. It shows the great history have occurred in the past that carry the Yogyakarta City is known as the City of Culture. The number of relics of the building as a valuable asset brings impact to the Yogyakarta City until now. Where the relics of these buildings, especially the four objects in this research as well as the main elements of the city that is easily recognized/memorized for the local communities and tourists. Another reason for the primacy of the four objects chosen because it is located in the heart of the city in Yogyakarta which becomes 'the breath of people's lives'.

This research uses mix-methods. Qualitative methods by conducted the survey and collected the literature descriptively. The field sur-vey included observation and interviews with resource persons (stakeholders) and distributed questionnaires of 327 respondents. Questionnaires were distributed to involves local communities perceptions of 260 respondents (170 were natives and 90 were newcomers, (with a ratio of 65:35)) and tourists were 67 respondents (62 were domestic tourists and 5 were foeign tourists) in Yogyakarta City. Comparison of the number of respondents between natives and newcomers is obtained from the information of Yogyakarta Government (Central Bureau of Statistics) which is then calculated from the magnitude of the population in Yogyakarta City, especially the sub-districts passed by the philosophy axis.

Questionnaires are data collection techniques to capture public perceptions towards questions related to the research by giving writ-ten questions to a select number of people and can be used as samples to represent the com-munities in the research area. In this research the distribution of questionnaires using stratified random sampling. Assessment of the perception of local communities and tourists con-ducted in this research with qualitative methods, which then analyzed descriptively by SPSS version 17. The level of significance/level of trust used in this research is 95% (0.05), which is generally conducted in social research. Therefore the data comes from the same variance (homogeneous), then it is done by using non-parametric test with a different test "Mann-Whitney U", which is included in quantitative methods.

Result and Dicussion

The Study Area

Four objects located on the philosophy axis in Yogyakarta City were chosen as the set-ting for the study. The objects research is Tugu, Malioboro, Zero Kilometer, and South Square (Figure 1). This research is in the first ring with delineation of natural elements that's the river and man-made elements that's the building. Delineation of in the east is Code Riv-er, in the west is Winongo River, in the north is Tugu, and in the south is Plengkung Gading (Figure 2).

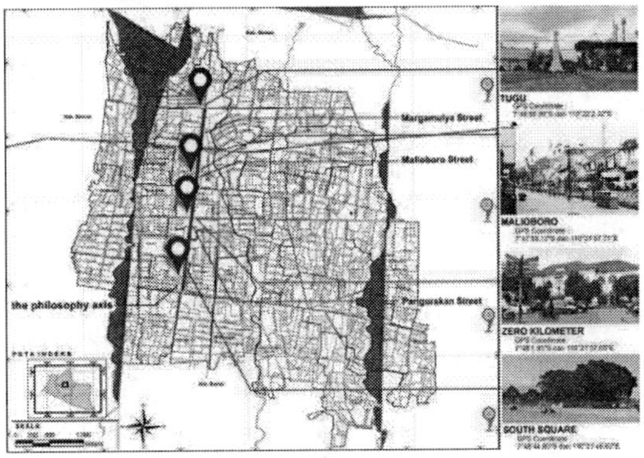

Fig. 1. Research Area on the Philosophy Axis.
Source: Department for Land and Spatial Planning, 2015, modified by Author 2017.

https://doi.org/10.1051/shsconf/20184104007

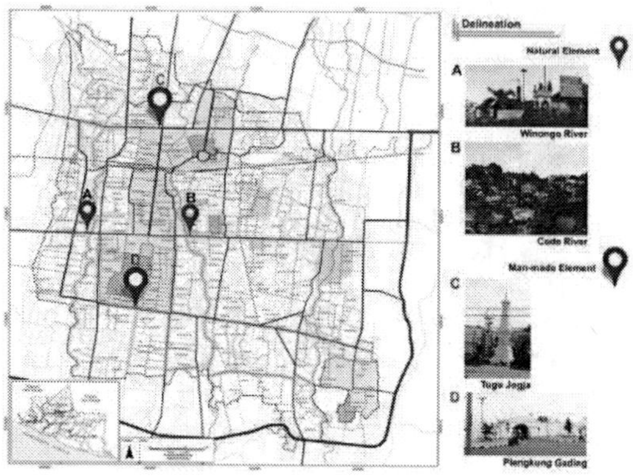

Fig. 2. Delineation of Research Area.
Source: Department for Land and Spatial Planning, 2015, modified by Author, 2017.

The Tugu was built by Sri Sultan Hamengku Buwana I shortly after Sultan occupied the Kraton (palace) in 1756. On June 10, 1867, earthquake happened to demolish the building of Tugu. The Tugu was rebuilt that was planned by Opzichte van Waterstaat JWS van Brussel. Under construction supervision by Patih Danurejo and inaugurated by Sri Sultan Hamengku Buwana VII accompanied by Resident Yogyakarta J. Mullenmeister on October 3, 1889. The Tugu itself located in Jetis Sub-district, where Jetis Sub-district has Indische Architectural Style (Figure 3). Conditions in Jetis Sub-district, now is a lot of high hotels, less adopt local culture, and buildings that do not interpretation 'elements of Jogja'.

Fig. 3. Architectural Style in Jetis Sub-district.

The Malioboro etymologically has to mean is 'be a wandering wali'. The function of Malioboro in 1755-1945 was ceremonial, politics, and economic, in 1945-1950s changed into politics, culture, and economic, and in 1950 until now there is a change of Malioboro function as economics and tourism (Septirina SN, et al., 2016). Malioboro condition was found many vegetation planted on the east side of the road, as well as the pedestrian area the east of the road is wider than on the west side of the road. In 2017, the revitalization of Malioboro has just been done, now Malioboro is equipped with furniture pedestrian, such as benches, dump on the east side of the road (Figure 4). While on the west side of the road is still the same as in previous years used for a particular lane, such as Andong, pedicab, and bicycle (Figure 4). The building style in Malioboro Area is Indische Architectural Style, Chinese Architectural Style , Javanese Architectural Style, and mixing (Figure 4).

Fig. 4. Pedestrian and Architectural Style in Mali-oboro Street.

The Zero Kilometer located nodes of Margamulya Street and Pangurakan Street. The Zero Kilometers are important buildings among other of them is the Indonesian Post Office, BNI 46, Central Bank, Vredeburg Fortress, Presidential Palace, and others (Figure 5). The architectural style of the building in Zero Kilo-meter is the Indische/Transitional /Modern Co-lonial Architectural Style. In addition, many important events occur in Zero Kilometer which is a dispute between Indonesia, Netherlands, as well as Japan in seizing the territory.

Fig. 5. The Important Buildings in Zero Kilometer Area.

The South Square Area was originally used as a training ground for warrior and served as a procession lane in the funeral ceremony of a Sultan's who will be buried in Imogiri (tombs of kings and families) (Department for Culture, 2015). Today, South Square serves as a place visited by people and tourists to enjoy historical objects. The point that is still interesting is the existence of two Banyan Trees (Ficus benjamina), the attractions at night, and at the weekend there are events such as archery (in Javanese; jemparingan), bird competition, or something related to intangible cultural heritage. Not far from South Square there is Plengkung Gading, the buildings of the Kraton complex, such as Sitihinggil, Kedhaton yard, and residential buildings of the community with Javanese Building Style (Figure 6).

Fig. 6. Tangible Cultural Heritage in South Square Area.

Yogyakarta until now recognized as a locus (place) of Javanese culture which is widely known. Cultural heritage not only about tangible heritage, but also intangible heritage. Based on observations and literature, the intangible heritage still held by the Government, the Kraton, and the community of several things can be seen in Table 1.

Table 1. Intangible Heritage.

Items of Intangible Cultural Heritage	
Characterization	The mythical story, event or festival of culture, beliefs, music, song, performing arts, traditional culinary arts, ceremony, and fabric/cloth.
Photos (represent only)	

The Respondents

Contribute to local communities perception in assessing the important elements of historical objects can provide their views on the relics of tangible cultural heritage. This is very giving of information, especially in planning the Yogyakarta City in the future as a city known for its culture. Criteria of respondents selected in this study are natives and newcomers who live around the philosophy axis that is expected to give an explanation of how aspects of the place attachment, namely the local communities to the architectural heritage that is around. Data collection using questionnaires was distributed to 170 were natives and 90 were newcomers.

In this research, local communities respondents are natives aged between 20-30 years old as much as 40.6% and aged between 31-40 years old as much as 29.4%. While new-comers, the majority of the survey respondents aged between 20-30 years old were 67.8%. Gender for both natives and newcomers respondents was dominated by men who achieved ± 52%. The last level of education is mostly senior high school, it can be said that they are currently taking Diploma/Bachelor-1. The last education from natives respondents as much as 56.5% and newcomers respondents as much as 68.9% are senior high school.

The involvement of tourist perception in this research also becomes a part in giving the assessment to the heritage building which is in the philosophy axis as the icon of Yogyakarta City. Distribution of questionnaires given to the tourist's respondents as much as 67 people. Criteria of respondents selected were tourists who have visited/toured in four objects. A total of 67 respondents who were interested to fill out the questionnaire, 5 respondents came from foreign tourists who came from India, Japan, Singapore, and Spain which was at that moment in Malioboro Area and Zero Kilometer Area.

The contribution of tourists in filling the questionnaire is as part of the Yogyakarta City which is a favorite tourist destination objects for tourists because of the uniqueness of cultural tourism. Based on gender, female respondents were 64.2% and men respondents were 35.8%. Most of the travelers are aged be-tween 20-30 years old were 62.7% with their last education is Bachelor 1, where the purpose of their visit/tour is the release of the busy work and studying program of postgraduate. That matter seen from the questionnaire from their have employment as a private employee of 34.3% and as a college student of 26.9%. More than 50%, tourists who travel to the Yogyakarta City is not the first but has 5 to 10 times (especially domestic tourist).

Overall the criteria of all respondents above 17 years old and knew about the four objects of research. In this paper, the natives and newcomers were tested to prove there is a difference or not, while tourists to know they're the notion, and compare the results of the three categories of respondents. The results of this research are divided by three substances regarding historical objects, mental map elements of the city and the spirit of place can be seen in table 2.

Historical Objects

Historical objects are an important part of history, architecture, culture, and also tour-ism as a priceless relic because it can still be seen and felt to this day. The survey results through of the distribution of questionnaires are found that Malioboro Area received a very high rating by tourists than the Tugu Area, Zero Kilometer Area, and South Square Area. With mean value on historical objects of 4.24 which means tourists strongly agree that Malioboro Area is a historical object that has a very high historical value or 4.21. It is seen that Malioboro Area has a pedestrian in every right-left of Malioboro Street there is building style which is important in the historical event, and the existing culture can still be seen from the event, the attractions organized by the Government and the community. This is the main attraction for tourists, especially in terms of cultural tourism visible from the respondent's rate tourists at 4.45, which means very high.

Relates to the historical objects leaving important elements that are able to bring memories of person to remember of the city identity ever visited. Malioboro Area and Tugu Area are a strong image of the city for tourists. This looks very high against Malioboro with a value of 4.36 and Tugu with a value of 4.24. Of course, this is supported by good protection and management along Malioboro Street which still stands the heritage of the building with a fairly well-maintained of originality/authenticity where tourists rate of 4.12/agree and 4.09/agree with it. While Tugu is a landmark of the Yogyakarta City is judged by tourists have originality/authenticity with a value of 4.19/agree and have an important value in historical events be-cause Tugu is 'the silent witness' of the establishment of Ngayogyakarta Sultanate with a value of 4.09/agree.

The valuation of the local communities to historical objects that are near with them into doubt of the place attachment to a historical object spiritually. This is indicated began to gradually consciousness of local communities of the value of historical objects that are so precious. Based on observations, historical objects for some local communities are used solely for personal gain (financial gain). It is seen in a number of commercial areas, such as the center of souvenirs and culinary. In addition, the heritage buildings that have been assessed have style building of originality/authenticity sold to investors with the reason the cost of the land and building tax is rated high annually. This is the result of interviews with informers and local community leaders.

Mental Map Element of the City

The valuation of tourist respondents to mental map elements of the city have the same value between Tugu Area and Malioboro Area with 4.04/agree. Tourists assessing with a very high weight of 4.39 where Malioboro Area is a region the Yogyakarta City. The notion of tour-ists to Malioboro Area is a city element that is imageability and legibility as a center of cultural tourism with a value of 4.21 or very high. Malioboro Area is known by tourists as a center for traditional shopping, especially souvenirs along shops Malioboro Area. From that, tourists enjoy Malioboro by walking while looking at souvenirs, such as T-shirts, antique goods, as well as typical Jogja food such as Sate Kere, Pecel, Bakmi, and so on. This makes the tourist memory keep in mind of Malioboro with a strong corridor as an emphasis on the north-south axis, especially their memory of the Kraton which is on one philosophy axis with Malioboro. Tourists assessing Malioboro has a strong corridor/line of 4.12/agree. It almost very strong about Malioboro has a strong corridor/line.

The Malioboro Area gives a lot of a mil-lion meaning and impression for the tourists. This is seen from tourists assessing Malioboro has a physical design expression that reflects of identity the city a very strong meaningful culture and historical value in Yogyakarta City with a value of 4.12/agree. The same value that is 4.09/agree Malioboro Area can be used as a guide by giving direction when there are people who confused (ask/lost). With the same value, travelers rate agree with Malioboro Area as a reference/navigator of the city. The lowest value provided by tourists to the can be regarded as a divider of Yogyakarta City with a value of 3.60/agree. In fact, in Malioboro Area, there is Tugu Railway Station which has a railway line that can be said to be a breaker/cutter with an-other area.

The Tugu Yogyakarta is a physical de-sign expression that reflects of identity the city a very strong meaningful culture and historical value in Yogyakarta City. This is assessed by tourist respondents a very high of 4.28 about Tugu. There are five question items with values that are not significantly different. Tugu can be said as the nodes of Yogyakarta City with a value of 4.13, the elements of the city that is imageability and legibility with the value of 4.12, as the area of Yogyakarta City with a value of 4.12, can provide guidance as a reference with a value of 4.10, and as a landmark with a value of 4.10. Thus it can be concluded that Tugu is an icon of a city as a mental map (cognitive map) that attachment the memory of tourists about Tugu and surrounding areas. Furthermore, tourists assessing 4.04 where South Square Area is a physical design expression that reflects of identity the city meaningful culture and historical value in Yogyakarta City. It can be said that tourists have the allure (temptation) of its own with the meaning of South Square, such as the existence of two Banyan Tree (Ficus benjamina), Elephant cage, and buildings that are part of the Palace Complex with the Javanese Architecture that can still be seen by tourists.

The assessment of local communities towards the mental map element of the city as a whole shows doubt on the four research ob-jects. However of the four objects are shows that Tugu Area as an option for local communities as the mental map element in Yogyakarta City compared to other objects. The newcomers assessing of 4.06 that Tugu is a physical design expression that reflects of identity the city a very strong meaningful culture and historical value in Yogyakarta City. In this case, the natives also share the value of 3.89/agree with other objects. Public perception about Tugu has an important value in historical events and figures that is Sri Sultan Hamengku Buwana I as architect/visionary. In addition, newcomers also assessing of 3.90/agree that Tugu can be said as a region the Yogyakarta City. Not much different from the natives of 3.91/agree that Tugu can be said as a region the Yogyakarta City. Based on field survey, Tugu Area has a distinctive architectural style. Especially, Jetis Sub-district has Indische Architectural Style. For example is the building of the Ex Hotel Toegoe, Kedaulatan Rakyat (newspaper) Office, and much more.

Spirit of Place

The 2008 ICOMOS Quebec Declaration on the Preservation of the Spirit of Place is defined as the tangible (buildings, sites, landscapes, routes, objects) and the intangible elements (memories, narratives, written documents, rituals, festivals, traditional knowledge, values, textures, colors, odors, etc.), that is to say the physical and the spiritual elements that give meaning, value, emotion, and mystery to place (ICOMOS, 2008). As a city of cultural heritage, the Yogyakarta City has a privilege located on the philosophy axis. Along the philosophy axis, there are the main elements of historical objects that have public space to be enjoyed by the communities and tourists.

Based on the results of this study, Malioboro becomes a spirit of place for natives, newcomers, and tourists compared with other objects. For the natives, overall Malioboro got a value of 3.52 with the highest value of 3.88 which emphasizes that Malioboro is a corridor to meet, gather, and release the longing (com-munity life). It reminds that Malioboro can pro-vide space and facilities that meet them, such as wide space, availability of parking, etc. Also with the attractions that are served by the community such as music community in the evening. In addition, Malioboro got a value of 3.74 that Malioboro is the heart of the city, which means the inheritance of all things. This can be seen from the natives answer by choosing the street elements in Malioboro, such as lamps and benches are unique. The feature of street lamps owned in Yogyakarta City and not owned by other cities.

The newcomers assessing Malioboro on an item of the spirit of place got a value of 3.53, this average value is not much different from the assessment given by the natives. The highest value by natives stating that Malioboro is the heart of the city, which means the inheritance of all things with a value of 3.86. This means that newcomers depend on Malioboro as a place for them to enjoy the beauty of the city in preserving the culture in which their life now. The next valuation is given that Malioboro is the place to social interaction and release the longing with a value of 3.56. Based on inter-views to newcomers, Malioboro is fun and pleasure place to meet or gather with friends or families with the same village background. This case is found when the newcomers who have family status and live in Yogyakarta City to see/meet relatives of one hometown.

The valuation of tourist respondents in assessing Malioboro the value of 4.12 that Malioboro has the spirit of place. In contrast to the communities, tourists are more pressing Malioboro is public space to escape them-selves from the routine of daily life (fatigue) (urban well-being) the value of 4.06. 76.1% of the background of their education is Bachelor 1 with employment are private employees and postgraduate students. The purpose of their visit to Malioboro is to take a walk and see the panorama. Furthermore, travelers who assessing of 4.01 is Malioboro is the heart of the city, which means of all things. Reading in the tourist sites is cultural tourism is warm growing a new trend among tourists where the tendency to seek something unique and authentic to a culture. How precious the uniqueness, beauty, and cultural values become the tourist attraction in their visit to enjoy the Yogyakarta City, especially in Malioboro.

The perception of the local communities and tourists to Malioboro is the spirit of life for the urban citizens get a value that is close to significant. The natives assessing the value of 3.47/agree, newcomers the value of 3.40/agree, and tourists the value of 3.91/agree. Based on survey results in Malioboro Area, neither the local communities nor the tourists have yet to feel the quality of life for the welfare, especially for the local communities. Environmental conditions give effect when they want to enjoy Malioboro, lack of vegetation causes reluctance to enjoy Malioboro especially the local communities. Similarly, what is perceived by tourists, Malioboro has not been able to provide a cold (mind) environment. Allegedly, this is due to Malioboro has just completed the revitalization process, while the planting of trees in the Malioboro Street is in the process of growth. The planted trees are Gayam Tree (Inocarpus fagifer) (In Javanese; ayom) and Javanese Asem Tree (Tamarindus indica) (In Javanese, sengsem).

Different Test of "Mann-Whitney U"

In this research different tests were used to prove the differences of perceptions of natives and newcomers. Different tests function to know the relationship or meaningful difference between two variables (natives and new-comers). Hypothesis testing in this research is determined based on the result of prasarat test that is a test of data normality and homogeneity of data. From the results of a test of data normality and homogeneity of data, will be determined what test equipment is most suitable to use. Based on the results of preliminary test analysis, it is found that the majority of data is not normally distributed, but the data comes from the same variance (homogeneous).

With the Levene statistic method, it is seen that the significance value is greater than the alpha (>0.05) can be seen in Table 3. Where the level of significance/level of trust for social research is 95% (0.05). Therefore it can be assumed that the prasarat test is not fulfilled and for further analysis done by using non-parametric test by different test "Mann-Whitney U" can be seen in Table 4. Based on results obtained different test by using "Mann-Whitney U" from the result of the difference of perception between natives and newcomers, there is no difference.

Table 3. Test of Homogeneity of Variances.

Test of Homogeneity of Variances				
	Levene Statistic	df1	df2	Significance Value
Historical Objects	3.851	1	258	0.051
Mental Map Element of the City	0.760	1	258	0.384
Spirit of place	2.273	1	258	0.133

Conclusion

The research indicates the significance of the tangible heritage and intangible heritage attachment in shaping the heritage city. Cultural heritage is a great asset and a proud to be pre-served. The main elements of historical objects become of identity the city because of its has architectural style and historical background. Moreover, the supporting elements that sustain from the main elements should be protected and managed properly. Unfortunately, the proliferation of buildings not in the context of the environment can reduce the attractiveness. The soaring of the cost of the land and building tax is considered to be very burdensome for the owners. This needs cooperation between the Government, the owner of the building/land, and the whole society.

Based on the results of the distribution of questionnaires, both natives and newcomers in assessing the main elements of historical objects that also become mental map elements of the city above average or agree. Researchers think it "looks flat". While the spirit of place of the four objects selected by local communities based on the availability of public space that is equipped with soft elements, street elements, and adequate facilities such as parking area. Otherwise, tourists are more appreciated with the culture of Yogyakarta City. Original/authentic cultures and good protection and management are an attraction for them. Furthermore, tourists have memories with the Yogyakarta City as a strong the image of the city, imageability, and legibility

Malioboro as corridor/routes becomes an option for local communities and tourists. Pedestrian as public space of Malioboro has the meaning of the spirit of life and heart of the city, which means the inheritance of all things both tangle (buildings) and intangible (cultural attractions). Based on the preferences, many of them seek the atmosphere of the evening to enjoy the beauty of the city while nostalgic. The researcher suggests improving place attachment in all the main elements of historical objects to be distributed evenly and to all levels of local communities. Historical objects are crucial to being done in order to survive and increase as well as give a good impact on the local communities welfare (environmental psychology).

To get the predicate of World Heritage City by UNESCO is not easy. The local com-munities need to be spiritually motivated to support Yogyakarta as a Heritage City. Four objects in the research that Tugu as monuments, Malioboro as corridor/routes, Zero Kilometer Area as the groups of buildings, and South Square as sites are tangible elements as the spirit of place in Yogyakarta City. With the spirit of place in public space to gather and interact can provide awareness to continue to attend for cultural heritage. Researchers argue, through soft elements are expected to be input in the planning of public spaces in urban areas. It aims to be more in demand by local communities and tourists, such as the use of trees that have sense and meaning in the philosophy.

Acknowledgements

This paper presents part of magister re-search by the author with support from Bogor Agricultural University.

Table 4. Test Statistics with "Mann-Whitney U".

Test Statistics[2]			
	Historical Objects	Mental Map Element of the City	Spirit of place
Mann-Whitney U	7331.500	6959.500	7396.000
Z	-.561	-1.209	-.446
Asymp. Sig. (2-tailed)	.575	.227	.656

[2]Grouping Variable

Table 2. The Perception Score of Natives, Newcomers, and Tourists.

Items	Tugu Area (TG)			Malioboro Area (MB)			Zero Kilometer Area (ZK)			South Square Area (SS)		
	NT	NC	TR	NT	NC	TR	NT	NC	TR	NT	NC	TR
HISTORICAL OBJECTS												
TG/MB/ZK/SS has an important value in historical events, style of building, and culture **[outstanding universal value, historical value]**	3.91	3.94	4.09	3.82	3.80	4.21	3.74	3.67	3.48	3.86	3.88	4.04
TG/MB/ZK/SS has an originality/authenticity **[authenticity]**	3.79	3.84	4.19	3.72	3.69	4.09	3.61	3.34	3.30	3.82	3.77	4.10
TG/MB/ZK/SS has good protection and management **[heritage city management]**	3.70	3.68	3.93	3.71	3.78	4.12	3.56	3.52	3.37	3.66	3.47	3.78
TG/MB/ZK/SS as the image of the city **[the image of the city]**	3.86	3.76	4.24	3.77	3.67	4.36	3.64	3.42	3.34	3.73	3.60	3.88
TG/MB/ZK/SS has a tourist												

The Perception Score of Natives (NT), Newcomers (NC), and Tourists (TR)[1]

attraction **[tourist attraction]**	3.83	3.73	4.00	3.93	3.69	4.45	3.71	3.76	3.51	3.81	3.82	4.10
Mean value	3.81	3.79	4.09	3.79	3.78	4.24	3.65	3.54	3.40	3.78	3.69	3.98
MENTAL MAP ELEMENTS OF THE CITY												
TG/MB/ZK/SS is a physical design expression that reflects of identity the city with meaningful culture and historical value in Yogyakarta **[identity in urban]**	3.89	4.06	4.28	3.78	3.91	4.12	3.72	3.79	3.46	3.78	3.66	4.04
When someone is lost, you give directions TG/MB/ZK/SS as a reference **[navigator]**	3.71	3.63	4.10	3.72	3.54	4.09	3.68	3.32	3.16	3.54	3.20	3.82
TG/MB/ZK/SS is a city element that is imageability and legibility **[imageability and legibility]**	3.78	3.80	4.12	3.82	3.69	4.21	3.66	3.42	3.24	3.60	3.46	3.90
TG/MB/ZK/SS as landmark of the city **[landmark]**	3.81	3.69	4.09	3.71	3.63	4.09	3.70	3.47	3.40	3.58	3.42	3.85
TG/MB/ZK/SS Area can be regarded as district of the Yogyakarta City **[district]**	3.91	3.90	4.12	3.84	3.86	4.39	3.71	3.60	3.40	3.67	3.63	3.91
TG/MB/ZK/SS can be regarded as node of the Yogyakarta City **[node]**	3.74	3.58	4.13	3.54	3.23	3.69	3.63	3.30	3.33	3.48	3.16	3.81
TG/MB/ZK/SS can be regarded as edge of the Yogyakarta City **[edge]**	3.51	3.23	3.60	3.38	3.06	3.60	3.45	3.17	3.18	3.35	3.19	3.39
TG/MB/ZK/SS can be regarded as path of the Yogyakarta City **[path]**	3.77	3.64	3.87	3.70	3.34	4.12	3.67	3.38	3.46	3.49	3.39	3.70
Mean value	3.76	3.61	4.04	3.69	3.53	4.04	3.65	3.43	3.33	3.56	3.39	3.80
A SPIRIT OF PLACE												
TG/MB/ZK/SS is the heart of the city, means the inheritance of all things **[heart of the city]**	3.75	3.81	3.94	3.74	3.86	4.01	3.51	3.68	3.09	3.59	3.60	3.60
TG/MB/ZK/SS is meet, gather, and release the longing **[community life]**	3.05	3.21	3.10	3.88	3.56	3.94	3.14	3.38	3.10	3.37	3.49	3.61
TG/MB/ZK/SS is the spirit of place for urban residents **[spirit of the place]**	3.29	3.17	3.55	3.47	3.40	3.91	3.20	3.28	3.09	3.33	3.30	3.36

TG/MB/ZK/SS is public space to escape themselves from the daily routine (fatigue) **[urban well-being, city neighborhoods]**	3.05	3.02	3.06	3.45	3.43	4.06	3.19	3.31	3.10	3.43	3.50	3.61
Atmosphere is something that is sought after from TG/MB/ZK/SS **[atmosphere]**	3.21	3.19	3.33	3.56	3.43	3.96	3.28	3.24	3.10	3.44	3.51	3.75
Mean value	3.27	3.28	3.40	3.52	3.53	3.98	3.26	3.38	3.10	3.43	3.48	3.59

[1]Mean value;
1.00 – 1.80: Strongly Disagree, 1.80 – 2.60: Disagree, 2.60 – 3.40: Doubt, 3.40 – 4.20: Agree, dan
4.20 – 5.00: Strongly Agree. (reference: Umar, 2005)
Source: Author, 2017 (based on field survey from May-September)

References

Akkar, Z.M. (2005a). Questioning the 'publicness' of public spaces in post-industrial cities. *Traditional Dwellings and Settlements Review*, 16, 75-91.

Akkar, Z.M. (2005b). The changing 'publicness' of contemporary public spaces: a case study of the Grey's Monument Area, Newcastle upon Tyne. *Urban Design International*, 10, 95-113.

Alexander, C. (1977). *A Pattern Language. Towns-Buildings-Construction*. Oxford: Oxford University Press, Inc.

Aplin, G. (2002). *Heritage, identification, conservation and management*. United Kingdom: Oxford University Press.

Bourdieu, P. (1984). *Distinction: A Social Critique of the Judgement of Taste*. London: Routledge.

Carmona, M., Tiesdell, S., Heath, T., & O. Taner. (2003). *Public Places, Urban Spaces, The Dimensions of Urban Design*. Oxford: Architectural Press.

Childs, M.C. (2004). *Squares: A Public Place Design Guide for Urbanists*. Albuquerque: University of New Mexico Press.

Cooper-Marcus, C. & Francis, C. (1998). *People Places: Design Guidelines for Urban Open Space*. New York: John Wiley & Sons.

Department for Culture of Yogyakarta Government. (2015). *Yogyakarta City of Philosophy*. Yogyakarta: Department for Culture.

Duany, A., Plater-Zyberk, E., & Speck, J. (2000). *Suburban nation: The rise of sprawl and the decline of the American dream*. New York: North Point Press.

Gehl, J. (1987). *Life Between Buildings: Using Public Space*. Washington-Covelo-London: Island Press.

Handinoto. 2010. *Arsitektur dan Kota-kota di Jawa pada Masa Kolonial*. Yogyakarta: Graha Ilmu.

ICOMOS (International Council on Monuments and Sites). (2008). *Québec Declaration on the Preservation of the Spirit of Place*. Paris: ICOMOS. Retrieved from http://whc.unesco.org/uploads/activities/documents/activity-646-2.pdf.

Koseoglu, E. & Onder, D.E. (2011). Subjective and objective dimensions of spatial legibility. *Procedia Environment and Behavior Sciences*, 30, 1191-1195.

Lynch, K. (1960). *The image of the city*. Cambridge: MIT Press.

Lynch, K. (1981). *A Theory of Good City Form*. Cambridge: MIT Press.

Madanipour, A. (1995). Dimensions of urban public space: the case of the Metro Centre, Gateshead. *Urban Design Studies*, 1, 45-56.

Madanipour, A. (2003). *Public and Private Spaces of the City*. London: Routledge.

Németh, J. (2006). Conflict, exclusion, relocation: Skateboarding and public space. *Journal of Urban Design*, 11(3), 297-318.

Newman, O. (1980). Community of interest. *Society*, 18(1), 52-57.

Norberg-Schulz, C. (1980). *Genius Loci: Towards a phenomenology of architecture*. New York: Rizzoli.

Project for Public Spaces. (2000). *How to Turn a Place Around: A Handbook for Creating Successful Public Spaces*. New York: Project for Public Spaces.

Saeidi, S. & Oktay, D. (2012). Diversity for Better Quality of Community Life: Evaluations in Famagusta Neighbourhoods. *Procedia Environment and Behavior Sciences*, 35, 495-504.

Sandler, D. (2007). Place and process: Culture, urban planning, and social exclusion in Sao Paulo. *Social Identities*, 13, 471-493.

Septirina SN, Takeo O, & Satoru K. 2016. Conservation of historical architecture in Malioboro street, Yogyakarta city, Indonesia. *Procedia Social and Behavioral Sciences*, 225, 259-269.

Tavano Blessi, G., Grossi, E., Sacco, P.L., Pieretti, G., & Ferilli, G. (2016). The contribution of cultural participation to urban well-being. A comparative

* Corresponding author: rrpawitrasari@gmail.com

study in Bolzano/Bozen and Siracusa, Italy. *Journal of Cities*, 50, 216-226.

Tavano Blessi, G., Tremblay, D.G., Sandri, M., & Pilati, T. (2012). New trajectories in urban regeneration processes: Cultural capital as source of human and social capital accumulation-Evidence from the case of Tohu in Montreal. *Journal of Cities*, 29(6), 397-407.

Tibbalds, F. (1992). *Making People-Friendly Towns*. London: Longman.

Tweed, C. & Sutherland, M. (2007). Built cultural heritage and sustainable urban development. *Journal of Landscape and Urban Planning*, 83, 62-69.

Umar, H. 2005. *Riset Sumber Daya Manusia dalam Organisasi*. Jakarta: Gramedia, Pustaka Utama.

UNESCO (United Nations Educational, Scientific, and Cultural Organization). (2005). *Basic Texts of the 1972 World Heritage Convention*. Paris: UNESCO. Retrieved from http://whc.unesco.org/uploads/activities/documents/activity-562-4.pdf.

UNESCO (United Nations Educational, Scientific, and Cultural Organization). (2005). *Operational Guidelines for the Implementation of the World Heritage Convention*, 2 February. Paris: UNESCO. Retrieved from http://whc.unesco.org/archive/opguide05-en.pdf.

Van Melik, R., Van Aalst, I., & Van Weesep, J. (2007). Fear and fantasy in the public domain: the development of secured and themed urban space. *Journal of Urban Design*, 12(1), 25-42.

Whyte, I.B. (2003.) *Modernism and the Spirit of the City*. New York: Routledge.

Young, I.M. (1990). *Justice and the Politics of Difference*. Princeton: University Press.

Zakaria, K., Harun, N.Z., & Mansor M. (2015). Place Meaning of the Historic Square as Tourism Attraction and Community Leisure Space. *Procedia Environment and Behavior Sciences*, 202, 477-486.

Singosaren As Interactive Space In Kotagede Tourism Destination

Irwin Panjaitan[1,*], *Moh.* Prasetiyo[2]

[1]Magister student of Urban Design, Departement of Architecture; School of Architecture, Planning, and Policy Development, Institute Technology Bandung, Indonesia
[2]Lecturer, Departement of Architecture; School of Architecture, Planning, and Policy Development, Institute Technology Bandung, Indonesia

Abstract. The development of the area for tourism purposes should have good sensitivity related to the product to consumption aspects, where the interaction between the local community is seen as the supply (insider) produce the products, which can then be consumed by the visitor as demand (outsider). Such sensitivity can be seen in the space between the private realm and the public sphere called the interface space. Development of the area for tourism, will enhance the touristsaccessibility in exploring its uniqueness as a tourist attraction. Accessible tourist attractions can improve the degree of permeability, which can also affect the degree of privacy of local communities. Singosaren as an area that will be designed as part of the united Kotagede tourist destinations, Yogyakarta, should consider the influence of tourism activities on a daily basis of the community. Permeability study in the area with mapping method, is expected to connect local potency to the circulation pattern in the interface, by finding the street corridor that needs to be anticipated, to maintain the sustainability of tourism activities in Singosaren area.

Keywords: Permeability, interface, tourism destination, socio-culture

Intoduction

The tourism development as industry has the ability to encourage economic improvement for the surrounding community. Economic enhancement happens when the job opportunity available for the community, through the ability that can be a selling point. Yogyakarta with a large degree of Natural and Cultural diversity, potential to continue to be developed with a variety of alternative tourism destinations.

Tourism basically offers a unique experience, different from the daily life of the visitor. Kotagede as one of the alternative tourist destinations in Yogyakarta, has a very dense circulation intensity, which is not comfortable for tourists. Density occurs because the street dimension is not proportional to the high circulation intensity. Convenience achievement for travelers is the most important part of tourism, when one's experience of exploring an area will be influenced by what a visitor can access from his/her path (Dovey, 2016: 187). Visitor's space experience will provide a complete picture of the image of the area.

Singosaren located between Kotagede as tourist destinations and the Ringroad corridor of Yogyakarta as the primary arterial road. There is opportunity in Singosaren area to develop a new tourism access to the Kotagede tourist destination. The sustainability of Singosaren area development as a unity of Kotagede tourist destination is influenced by Howe the interaction of the inhabitants as the supply and the outsider as demand is formed. This paper attempt to identify the influence of the new tourism access development in the Singosaren area.

Methods

Data collection method is done by doing direct observation such as recording using camera, taking notes, and also sketching in the study area.
Informal interviews also conducted to some locals and tourists to get information related to daily activities on the area and the tourism activities.

Qualitative descriptive analysis method used to describe a certain character in a systematic, actual, and accurately. (Ariknto, 1998). The paper attempts to explain the relationship between theory and the existing conditions found in the study area. Mapping technique is used to facilitate the understanding of the relationship of activity patterns on space.

Understanding The Permeability study

Moughtin (2003) attempted to map The Townscape analysis by Cullen (1966) into three parts: perceptual structure by prioritizing the legibility of urban structures, how easily a user accepts, understands, and responds the environment; Permeability study talks about the degree

* Corresponding author: irwinsmart23@gmail.com

of choice that users have when doing a movement within the area; and Visual study is how a user can recognize the area uniqueness that can be the meaning of a place. Urban design studies specialized in public sphere as its study area, and attempt to accommodate both private and public needs simultaneously. Carr (1992) explained that urban design should be able to realize three main values as a determinant quality in public space, such as responsive, democratic, and meaningful.

Urban design is strongly influenced by the movement that can connect the activities in the area (urban connectivity) by dividing the area into blocks of building mass and open space for circulation needs, which then form the structure of the spaces in the area (Lynch, 1960; Carmona, 2010; Dovey, 2016). In designing tourism area, providing a unique experience that visitors are looking for will depends on how the area is formed (product for consumption). Making the area as a product for tourism consumption means being able to bring together the expectation of the visitors with the experience received in the area. Tourism activities as an industry is the most effective way of learning a new and old area (Dovey, 2016: 185). There is a clear distinction of seeing tourism from the perspective of the residents (insiders) and visitors (outsider), where the visitors should be able to capture the image through the area morphology.

- The text should be set to single line spacing.
- Paragraphs should be justified.
- The first paragraph after a section or subsection should not be indented; subsequent paragraphs should be indented by 5 mm.

The use of sections to divide the text of the paper is optional and left as a decision for the author. Where the author wishes to divide the paper into sections the formatting shown in Table 2 should be used.

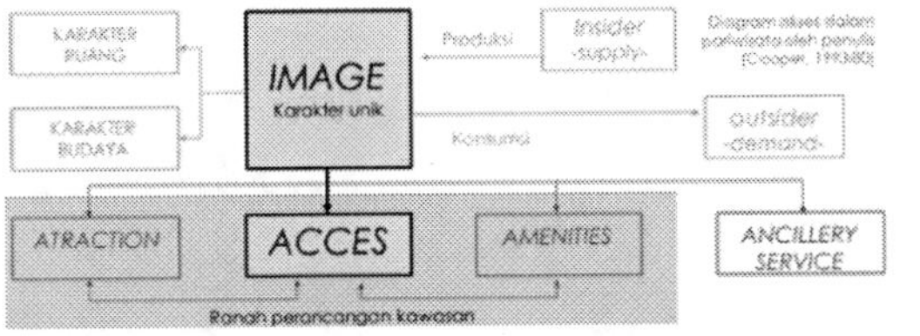

Fig. 1. The Relationship Between Access and Tourism Destination Image.

Dovey (2016) explained that flows of pedestrian traffic together with the traffic ideas, identities, images, & products, are at the heart of what makes cities urban, in what ways do they enable and constrain urban connectivity. The design of Singosaren area as part of Kotagede tourism destination will try to increase circulation intensity to the tourism attraction in the area. This improvement will shown in the connected corridors(streets), where the visitors movement can make a space experience, then form an image of the area that can be accepted by visitors.

Moughtin (2003) explained that the design of the interface between the public realm and the more private areas of individual properties is a legitimate concern of urban space. The statement strengthen the needs of designer's sensitivity in responding to the presence of visitors in the Singosaren, which will greatly affect the space between the private and public sphere (interface spaces), when visitors have the ability to access the formed streets. There is a clear connection between the degree of accessibility to privacy, where access leads the freedom of visitors to move, and privacy leads to the sense of security that residents require.

Bentley et al (1985) stated that both physical and visual permeability depends on how the network of public space divides the environmental into blocks area: areas of land entirely surrounded by public routes. The statement explain that the degree of permeability has a strong relationship between the street layout and the building blocks formed on the area structure. The higher the accessibility of the area, the privacy of the occupants will decrease, and vice versa. The Singosaren area sustainability as a tourist area is determined by how the designer's sensitivity sees the influence of tourist circulation in accessing the area, especially interface space. Mapping technique and direct observation (photo recording and skecthing) were being used to identify the permeability in the area.

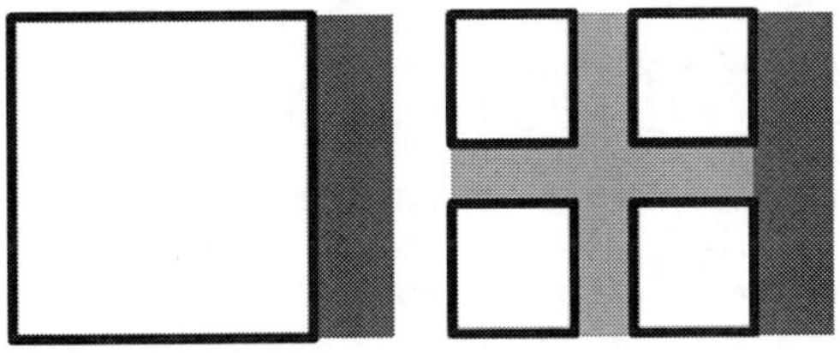

Fig. 2. Diagram of Relationship Between the Increasing Access and Permeability on the Same Spatial Scope.

Typology of interface space

The quality of the interface space formed in the private sphere and the public sphere can determine the life activity that occurs in the street space as a circulation. Jacobs (1961, 39) tried to promote the slogan *'eyes on the street'* as a benchmark in assessing the contribution of the interface space to the user's sense of security, as well as the interactions that occur in the space. Gehl classified public/private interface morphology as a long series between *'soft'/active'* that are social and permeable and the *'hard'/passive'* that are antisocial and impermeable, which both contradicted. This classification is based on social interaction. The influenced factors are ***transparency, activity, diversity, design quality, and building dimensions*** (Dovey, 2015).

Typology of interface space is made to facilitate the designer in designing a livable and attractive street. Dovey & Wood then developed a more detailed public/private interface typology, with criteria including: ***permeability***, related to the pedestrians ability to access the interface space; ***transparency***, related to the media in interface space that limits the public gaze into private space; ***setback***, the distance between the private and public spaces in the private sphere, which makes it publicly accessible; ***Car dependency***, the object that dominates access to the building facade (interface) by people or vehicles. (Dovey 2016, 59).

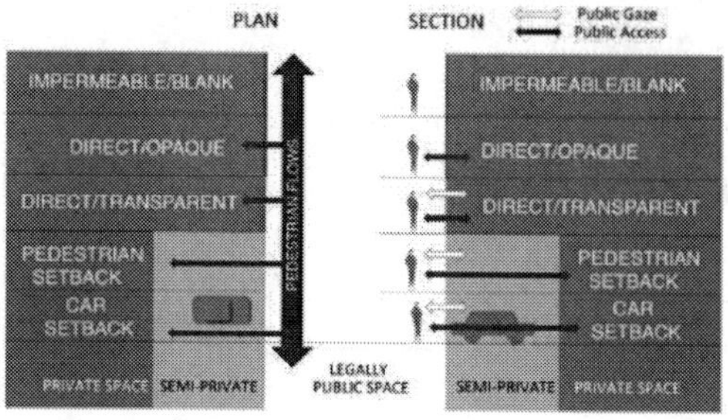

Fig. 3. Interface Tipology.
Source: Dovey & Wood, 2015.

These variables form the basis of an interface typology consisting of 5 types (Dovey, 2016): impermeable / blank (type A), direct / opaque (type B), direct / transparent (type C), pedestrian setback (type D) and car setback (type E).

Type A is an interface space in a building that does not allow spatial and visual access. This model is usually appears as building walls or fences without access, directly adjacent to the street, which is often used as billboards or street art.

Type B is a building interface space adjacent to the street space, allowing pedestrian access into the private space within the building directly, but does not provide visual access because no openings or the walls are less transparent. This type appears as building wall with door access, has a large window openings, or an accessible building fence, usually seen in apartment buildings, offices, etc.

Fig. 4. The Examples of Interface Tipology
Source: Dovey & Wood, 2015.

Type C is an interface space in buildings that allows pedestrian and visual access. The building walls or fences are dominated by openings. This type can generally be found in commercial buildings, and there's possibility of this typeto be adapted into other functions.

Type D is an interface space in the private realm with buildings that have a setback between private buildings and public street spaces, as semi-private spaces. Building walls or fences can be directly accessed spatially and visually, without any obstacles.

Type E is an interface space that has a similarity to type D. The difference is the presence of interference or obstacles in the semi-private space to be accessed spatially and visually. The interface space is often used as parking area for vehicles.

The diversity of these typologies can indicate the relationship between interface space's function alteration with the degree of circulation intensity within the space. Singosaren area planning as part of Kotagede tourism destination, needs to consider the possible level of change in the space of its interface

The Description of Singosaren Area

To study the Singosaren spatial pattern, it is necessary to describe the relationship of the surrounding area. Singosaren area is located between the center of the tourist destination Kotagede, and South Ring Road of Yogyakarta. The opportunity to open new tourist access that can connect the southern ring road of Yogyakarta to the center of tourist destination Kotagede is discovered, which is passing the Singosaren area. The new access will have an influence on the development of Singosaren area, so the arrangement of Singosaren Area needs to be done carefully to support the creation of he united Kotagede tourist destination.

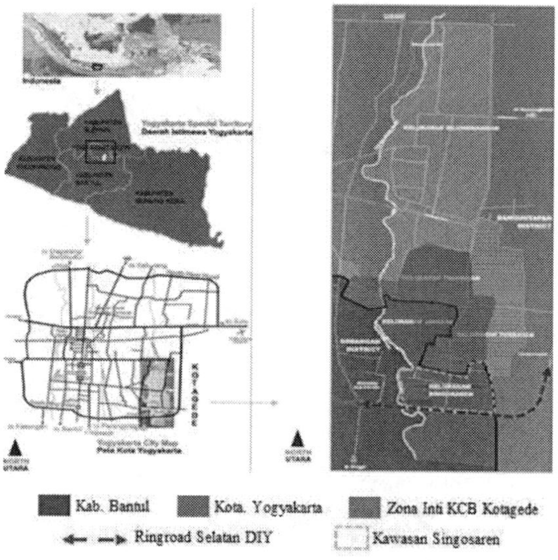

Fig. 5. Singosaren in Yogyakarta.
Source: Remade by Panjaitan, 2017, from Kotagede Pusaka Area, JRF, 2007.

 https://doi.org/10.1051/shsconf/20184104008

The Singosaren area as a designed area has territorial boundaries, such as: the north side by the administrative boundaries of Jagalan and Purbayan districts; the southern side by the southern ring road of Yogyakarta; the west side by the Gajahwong River; and the eastern side by the Purbayan street.

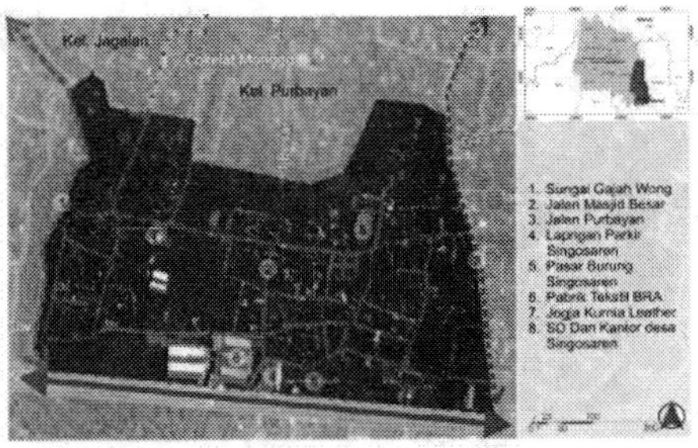

Fig. 6. Designed Area Boundaries.

To facilitate the understanding of the Singosaren area exsisting conditions, the pilot study is done based on the urban design elements, adapted to the movement within the area. The urban design elements by Shirvani (1985) can be adapted to the needs of the designed area context, which is the need for interaction between visitors and occupants in the interface space along the space of movement/circulation. The design elements used in this permeability study are: space utilization (landuse & activity support), to see the relation of spatial function influence the movement within the area; Circulation and parking, to see the relationship between street classification, transportation modes and the placement of parking nodes within the area; building masses arragement (Building form and massing), as the forming scope of street spaces which can encourage interaction; Open space, intervened space for tourism activities.

Function Of Space

there are several functions in the Singosaren that can affect the movement within the area. Based on the observation, the existing textile factory as industrial function at the main gate of Singosaren, has the greatest influence. Large factory have many employees, increasing the neccesity of residential function for the workers who are dominantly come from outside the area, impacted in exsisting residential houses turned into a boarding house. In addition, Singosaren parking lot, empty buildings, and open spaces surround the factory utilized by the local community as a parking lot for employees.

Fig. 7. Example of Space Utilization in the Area.

Some functions can be the driving forces in the Singosaren, such as: three mosques in the area, two clinics,the presence of elementary school, and district office in the designed area. The presence of car parking for Kotagede tourist in the Singosaren region, generating supported activities such as kiosks and the afternoon market, to meet the tourists needs. The building functions along the collector road class become a commercial activities. Dominant functions of the existing commercial use come in food stalls, grocery kiosks, salons, and some textile and invitation cards production offices. There are also several functions of the home industry like traditional culinary 'yangko', puppet handicraft, and silver handicraft. Some leather craftsmen also exist as wholesalers and warehouses.

In addition, in the area are found some unique cultural activities, such as the discovery of leather puppets craftsmen, and silver craftsmen, who work on their homes on the basis by orders. The center for art training activities for various art communities in the area Singosaren were found, on Singosaren art studio building. Lastly, there area also found the existence of pigeon lovers community activities in the open space of rice fields.

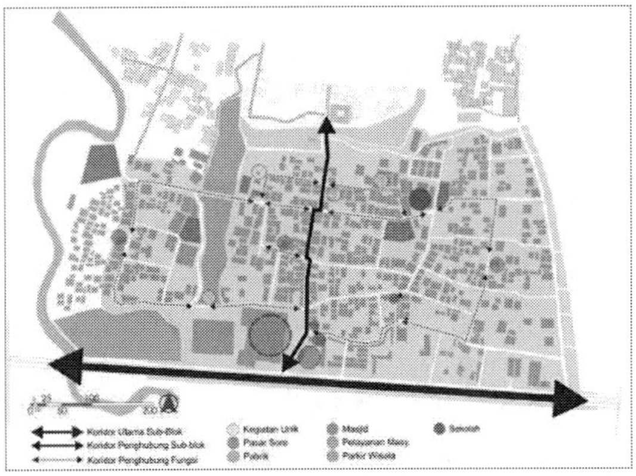

Fig. 8. Mapping of Space Utilization on Singosaren.

Circulation and parking

The movement in the Singosaren Area is centered on the Singosaren corridor from the center of Kotagede to the southern Ringroad of Yogyakarta. Thick and thin lines on the map illustrate the dimensions of the road in the Singosaren area. The existing road hierarchy is formed based on the road width. To reach the Singosaren whole area can be done by walk because it is within a 500 meters (walking distance). The street dimensions are diversed and mainly have two directions, often causing traffic jam. The singosaren corridor is also used by locals as the main route to reach the functions within the area. The intensity of two-wheeled and four-wheeled vehicles dominates the area circulation, while the pedestrians intensity is very low, and the area is unreachable by public transport such as buses.

 https://doi.org/10.1051/shsconf/20184104008

Fig. 9. Example of Circulation and Parking in the Area.

The limitations of street space in the area, making some houses within the area cannot be accessed by four-wheeled vehicles, and not allow for on-street parking. The available parking area in Singosaren usually use the open space in private plots located right on the side of the road, used as shared vehicle parking.

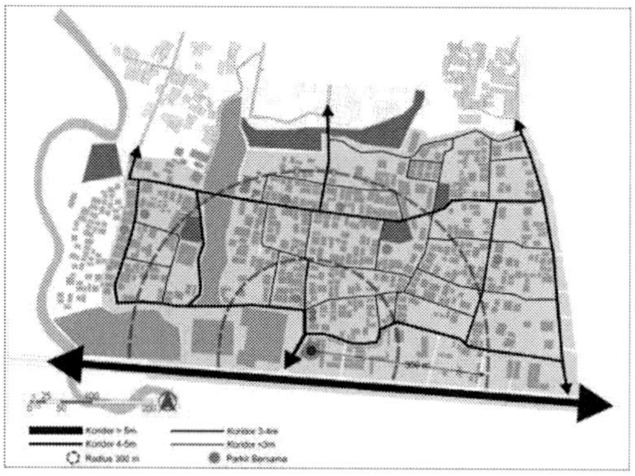

Fig. 10. Circulation and Parking Area Map in Singosaren.

Building Layout

In general, the buildings are dominated by one-floor residence, with a neo-vernacular style from the strong Javanese character with sloping roof. Some buildings are not maintained because it's not vacant, and the other reason is because the building is still semi-permanent. The buildings are oriented facing the main road as access to enter the building. Space in buildings adjacent to roads often turns into commercial activities. Meanwhile, the increasing occupancy needs in the area, making some buildings increase the intensity.

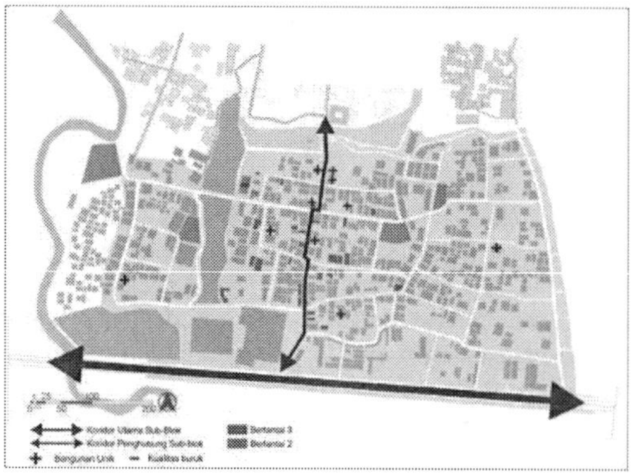

Fig. 11 Building Quality Mapping in Singosaren.

The existing building mass in the designed area, especially the residential area, generally consists of 1-2 floors buildings where the dominant building sizes are 50-200 square meters. Residential are generally has a private open space on the front or rear, there are even some who have setbacks on four sides.

Fig. 12. Example of Building Quality in the Area.

The fence as element in the area acted just as a barrier between private space and public space. Fences in general have a height that can still be accessed at the height of the human eye, which can be accessed visually. Some buildings with function as boarding houses with modern style, have a high fence and formed in more massive and solid built. Modern boarding house buildings have minimalist style with less Javanese character, either in shape or its building facade.

Open space

In the area, no open spaces are planned for social community activities. Lots of open space in private plots were found and often used as social space of the community. The open space were used for interaction space, children playground, community meetings, shared parking, etc. Also, visitors were allowed to access private land for public access, utilizing the setback space on the fence.

Fig. 13. Example of Open Space in the Area.

Open space such as streets and courts, formed by building fence or directly adjoining by building wall. The dimension difference between street width and the building height is 1:1 ratio, so the formed space is very strong. Some parts of open spaces that do not have a building facade, serve as a creative medium for mural activity. Some open space also has special functions such as rice fields and cemetery.

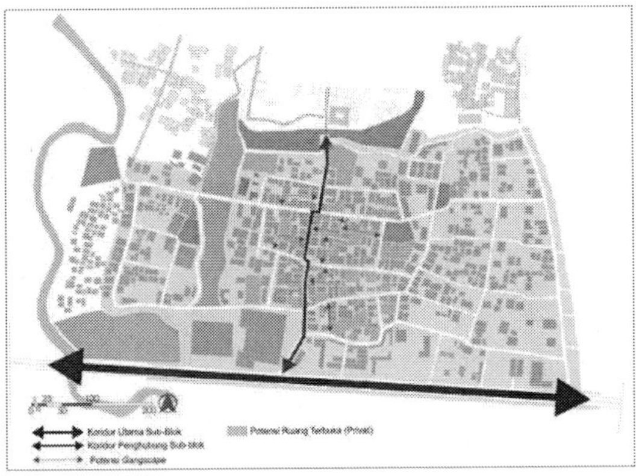

Fig. 14. Mapping of Open Space on Singosaren.

Interface on the Singosaren area

Singosaren region has a variety of interface conditions. In general, the function of buildings in the area are residential and commercial functions, with main entrance is oriented to the road. Very limited street space dimensions with an average width of two to five meters, making public visual access (public gaze) easy to reach space in community-owned buildings. This condition does not interfere with the privacy of the community because the community has strong social ties so that frequent interactions between local communities occur in the interface space. Here are some interface types found in the Singosaren region:

Impermeable/ blank Interface

Generally found in buildings located at crossroads without visual and spatial access. The orientation of access leads to the main road, making the other side of the building has a wall without the facade. In addition, some case happen to avoid unexpected views such as funerals. The sides of the building without setback have the lowest class streets, some of them are decorated with street art, and the interaction considered minimal. This condition is not related to the function of the building.

Fig. 15. Example of Impermeable/ Blank Interface in the Area.

Direct / Opaque Interface

The building has no setback on the road, which can only be accessed spatially. Those generally are found in buildings with boarding house function or commercial like kiosk. Interaction occurs on the street space which facing the entrance access of the building. The distribution of this interface is influenced by building functions and the limitation of building lot in the area.

Fig. 16. Example of Direct/Opaque Interface in the Area.

Direct / Transparant Interface

This Interface space is commonly found in the public buildings and commercial buildings in the area. The openings have dominated the walls of buildings. The interaction in this typical interface space is very high, due to the easy visual and spatial access from the street space into the building space. This model is often found in high circulation intensity street.

Fig. 17. Example of Direct/Transparent Interface in the Area.

Pedestrian Setback Interface

Buildings with setback are found in buildings with residential functions. The barrier as element between private plot boundary and public street space formed as wall fence. Fences, in general, have a good transparency, a lot of house fence that is not solid or discontinuous, easily accessed spatially. This can indicate the strong social interaction of the local community in the interface space.

Fig. 18. Example of Pedestrian Setbak in the Area.

Vehicle Setback Interface

Having a setback that is not much different from the typical pedestrian setback, the difference is the discovery of obstacles in public space when accessing visually or spatially to the private space within the building. Generally, the disturbance is caused by the use of parking vehicles, and also some because of the trees. The interaction between residents and visitors is very low. Some types of these interfaces occur in buildings that are already uninhabited, and which has a fairly wide setback.

https://doi.org/10.1051/shsconf/20184104008

Fig. 19. Example of Vehicle Setback Interface in the Area.

Design Guidelines

From the observations made in the designed area, has found some potency and problems that can be used as a guidelines in the designing Singosaren area as part of Kotagede tourist destinations.

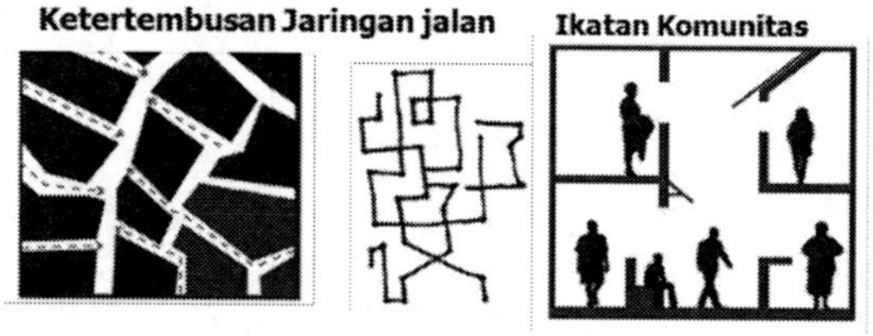

Fig. 20. Output Sketch in the Area.

Problem

The building style in the Singosaren area generally has no conformity to the typical character of Kotagede with the typology of Javanese buildings, where the building has a roof slope. Building function change into a boarding house and commercial, impacted to a new modern style building with a flat roof.

The unstructured pedestrian path quality, which does not have clear articulation according to the mode of transportation it carries. The material used has not been able to provide direction of movement and the movement speed of the users.

The incompatibility of the quality of the visual range between the spatial scale to the speed of movement, affecting the ability to memorize of the users.

Lack of open space quality for social activities, which does not support the diversity of Singosaren community cultural activities especially to support tourism activities.

Potential

Strong community social interaction in conducting activities together, also in daily life, can be seen through the building elements on the interface space, which allows interaction.

Utilization of private space for the public in the area can show the level of flexibility of the community to adapt to the new things they receives

Potential diverse cultural activities for tourist attraction in the Singosaren, used as circulation orientation , to show the possibility of appropriate building functions changes along the street.

Area ranges within a pedestrian radius (500m), can be used as a pedestrian priority area, encourage tourists to explore the area freely and discover a variety of uniqueness within.

From the potential and the problems, the design guidelines try to be compiled based on the elements of the design of the area, which connects the enhancement in the movement circulation to the degree of permeability of the area. The discussion focuses on the interface space along the road in the Singosaren area as part of the Kotagede tourist destination.

Function of space

In general, the land use in the Singosaren does not need to undergo significant changes, remaining as medium-density dwelling areas. Guidelines for functional change should be made in some street corridors for tourists circulation as a planned pedestrian priority. Changes in the function need to be done on the space within the building that is bounded directly on the public road. The function of the space is a semi-private function. In residential buildings, the appropriate functions are living room, family room, etc. Changes in semi-private rooms may also be directed for commercial functionss. On street corridors that accommodate the circulation of motor vehicles, building functions can be directed to commercial functions become commercial corridors. The road setback that lies along the commercial corridor does not use fences, encourages continuity of movement and allows for parking space activities.

In addition, street setback can also be used for supporting activities that can improve the social life interaction of the community such as guard posts, or encourage economic activities such as 'angkringan' (hawker), as well as place for road amenities. Changes in building functions that can be accessed directly from the main corridor of Singosaren, especially in the corridor with alley typology, can be directed to accommodate the tourists with guest house building typology.

Circulation and parking

The main parking area is in Singosaren square, often used by the community as a space for entertainment activities at night or non holiday season, and also for tourist's parking to Kotagede. The land use needs to be re-design, especially the parking lot layout, to optimized the functions as green open space which can also be utilized as a space of social activity (multiuse space). The parking spots are also directed transit spaces with varied modes for tourists, which are still accessible to the attractions within the area.

Some of the corridor's circulation within the Singosaren area are still within a pedestrian distance radius, and others for vehicle circulation, should be supported by the use of materials, textures, colors that can clarify the articulations according to the movement modes. The main corridors connecting the southern ring roads of Yogyakarta and the Kotagede destination are prioritized for pedestrians, while circulation by vehicles circles the periphery of the area. This intends to support the equity of circulation within the Singosaren region.

To ensure the safety of each mode of transport between the pedestrian and the vehicle, it is necessary to maintain the speed of movement within the area with traffic calming, especially on the nodes of pedestrian encounters and riders. Traffic calming can be done with different color, texture, bumper and also strengthening the fence with the height of the human eye at the corner meeting.

Building Layout

The intimate scale of the road space has been formed on the comparison of road space and ground floor of the building. An intimate scale needs to be guarded by directing a roof-tiled building, and also adjusting the height of the building such as adding a floor of a building on the ground floor should use a podium system. To support tourism activities, building facades should be designed to allow for interaction between visitors and residents. The design can be done by placing openings on the walls of the building facing to public access.

The degree of transparency on the walls of the building adjusts the degree of privacy of space functions, and the setback on the side of the road corridor should not be used for permanent parking. Fences became the most typical element of the current Singosaren region. In the fact, the fence is only a barrier element between plots, which is also used as a barrier element between private and public space. To support tourism activities, fencing in commercial building functions needs to be eliminated in order to increase spatial continuity. While in the building of residential function, the fence elements need to be strengthened so as not easily accessed spatially, while maintaining wall transparency or according to the height of the human eye.

Open space

Creating an open space for public activities and as a space of interaction between visitors and residents, can make varietes of attractive activities. The high level of community flexibility can be utilized by plotting private plots that are not used as a space of public activity, becoming a tourist attraction in thearea. Potential utilization of connected open space on private plots become new publicly accessible corridors.

The corridor between private plots can be a typical street with the concept of the alley, which connects visitors to the attractions in the area. The typical corridors of the selected alley are those that have a direct spatial connection to the main corridor of Singosaren. In addition, to increase the quality of pedestrian comfort during interaction or activity, open space along the road corridor in the form of setback plot, can be done by adding the number of trees along the Singosaren corridor.

Conclusion

The movement pattern has an important role in conducting the design of the area as a tourist area. The local potency used as a tourist attraction, can stimulate the connection through new access. Growing more access will further increase the permeability of the area, which can indirectly decrease the privacy of the residents. The design of the interface space is important, in order to support the interaction space between the visitor (outsider) and the inhabitant (insider), where the occupants remain their privacy and still increase the route options for visitors. The new access that is formed is influenced by the placement of tourist attraction nodes and the existence of open space that allows design intervention.

In exsisting area are found that strong factors affect the quality of the interface space as interaction space, just as what Dovey had found: Permeability in the area is formed due to the tourist attractions that need to be connected to tourists; transparency in the region can encourage interaction either through wall fences or on the walls of buildings; The setback on private land can be utilized to support the convenience of interaction; car-dependency can inhibit visual and spatial access from the path space to the inner building space limiting the interaction.

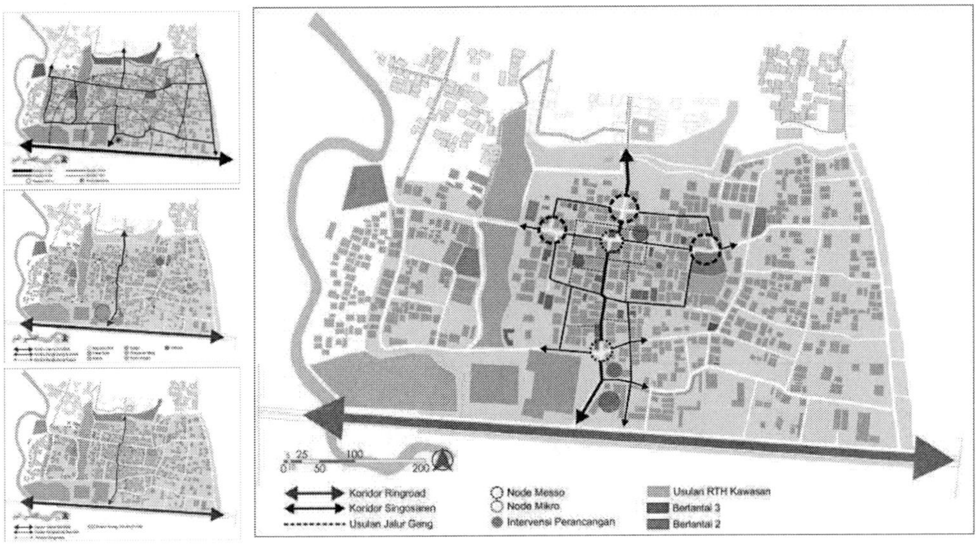

Fig. 21. Mapping of permeability analysis results on Singosaren.

https://doi.org/10.1051/shsconf/20184104008

References

Bentley, I., Alcock, A., Murrain, P., McGlynn, S., & Smith, G. (1985). Responsive Environments a Manual For Designers. London: Architectural Press.

Carmona, M., T. H., OC, T., & S. T. (2010). Public Places Urban Spaces: The Dimension of Urban Design. Oxford: Architectural Press.

Carr, S., Francis, M., Rivlin, L. G., & Stone, A. M. (1992). Public Space. New York: Cambridge University Press.

Cooper, C., (1993). Tourism: Principles & Practice. London: Pitman Publishing.

Dovey K. & Wood, S.,2011, Public/private Interfaces in the Innercity: Types, Adaptations, Assemblages, State of Australian Cities Research Network, viewed 04 May 2017

Dovey, K. & Symons, F. 2013. Density without intensity and what to do about it: reassembling

Understanding the Importance of Time in Interior Architectural Design Method

Anak Agung Ayu Suci Warakanyaka[1,*], *Yandi* Andri Yatmo[1]

[1]Department of Architecture, Universitas Indonesia, Indonesia

Abstract. The presence of time holds an important position in interior architectural practice and education. On theother hand, the presence of time challenges the stability and steadiness that framed the discipline. Furthermore, in interior architectural practice it is usually seen as a threat that should be either eliminatedor restricted. Rather than establishing defense mechanism against it, this paper argues that interiorarchitectural practice should be able to progress with time. By looking through undergraduate designstudio projects in Interior Architecture Program, Department of Architecture, Universitas Indonesia, this paper aims to addressed how the presence of time might affect, transform and even generate context specific interior architectural design methods that enables several dynamic forms of inhabitation. The out come of this study provides the opportunity for interior discipline to switch its focus, from the discipline that focuses on the aesthetic and constructional aspects, to the one that embraces the temporal aspectsof sociocultural conditions to enhance the wellbeing of its inhabitant.

Keywords: time, transformative, interior architecture, design methods

Introduction

Architecture, and also the interior, has a quirky relationship with time. In both disciplines, time is seen more as a threat rather than as a benefactor. Time, with its ability to alter and transform space, is seen as the genesis of chaos and disorder, which is very much what both disciplines try to avoid. It becomes an intimidating terror that should be abolished or at least control in architecture and interior architectural practice. This dynamic quality of time is conflicting with the obsession of architecture and the interior in immortalising purity in space (Till, 2009). Especially with the discipline of the interior, which has long been responsible for controlling the physical forms of the space, is mostly threatened by the offensive presence of time that introduces the elements of change, unruly, and weariness. It has led the discipline of the interior to cast-out the influence of any temporal aspects and contingency in its practice and to developed a barricade mechanism to restrain it (Till, 2009).

Contrary to this argument, the interior architectural practice has no authority nor the ability to control time. Interior would never win the battle against time because the nature of the interior itself is ephemeral. Interior is a process of enabling inhabitation that always involves complicated spatial-temporal compositions. It is a transient process of becoming before it transforms into the rigid spatial entity (Attiwill, 2012). It is a being, rather than a setting, that presence largely depends on the way its inhabitant perceive space. The interior should be able to progress with time, not against it if it still wants to be able to accommodate the ever-changing needs of its inhabitant.

This paper aims to explore the importance of time in interior architectural practice and variety of design strategies it generates by looking through several undergraduate design studio projects in Interior Architecture Program, Department of Architecture, Universitas Indonesia. These projects addressed how the presence of time might affect, transform and even generate context-specific interior architectural design methods. It investigated the role of temporal elements, which are presented in the existing context, such as the layer of inhabitations, the everyday routine of the inhabitants and the sociocultural conditions in the context, that would act as the basis to produce time-based interior architectural design solutions.

The outcome of this exploration suggests that by acknowledging the presence of time, interior architecture could develop a ready in hand design tools that would subsequently produce responsive interior settings that would accommodate the changing needs of the inhabitants and also the environments. This study provides the opportunity for interior architecture discipline to widen the range of focus within its discipline, from the one that specifically concerns to the aesthetic and constructional aspects of the space, to the one that embraces the temporal aspects of sociocultural conditions in order to enhance the wellbeing of its inhabitant. This study could also contribute to enrich the theoretical tools in the interior architectural education

* Corresponding author: ayuwarakanyaka@ui.ac.id

https://doi.org/10.1051/shsconf/20184104009

that often suffered from its image as an intellectually lightweight discipline.

The Power of Time

Time, like the "outside", is a dreadful concept for the traditional interior architectural practice. Time, and so the "outside", is occupied by the immeasurable and the arbitrary (Bachelard, 2006), a concept in which the interior, the practice that was only operating from the "inside", could not fathom. It is the immediate threat to the value that the interior holds dear: control, order and familiarity. Time has the ability to produce temporal conditions such as weathering, programmatic changes, and accidents that would challenge the immutable authority of the interior architectural practice (Till, 1996). This correlation makes the practice of the interior develop certain attitude towards time. It either eliminates it all together or attempts to restrain any temporary possible variables in its design process.

Since modernity, the practice of the interior has been in a long battle to overpower time. It employs various strategies to help them reach a certain level of "purity", where they imagine that space is always in its perfect condition, regardless how harsh the reality is. It attempts to "freeze time, to hold onto that perfected moment of the completion of the building for as long as possible before and after the event" (Till, 1996, p.2). It utilises technology to banish any trace of time and generates the flawless, shiny and lasting image of the interior space where there is no dirt, accident and weather (Till, 1996). The architects and the designers believe that they have the full authority to set a certain limit to time to meet the terms of their design. This understanding has led the interior architectural practice to progress in the fairy tale, where it far removes from the contingent world which space will eventually occupy (Till, 2009). The interior believe that time is just an unproblematic universal, where they could get hold on to it on formal grounds (Till, 2009). It introduces time as aesthetic criteria, where it could be freeze and fold as a form or ornament. What they often do not realise is that this control is only achieved representationally and could never be applied in real life (Till, 1996). In reality, interior, as well as architecture, is a very dependent subject that affairs are not only determined by the architects as its "creator" but also contingent upon the spatial-temporal conditions of its inhabitant and its context.

The design of interior space is always vulnerable to the authority of the time. It exists IN time, not the other way around (Till, 1996). Acknowledging the importance of time does not mean that the interior practice should imitate the quality of time, let alone reproduce it, but to try to make sense of it "so as to know how to operate within them" (Till, 1996, p.5). By accepting that the interior is at the mercy of time, the practice of the interior should be conscious and progressing with the dynamic implications of time, such as programmatic and physical changes of the space. For this to happen, the interior architectural practice should shift its attention from the "superior cultural narratives", creating the perfect setting, to shift into the world of the everyday (Till, 1996). The everyday is the site of the ephemeral and it is by addressing the everyday, the interior architectural practice could generate an interior space that could accommodate the ever-changing needs of its inhabitant (Till, 1996).

The Temporal Nature of Interior Practice

The traditional practice of the interior, with it excessive fear of time, proposed that interior is a substantive being, which has a separate entity with the rest of its environment (Attiwill, 2012). But the recent theoretical progression of the discipline shown that the practice of the interior is shifting "from foregrounding things to processes, from the individual to process of individuation, from form to information, (and this is our highlight) from space to time and movement" (Attiwill, 2012, p.1). Interior space is generated in the midst of flux and the product of the complex temporal-spatial compositions of its inhabitant and its context. It acknowledges that time has a crucial role in the production of the interior space in to enable inhabitation. The shift in the discipline is based on the thorough look on how our modern society inhabits space. The way we inhabit space changes from time to time according to the shift of sociopolitical affairs, technological advancement and the progression of lifestyle. At the moment, our society lives in the state of "atopia" where we are belonged nowhere except exactly in permanent "in-transit-condition" (Verschaffel, 2012). This makes the interior, as the discipline that is supposed to provide inhabitation should be able to accommodate our nomadic ways of living. It should not be pre-determined for our mode of inhabitation that involves constant changes requires continuous adjustment.

By nature, the interior is dealing with the transformation of a given space; it is dealing with a situation that is continuously made and re-made (Brooker and Stone, 2007 and Attiwill, 2012). Interior space "'is not made just once, but is made and remade over and over again each time it is represented through another medium, each time its surrounding change, each time different people experience it" (Forty in Borden, 2006, p.50). This statement reflects that adjustment and temporality are the essence of the interior space in providing the physical and emotional wellbeing of its inhabitants. This adaptive nature of the interior is supported by a permeable boundary of the practice. Interior space is no longer just the room inside of the building; rather its territory is highly depending on how the body manoeuvre between the objects and space (Borden, 2006). It is not restricted to any physical form or placement; rather it is depending on its inhabitant state of affairs at the particular moment in time. Thus, interior should not be seen as a substantive being but as a process of becoming, as a product of a complicated relationship between space and time that enables inhabitation (Attiwill, 2012).

This made the interior architectural practice should develop a slightly different attitude towards time compared to the architectural practice. Dealing with an existing system makes the interior has to be a fleeting condition, where it supposed to be a set of "constantly shifting relationships between space and its users" (Lim, 2011, p.134). This is supported by McCarthy (2005) who stated that the practice of the interior is based on the interiority where it is not an absolute condition, rather it is a mobile and promiscuous situation, that generates the art of relationship where the limits is always redefined and renegotiated. It makes the interior not deal with the rigid layers of existing architectural space in what Frank Duffy called as "shearing layers of change" (in Brand, 1994). Instead, the interior is dealing with the most mundane, temporal and high-frequency layers of change such as the layer of service, space plan, stuff and soul (Brand, 1994). By addressing these layers of change in its design process, the interior architectural practice would generate an adaptive and responsive interior system that is able to progress with the dynamic needs of its inhabitants. This system adopts the system of nature where there is a continuous feedback loop of change that makes the inhabitation setting get to be harmonious (Brand, 1994). This understanding of time gives the possibility of the interior space to learn and progress together with its inhabitants.

There is a need to introduce interior architectural design method which includes tools, strategy and tactic that would focus on how to progressing along with the temporal aspect of inhabitation and also to cope with the unpredictability of the context. Moving away from the traditional conception of the interior, Attiwill (2013) suggests a proposition of "interiorizt" which makes the interior should not be seen as an object, entity or even space, but as a product of interiorization, "a making habitable through a process of selection and arrangement to produce an intensive space; a temporal consistency, a 'fabrication of space' which enables inhabitation" (Attiwill, 2012, p.5).

The idea of interiorizt suggests that the practice of the interior should not tackle the challenging nature of time, but instead should address it as "a creative problematic through design" (Attiwill, 2013, p.107). It makes the art of the interior is distinct from architectural tectonics where movement is not going to be frozen to generate certain ornamental form. But rather, movement, as one of the powerful components of time, becomes the main resources that reorient the way the interior practice organises, operates and dynamises its space (Attiwill, 2013). "An interior making through the production of a temporal consistency... becomes a composition of movement - "a rhythm" - that creates a space-time that enables inhabitation physically and mentally" (Attiwill, 2013, p.114).

Proposing a design method that is generated based on time and movement is a challenge to the conventional practice of the interior where the spatial programming and form finding is based on the acceptance of the inherited situations and its intolerable results. It opens up alternative ways of design thinking and address the interior as a discipline that depends on the dynamic nature of the physical and temporal situations of its environments. "In twenty-first-century design, a shift can be noted from a spatial emphasis to one of temporality in which the concept of the gesamtkunstwerk as a total work of art continues but become one of "temporary totalities, nomadic encampments or natural environments composed of aggregated assemblies"" (Attiwill, 2013, p.116). With this design proposition, it would be possible for the contemporary interior architectural practice to "achieves a sense of unity and cohesion that is provisional and dynamic, enabling engagement with contingency, temporality, and change" (Attiwill, 2013, p.116)

The key to creating the adaptive and responsive design is by putting certain levels of flexibility and allowing the design to be "wrong". Interior is the process that is made by displacements, that is why being "wrong" is the process of becoming. One of the tools to create the adaptive design is called the scenario planning, where it would not produce design plans but strategies (Brand, 1994, p.178). The differences are "where a plan is based on prediction, a strategy is designed to encompass unforeseeably changing conditions" (Brand, 1994, p.178). This creates rooms for possibilities where the inhabitants could manoeuvre according to their spatiotemporal needs, not the predefined interior. Although the regular spatial programming covers almost all the needs to create a responsive space, it is "leaving future users out of the picture, making the building all too optimal to the present and maladaptive for the future" (Brand, 1994, p.181). It adopts the law of ecology where "the more adapted an organism to present conditions, the less adaptable it can be to unknown future conditions" (Brand, 1994, p.181).

The scenario thinking allows architects and designers to start thinking the unthinkable (Brand, 1994). It develops "robust" strategy that would be viable in the variety of future (Brand, 1994). The scenario planning suggests a bottom-up approach that would let the interior space learns from its environment (Brand, 1994). By inviting and exploiting the diversity aspect of the everyday, the mundane and the ordinary scenario would emerge as a various combination of driving forces and organisation strategies that befitting the dynamic needs of its inhabitants (Brand, 1994). By allowing time to immerse in the design process, we are allowing for the emergence of the new modes of inhabitation. It also turns the occupants into an active collaborator of the interior space, where it would generate the feeling of intimacy, control and belonging.

Interior Architectural Design Practice in Time

The following chapter will explore the influence of time in generating design strategies in interior architectural design practice, by looking through two undergraduate design studio projects in Interior Architecture Program, Department of Architecture, Universitas Indonesia. All of these projects implemented the scenario planning with its bottom-up approach to address a particular design issue. These projects investigate the role of temporal elements in the existing context, such as the layers of inhabitations, the everyday routine of the inhabitants and the sociocultural context.

By exploring how these projects generate their design methods, we could investigate various adaptive and responsive interior architectural design strategies that would act as the tools to produce a time-based interior space that would accommodate the ever-changing needs of our contemporary society. The highlight of using the "robust" design strategies, as shown in these projects, is the ability of the interior space to respond to any immediate change that is initiated by its inhabitant and its environment. The scenario planning could prepare the interior space to cope with such changes.

Transforming existing issue into design scenarios

This project explores the issue of the waste distributions in Lokasari neighbourhood, Central Jakarta. This project started with investigating the mundane aspects of the neighbourhood, such as: the spatial programming of the street, the traffic flow, the habit and ritual of the actors, the atmosphere and the urban calendar.

This site is famous for its contrast atmosphere of daytime and nighttime. In daytime, the neighbourhood is mostly used as a site of formal commercial activities and parking lot. At night, the neighbourhood is famous for its entertainment and informal commercial activities, such as street vendors and street musicians. This project employed both behavioral and spatial mapping as a method to unfold the emerging issues on site. The mapping addresses all the variables that are responsible for the quality of the spatial experience in general.

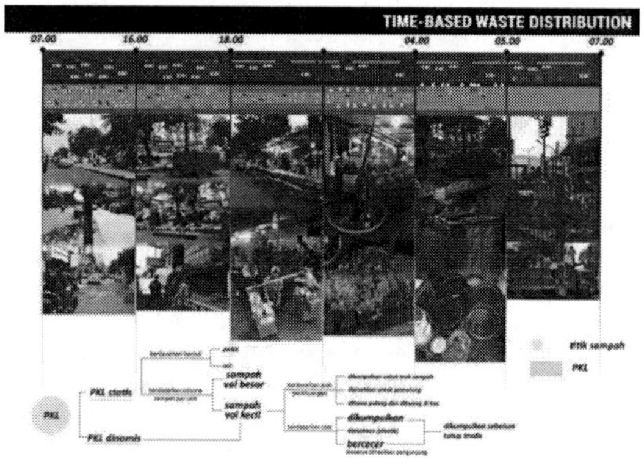

Fig. 1. Site Mapping.
Source: Kristofani, 2016.

One of the core issues was the distribution of liquid waste that affected all the activities in this neighbourhood. The production and distribution of the liquid waste involved all of the actors on site, such as: the street vendor, street musician, pedestrian, beggar and even the night escort. The mapping also found, because of the differences of activities in daytime and nighttime, some of the actors already employed the adaptive place making. For example, the street vendors use portable and collapsible tent/truck for their night commercial activities.

Fig. 2. Design Scenario
Source: Kristofani, 2016.

After thorough investigation, it is found that the issue could be transformed into design solution. All of the actors produced various types of liquid waste based on their consumption activities. The project attempts to exploit those consumption habit and tries to create an urban interior intervention that would not only solve the liquid waste issues, but also bring benefit and add value for the neighbourhood in general. It tries to cater multiple needs of the Lokasari's inhabitant at once.

Interior "deal with transformation of a given space" (Brooker and Stone, 2007, p.125). This argument is manifested in this project by trying to exploit the existing system to generate and support the new interior intervention. Utilizing the existing gutter and the water utility system, this project tried to address the issue of the liquid waste distribution by creating a water recycle system that could accommodate several activities surrounding water utilization. By creating scenario planning (Brand, 1994), this project came up with several possibilities on how this place would be used in both present time and also in the future (figure 2).

Fig. 3. Final Design.
Source: Kristofani, 2016.

The final design generate from this scenario planning was not substantially temporal or collapsible that could perform immediate response of physical changes. Yet, it is incorporating time in its design process that enables it to cope and adapt with the possible changes of the activities made by the inhabitants. It develops design solution with "robust" strategies to tackle on the heart of the problem and make use of technology, not to get rid of the waste or dirt, but to embrace it and transform it into design tools. This design strategy gives possibilities for the inhabitants to become an active dweller where they would keep the feedback loop and grow together with the interior intervention.

Generating transformative urban interior setting

Although located on the same site, this second project addresses a whole different issue from the first project: the issue of locality and cultural identity. Quite similar with the first project, it employed behavorial and spatial mapping in investigating the unseen sociocultural issues that might affected the quality of spatial experiences in the context. The mapping suggested that the context possessed an interesting history in the entertainment culture that was rooted far before the Indonesian independence.

Yet this locality was concealed by the current nightlife activities, which was only intended to and attracted particular target group and the presence of the street vendors, who contributed to crowdedness, waste, commodity exchange and lively atmosphere in the neighbourhoood. The site is currently developing an unpleasant image and there was almost nothing left from its former entertainment hub glory.

Fig. 4. User Behavorial Mapping.
Source: Kosasih, 2016.

The design solution generated from this issue was providing an urban interior setting that would accommodate all types of entertainment activities to claim back the image of Lokasari as Jakarta's entertainment hub without disturbing any existing commercial activities. By exploring the operation of how the street vendors organize and reorganize their stalls, this project proposes to create a transformative urban interior setting, therefore its forms, location and arrangement really depend on the demands of the inhabitants. In this project, the urban interior setting was generated through a series of modules and transmodules that would allow the interior setting to adapt with the immediate change of activities and accommodate various needs of entertainment activities.

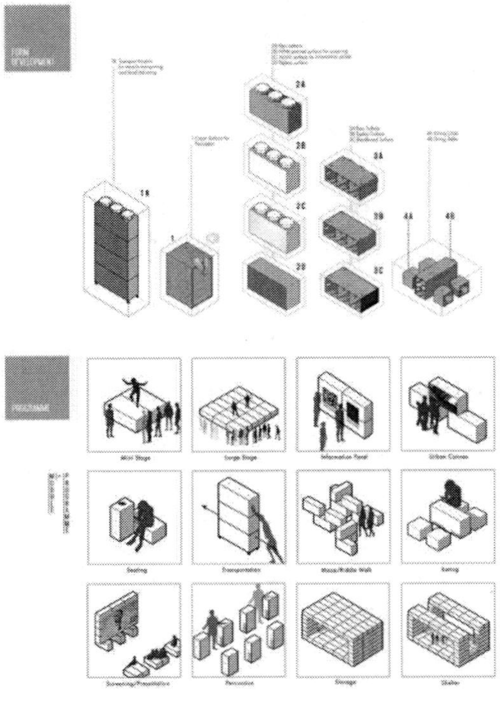

Fig.5. Design Scenario Diagram.
Source: Kosasih, 2016.

Aside from being able to generate various forms of temporary entertainment hub in any leftover spaces, these modules could be utilized to support some of the existing commercial activities in Lokasari. The details of scenario activities and transformative nature of the interior setting could be seen in Figure 5. Employing several features of collapsibility such as merging, adding, and retracting, these modules allow the inhabitants to freely adjust the arrangement of the modules to fit their spatiotemporal needs.

Rather than generating an exclusive interior setting to fulfill its purpose, this project brought the locality of the context - the entertainment culture, to the street where the activities actually happen. The interiorization of the street was made possible by employing the transformative spatial components that could adapt with the dynamic nature of the street. Rather than designing a rigid and high performative interior space, this project focused on exploring the ephemeral aspects of the interior to respond to the dynamic needs of the inhabitants.

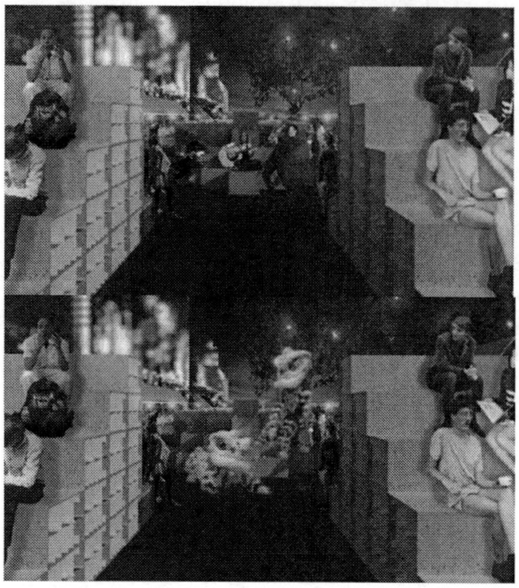

Fig. 6. Transformative Urban Interior Setting. Kosasih, 2016.

Both of these projects incorporates time on their design process and generate design strategies that is able to cope with the dynamic forces of the context. The first project use the bottom up approach to transform the existing issues into design solutions. It generates several design scenarios that would be able to catter multiple needs of the inhabitants. The second project use behavorial as well as spatial mapping to investigate the dynamic spatial operation that already existed in the context. The findings of this investigation is used to produce transformative urban interior setting that forms, location and arrangement responds the spatial-temporal situations of its inhabitants.

Conlusions

This paper argues that the presence of time is a crucial aspect in the progression of the interior architectural education. Rather than seeing time as intimidating forces, interior architecture should see it as the potential agency that would support the practice of interior architecture in enabling inhabitation. By incorporating time and its features in the design process, interior might be able to find suitable design tools to accommodate the ever-changing needs of the inhabitants. It might also generate context-specific interior architectural intervention, that not only addresses the current issues but is also well prepared for the future challenge of upcoming inhabitation. This exploration in design studio projects, illustrate that by employing scenario planning, interior architectural practice could generate adaptive and responsive spaces that could cope well with movement and changes. This study contributes to the enrichment of theoretical tools in the interior discipline that is aimed to provide inhabitation that enhances the physical and psychological wellbeing.

Acknowledgements

The projects and diagrams included in this paper are student works in Interior Architecture Design Studio 4 - 2017.

References

Attiwill, S. (2012). Beyond building: interior designs. IDEA Symposium 2012, (Interior: A State of Becoming), 1–7.

Attiwill, S. (2013). Interiorizt. In G. Brooker & L. Weinthal (Eds.), The Handbook of Interior Architecture and Design (1st ed., pp. 107–116). London: Bloomsbury Academic.

Bachelard, G. (2006). The Dialectics of Outside and Inside. In M. Taylor & J. Preston (Eds.), Intimus: Interior Design Theory Reader (pp. 22–25). Chichester: John Wiley & Sons Ltd.

Borden, I. (2006). Thick Edge: Architectural Boundaries and Spatial Flows. In M. Taylor & J. Preston (Eds.), Intimus: Interior Design Theory Reader (pp. 49–55). Chichester: John Wiley & Sons Ltd.

Brand, S. (1994). How Building Learn: What Happen After They're Built. New York: Viking.

Brooker, G., & Stone, S. (2007). From Organisation to Decoration. In Thinking Inside the Box: A Reader in Interior Design for the 21st Century. London: Middlesex University Press.

Lim, S. (2011). Tactics of The Peripatetic: On Vendors' Making an Interior of The Street. In Interior Tools Interior Tactics: Debates in Interiors Theory and Practice. Faringdon, Oxfordshire: Libri Pub.

McCarthy, C. (2005). Toward a Definition of Interiority. Space and Culture, Vol. 8 No. 2, 112–125

Till, J. (1996). Architecture in Space, Time. Architecture and Anthropology, (Academy Editions), 1–6.

Till, J. (2009). Architecture depends. Cambridge, Mass: MIT Press.

Verschaffel, B. (2012). The Meaning of Domesticity. In C. Briganti & K. Mezei (Eds.), the Domestic Space Reader (p. 156). University of Toronto Press.

Reading Architecture: Explorative *Text*-ual Reception for Architectural Education

M. Mirza Y. Harahap[1,*], *Yandi* Andri Yatmo[1], *Paramita* Atmodiwirjo[1]

[1]Department of Architecture, Universitas Indonesia, Indonesia

Abstract. This paper discusses how reading is central to the architectural practice and how it carries an immense potential for design method, it is further argued that an act of reading is important to be introduced and learned in the architectural education institution within which students learn and develop various design methods. Accordingly, to this importance, this paper proposes a discussion toward various examinations of reading of various types of text. Stressing the discussion toward the conception of reception, interpretation, and representation, this paper aims to disclose the potential of the textual reading for design development by dismantling a range of exploratory reading acts, from literal act of reading to a developed act of reading, each of which positioned text as an event rather than an object. Each of the acts of reading ex Amined in the study takes a role as a process of creative analysis and interpretation which results in a representation of received knowledge.

Keywords: reading, text, method

Introduction

In architecture we discuss human, space and inhabitation. We study a broad range of discipline, from human behaviour to engineering. We design, learn how to make a good design and along the way, we also learn about design itself, continuously developing our effort to improve well-being. Within those processes, we perform careful yet deep consideration and calculation so that our design could successfully respond the uprising spatial issue(s). Consequently, the way we respond the issues arise in the environment is argued to be an important aspect to be paid attention to.

Understanding the world with every single phenomenon exist where there are continuous "Shifts in weather, changes in the behavior and needs of people" (Meagher, 2015, p. 160), is believed to be central and important in the overall design process. The needs to produce a flexible and adaptable design which responds to the ever-changing and shifting environment is then unavoidable (Meagher, 2015). Consequently this suggests the importance of reading the environment which would enable us to understand the environment itself as thorough as possible. Hence, the thorough understanding would enable one to come up with an innovation which responds to the uprising issue(s) thus offering the world as a better place to be inhabit.

Accordingly to this importance, this paper tries to explore the idea of reading as a means to understand the environment. Conducting an act of interpretation as the core of reading, the study particularly proposes reading as an explorative interpretational method for architectural design research and design development process.

Reading in Architecture

In architecture, to read means to uniquely understand and thoroughly grasp toward, but not limited to, certain phenomena exist in the environment. An ability to read is a quality an architect has to adequately possess. Particularly for an interior architect, reading, as Brooker & Sally (2007) explicitly suggested in their book titled Basic Interior Architecture 01: Form + Structure, is especially important to perform an adaptation or remodelling toward an existing setting. Looking at this importance, we would like to invite the readers to take a heed toward what is reading.

It is believed that reading, as a distinctive manner of knowing and to achieve knowledge, to some extent could be described as an effort to construct particular meaning (Tierney & Pearson, 1983; McGinley & Tierney, 1988). Thus, reading —which particularly referred in this study, not as a literal act of reading such in reading a newspaper for example— can be considered as an act of extracting the meaning of a text.

However, the function of reading is not limited to an act of constructing the meaning of a text. Instead, this paper intends to explore how reading can be developed as an exploratory design method and be used in architectural design development. Considering that reading, as previously stated, is not limited to the literal act of reading and so is the text, this means reading possesses a wide range of possibilities to understand the

* Corresponding author: mmyharahap@gmail.com

environment with all of the phenomena occurred within. As reading is practically potential for architectural design, it is consequently then argued that reading needs our attention in the academic level, to be learnt and developed. McGinley & Tierney (1988) argue that developing reading into a learning situation, or education in general, would particularly influence thinking and learning once reading is strategically applied in various forms. Particularly for architectural education, this implies that reading might be used as a way to understand environmental phenomena as a means to have a better understanding about the uprising spatial issues in the environment; as a design method.

Considering the above explanation, one may look at reading eminent with either hermeneutical or phenomenological conception. However we invite the reader to explore reading through the conception of reception instead. Proposing the reception as the corridor of the discussion, we try to explore reading as an exploratory act of interpretation which constitute critical analysis and creative thinking.

Reception: Interpretation as a Means to Understand

The discussion toward reading in this paper is proposed by looking at the idea of reception within which interpretation is performed, aiming an understanding toward particular text being read, indicating a process of acquiring knowledge. Stead & Graduno (2013) argue that reception is distinct to the two conception which influenced it that are hermeneutics and phenomenology. It argues that the meaning of any particular phenomena which one tries to understand does not indwell within the text (Stead & Graduno, 2013). It is rather "...focuses on practices: embodied, performed, and informed by a specific and unique set of circumstances. It demonstrates that the meaning of a text is always a dynamic negotiation, that texts and contexts can never really be distinguished as one can never know a text outside of its specific reception processes" (Stead & Graduno, 2013, p. 269). This clearly informs that an act of operating the text in order to disclose the meaning lies within should be performed. This operating act is no other than reading itself. Stead and Graduno's argument also suggests two things: 1) In reception theory, reading and looking up at a particular meaning as a mean to understand is a dynamic process; and 2) Text is indistinguishable from context. The two hence indicate that the reception of knowledge from an understanding or extracted meaning about a particular text is a dynamic process of interpretation.

Extending the discussion, an interesting argument is suggested by Gough (2013) regarding the conception of reception is worth to be paid attention to. He stated that:

"The central tenet of any reception theory of architecture would have been the following call: collapse the subject–object divide. A literary text, it was said, is never—properly thought— an object, except as an after-effect. A literary text is rather an event—the event of reading that occurs in the interplay between the subject (the reader) and the object (the text)." (Gough, 2013, p. 280).

As argued by Gough (2013), reading text as an event rather than object becomes necessary since reading has a sense of give-and-take between us and what we are reading. This characteristics of reading, hence, needs to be our concern especially when reading is performed to understand the environment. This is also important especially when we consider how in architecture the relation between us –the inhabitant; the reader– and the space and everything exists within –the environment; the text– is extensively discussed.

Iser (2005) suggested that interpretation "... has always been: an act of translation" (Iser, 2000, p. 5), and we argue that it reflects how interpretation relates to representation. The translation here could be defined as an act of transposing the process of a thing into another form (Iser, 2000) which is conducted through representation. Here, interpretation and representation are seen as two aspects that affect one another. A form of interpretation, in a reading process, needs to be delivered in particular form of representation. On the other hand, each representation that is delivered in whatever form the reader chose, further performs the interpretation itself, leading to a potentially infinite process. Therefore, one must underline that the interpretation and representation are two inseparable aspects of reception. Schleiermacher in Iser (2000) further extended on how interpretation might be understand. He argued that interpretation is "... a rigorous practice of discovering and elucidating the ramified conditionality of how understanding comes about" (Iser, 2000, p. 41). Thus, we can consider that in architecture, an act of interpretation, within which representation is involved, exists as a process of translating what exists in the environment, as the text, so that we can understand the environment.

Although reading process as a way to acquire knowledge is notable in many fields, from literature to art and architecture, what we really need to underline is that we have to embrace the transposition of its abstraction which allows us to assemble the very meaning of the text (Gough, 2013). And this could be accomplished by applying certain interpretation approach.

Schleiermacher in Iser (2000) explained that there are two methods of interpretation which are known: the divinatory method and the comparative method. The two methods are respectively defined as a method which "... seeks to gain an immediate comprehension of the author as an individual..." (Iser, 2000, p. 50) and "... proceeds by subsuming the author under a general type" (p. 50). The two methods will be discuss further in the following part of this paper. However, we need to note that these two interpretation methods are have a very close relationship between one another and that they are claimed to be the process that possibly ends up in the reception of knowledge (Iser, 2000).

The Two Interpretation Methods in *Reading* of Architecture

Building an interpretation and delivering it through representation might be a challenge since it takes a reading which encounters a critical thinking and creativity in the process. Both divinatory and comparative methods require an effort of making each bears its own strength characterised from the representation itself. As previously described, the divinatory method incorporates an in-depth operation to understand the meaning while the comparative method is practically performed through an act of comparing.

In this part of the paper, we invite the reader to take a look at how reading through each interpretation methods might be performed. Each of the methods will be discussed separately. Then we will discuss how these two methods are related, by also referring to another example that combines these two methods.

The Divinatory Method: An Exposition Toward Elliot the Elephant

Conducting a reading which is performed through divinatory method might be a challenge since one needs to bravely take an in-depth translation toward what is offered in the text. Even reading a literal form of text, as what we will discuss, could be challenging yet compromising a distinct understanding of the discussed text. Elliot the Elephant: Loses his Voice, written by Christian Swift in 1998, might be just an ordinary children book. However, reading the content of this book as an event rather than merely just a book, an object, is another thing. Gough (2013) further suggested that "There is a meaning to the literary text, but this meaning is not (or not merely) passed across from the text to the reader, rather 'the meaning of the text is something that he has to assemble' (Gough, 2013, p. 280). Hence, in order to be able to assemble the meaning of a text, the challenge to overcome is how to see the text itself as an event which consists of what might construct and develop the meaning of the text itself. In the paper titled Text-Extracting Process: A Method in Developing a Story, Harahap & Yatmo (2014) comprehensively explore the process of understanding how the content of the book develop its story. Through an extraction process, Harahap & Yatmo attempted to explain how a literal text builds its story through the composition of what constructed a narrative text, which are the settings and the plot, and use the extraction for design purpose.

Reading the book seems reception of preliminary understanding that the book has certain aspects which could be used to further understand what is hidden beneath what is shown on the surface of the book. Settings and plot are firstly operated then the connectivity to one another is developed, which in the process is interpreted and delivered into certain form of representation.

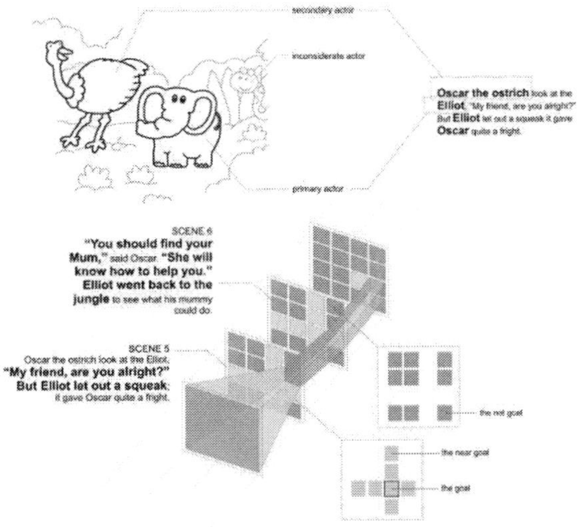

Fig. 1. The Operation of Settings.
Source: Harahap & Yatmo, 2014.

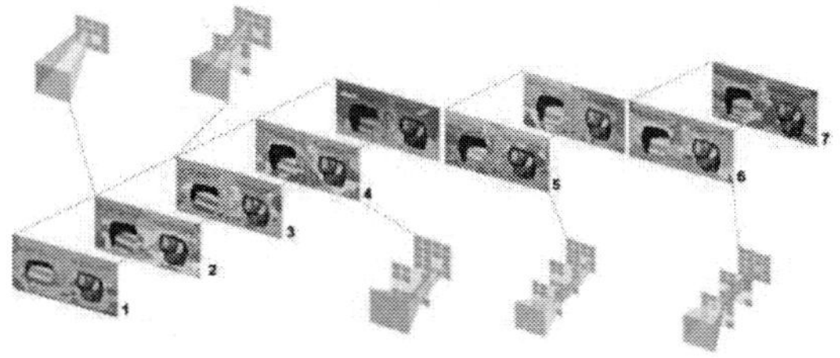

Fig. 2. The Operation of Settings.
Source: Harahap & Yatmo, 2014.

Figure 1 shows how this operation was conducted. Diagrams were used as the media to deliver the representation of the interpretation toward the setting. We suggest that it is a part of the reading itself, the extracting process. The diagrammatical representation performed particularly gives a chance to deeply interpret to what is exposed from the book and at the same time opens a possibility to make further interpretation. This is only possible as the authors deliver their interpretation into a developable and transformable diagram which respond to their needs on extracting the settings and plot, hence an understanding of the book is partly acquired. Overall, the process indicates a creative and critical thinking, a compromising preliminary act of reading of the setting.

What had been done as shown in Figure 1 was then applied to the whole pages of the book, resulting in a diagram as shown in Figure 2. This diagram particularly shows us how the interpretation which has been previously performed is conducted for other parts of the book. This is would then help the overall understanding of the book.

As an example of reading of which a divinatory approach for interpretation is applied, we argue that Harahap & Yatmo (2014) has attempted to conduct a reading which ought to be an act of understanding text as

https://doi.org/10.1051/shsconf/20184104010

event rather than object although the book is in a form of literal text.

The Comparative Method: The Melting Layers Project

The second interpretation method to be discussed in this paper offers a different approach which is conducted through a different representation yet a unique, if not even, quality of understanding. The discussion is conducted toward a set of text that appears as frozen liquids. However, this project is alike with how the previous interpretation method respond to the text, which is by not seeing the text as what it appears, the frozen liquids, but as an event of melting process instead.

Fig. 3. Data of Melting Frozen Liquids.
Source: Harahap, 2012.

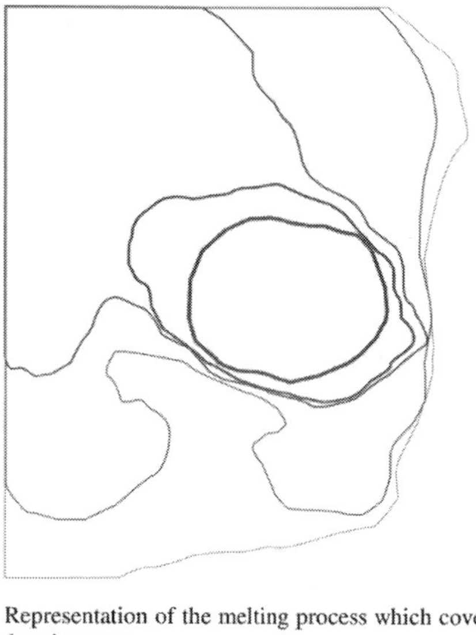

Fig. 4. Representation of the melting process which covers the two explored aspects.
Source: Harahap, 2015, development from Harahap, 2012.

As the name of the method suggests, the interpretation process to understand how the frozen liquids were melting is conducted by comparing the melting process from one liquid to another. Each melting process firstly went through an interpretation and delivered into certain representation. The Melting Layers project (see Figure 3), as explained by Harahap (2012) in the article which titled after the project, was done to explore how the frozen liquids melt by looking at three aspects: the dispersion, the shifting and the shrinkage of the melting frozen liquids. These three aspects are represented in the form of diagram. Basically, the reading conducted here is by interpreting and tracing the three mentioned aspects of a frozen liquid into a two-dimensional diagram, although in this paper, only two are discussed (dispersion and shifting position) out of three aspects as there is a limitation to the access of the data.

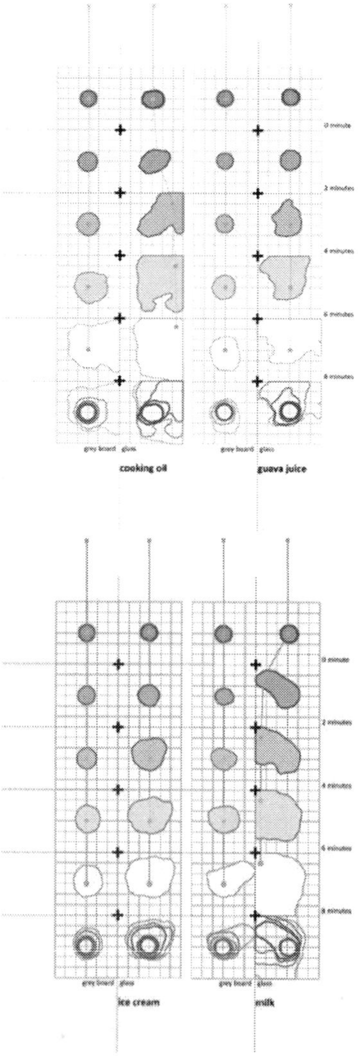

Fig. 5. Data of Melting Frozen Liquids.
Source: Harahap, 2015, development from Harahap, 2012.

The reception resulted from the observing of the melting process shown in Figure 3 determines the further interpretation and representation act. Particularly in this project, Harahap (2012) seems to look at the strong effect of time in the overall process, hence develop a form of representation which is correlated with how the data is recorded in a timely manner. Figure 4 shows the first stage of the diagrammatical representation which is more literal compared to the representation diagram of the first stage on reading Elliot the Elephant: Loses His Voice. The diagram shows the set of data of the melting liquid in a timely manner, performs the sequence of the melting process and, with the help of the additional grid, indicates how the two observed aspects are occurred.

This is then followed by applying the tracing process to the other frozen liquids as shown in Figure 5. From Figure 5, we can see that once the representation is applied to all of the frozen liquids, the comparison revealed the differences on how each of the melting processes occur. Moreover, the comparison also gives us an information toward how each frozen liquid is distinct to one another on how they melt, how their condition changes over time and how they later construct the understanding about the melting process. Moreover, the diagram represents the details of two out of the three explored aspects of the melting liquids which are the traceable changes of position and the measurable dispersion occur during the melting process. Accordingly to those explored aspects, thus through the conducted reading, Harahap (2012) then acquired a specific understanding toward melting process itself.

The Combination of The Two Interpretation Methods for *Reading* a Rather Complex *Text*

A *reading* process demonstrated by Atmodiwirjo et al. (2015) in the paper titled *Dwelling Spatial Arrangement, Active Body Movement and Health,* proposes a further approach to how one could interpret a text. This example demonstrated how a reading is conducted toward a more complex text; compared to the two examples which have been discussed previously. From their reading, we find a sense of combination of the two interpretation methods, each of which of course, for a specific purpose yet in the end of the reading process. Through this combination they could then acquire a whole understanding toward what they wanted to disclose.

Exploring the hodological space conception through an in-depth and comparative operation of particular dwelling context, they are able to pull out a rather important proposition for both architectural knowledge and for the importance of architectural design development as their proposition could be used for design justification as well.

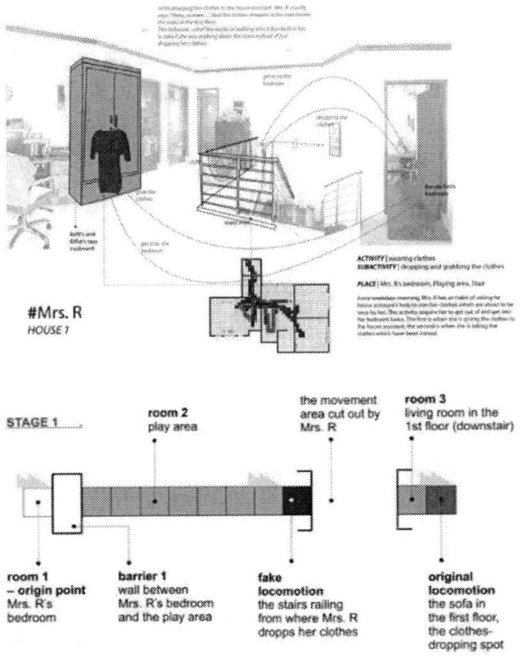

Fig. 6. Reading Dwelling Performed.
Source: Atmodiwirjo et al, 2015.

This example specifically indicates that reading is indeed could be used for architectural design importance, both theoretically and practically. It also indicates that reading is applicable even to a complex text. As long as a process of interpretation and representation is continuously and systematically conducted, we could understand and acquire knowledge from any kind of text. Up to this point, what we must further address is how far we can develop and apply the practice of reading in the academic environment. The three discussed examples clearly show how reading would be potential to be developed as a design method, yet, there is a challenge of how much we would concern ourselves with this potential? Architect, as well as interior architect and other design professions, does need reading to understand the uprising spatial issues, but learning and get used to reading is another thing. It is rather important to see reading as a necessary knowledge, especially in architectural education within which we try to understand human, inhabitation and space as what construct the world. Yet, we still rarely see the practice of reading is specifically cultivated among the students in architectural education institutions.

Reading for Architectural Design Method

As previously discussed, the first two project examples demonstrate how the two interpretation methods could be conducted and inform us three things: First, reading may or may not include a literal act of translation with no regard to when it is conducted in the overall process of interpretation and representation as it depends on how one chooses to interpret what he/she has been preliminary received; Second, as argued by Schleiermacher in Iser (2000) which suggested that there is a sense of correlation between the first two interpretation methods, the two discussed reading examples are each positioned the in-depth and the comparative approaches support each other in the overall effort to understand the text, although one will more likely to emerge stronger than the other in a single reading act. Consequently, the two interpretation methods show the very potential of reading itself.

Third, the last example which conducts how the two interpretation methods could be used simultaneously in one process of reading further expand what we have learn from the first two projects. The literal translation which may be performed as part of the overall interpretation could be repressed on its effect, thus, reducing the possibility of conducting an interpretation which could only acquire an outer understanding rather than an in-depth one. However the two methods hold their own strength for the interpretation importance, consciously combining them in a whole reading process is proven useful when the text appears in a more complex state.

These findings then further emphasise Iser's suggestion that "The two forces that Schleiermacher perceived in interpretation and in human thinking generally are versions of two processes that are indeed comprised in every realm of thought that can lay claim to knowledge" (Iser, 2000, p. 50). As reading is performed as an effort to understand a particular thing through a process of critical interpretation whether it is through divinatory method, comparative method or the combination of the two methods, the result apparently is full of potential with various, if not unlimited, ways to receive and claim knowledge. Related to architectural practice where an understanding toward any phenomena occur in the environment is necessary so that one could make a suitable respond to the uprising issue(s), the potential particularly indicates that reading might be useful to be developed to support, if not as, architectural design method as well.

Looking at reading as a method, the authors argue that it could be develop in the design process for various usage. The result of the reading conducted by Harahap & Yatmo (2014) as discussed earlier, were used to construct the programming idea of the architectural design proposed in their project. The result of reading in the research project conducted by Paramita et al. (2015), as previously discussed, could be used as a suggestion for design development. The two results also suggest the importance of design research which broadly learned in the architectural education institutions.

Conclusion

Reading, based on our discussion through various examples, is considerably and arguably potential to achieve particular understanding and claim certain knowledge on something. The variety of text which exists in the world, through a process of interpretation which incorporates representation within, gives us a chance and possibility to receive as much knowledge as we can. The two methods, that are the divinatory and comparative method, are proven effective for reading purpose. While each of those methods is exceptionally distinct in their own way, yet the two actually are interconnected one another in one process of reading.

All of the discussions toward reading itself suggest an urge to further explore reading, particularly for architectural education. Its flexibility and possibility make reading a rather important discourse to be developed. As the urge to deeply understand context so that we could have a better proposition to what kind of response we need to design to tackle the uprising spatial issues are getting significant, the urge to learn how to read is getting significant as well. Specifically for this matter, we argue that reading, in whatever form it is delivered in an architectural education, must always in an open state, meaning, that as reading is flexible and highly subjective in its nature and text is always changes over time, thus reading will continuously develop.

Looking at reading as a method also suggests that reading is essential for architectural design education. Reading generally enables us to understand the environment, thus, gives us a better chance for responding the arising issues. Reading as method is also specifically useful as a part of design research, covering both design development importance and the architectural research.

References

Atmodiwirjo, P., Harahap, M. M., & Yatmo, Y. A. (2015). Dwelling Spatial Arrangement, Active Body Movement and Health. i-dwell International Conference on Dwelling Form 2015, (pp. 29-44). Mataram.

Brooker, G., & Stone, S. (2007). Basics Interior Architecture 01: Form and Structure: The Organisation of Interior Space. Lausanne: Bloomsbury Publishing PLC.

Gough, T. (2013). Reception Theory of Architecture: Its Pre-History and Afterlife. Architectural Theory Review, 18(3), 279-292.

Harahap, M. M. (2012). The Melting Layers. arsitektur.net Jurnal Teori dan Desain Arsitektur, 6(2), 34-37.

Harahap, M. M., & Yatmo, A. Y. (2014). Text-Extracting Process: A Method in Developing A Story. [in]arch International Conference 2014 (pp. 139-148). Depok: Department of Architecture, Faculty of Engineering, Universitas Indonesia.

Iser, W. (2000). The range of interpretation. New York: Columbia University Press.

McGinley, W., & Tierney, R. J. (1988). Reading and Writing as Ways of Knowing. Champaign: Bolt Beranek and Newman Inc.

Meagher, M. (2015). Designing for change: The poetic potential of responsive architecture. Frontiers of Architectural Research(4), 159-165.

Stead, N., & Garduno, F. C. (2013). Architecture and "The Act of Receiving, or the Fact of Being Received": Introduction to a Special Issue on Reception. Architectural Theory Review, 18(3), 267-271.

Tierney, R. J., & Pearson, P. D. (1983). Toward A Composing Model of Reading. Champaign: Bolt Beranek and Newman Inc.

Cultural Assemblage as Genius Loci: Character Analysis of Medan City Center District

Hari Hajaruddin Siregar[1,*], *Petrus* Natalivan[2], *Agus* Suharjono Ekomadyo[3]

[1]Student of Magister Urban Design, School of Architecture, Planning and Policy Development, Institute Technology Bandung, Indonesia
[2]Assistant Professor in Urban Planning and Design Research Group, School of Architecture, Planning and Policy Development, Institute Technology Bandung, Indonesia
[3]Associate Professor in Architectural Design Research Group, School of Architecture, Planning and Policy Development, Institute Technology Bandung, Indonesia

Abstract. The city of Medan was formed from a rapidly growing plantation industry in the 1800s. The area that was originally only a village called Medan Putri with a population of about 200 people slowly changed since the Dutch investors saw the prospect of tobacco plantations in this region (Sinar, 2006). The amount of manpower needed to manage the plantation resulted in the investors bringing labor from Java, China and also Tamil. Moving the central government of the Deli Sultanate to Medan in 1891 increasingly crowded Medan at that time. The Arabs, Mining, Mandailing, and Aceh began to arrive for trading purposes as Medan began to grow and become more crowded. The study focused on locating the genius loci of Medan City through tracing the historical meaning by adapting the method undertaken by Norberg Schultz in tracing the spirit of the place and genius loci. The result of the analysis shows the role of culture and economic background that plays a major role in the formation of the character of Medan City center. The city is formed from the history of the plantation industry as well as the diverse cultures that share the same attachment and goals in the economic field.

Keywords: culture, character, genius loci, spirit of place, urban Reading in Architecture: Explorative Textual

Introduction

As the third largest metropolitan city, after Jakarta and Surabaya, with 2,210 million of population and and have about 265.10 km2 land area (Pemko Medan, 2013), Medan is now home for working people from various ethnicities and different cultures backgrounds. This cultural diversity will indirectly bring a unique social interaction between communities in accordance with culture and the culture itself and essentially require public space to perform such activities (Tanjung, 2015).

The city of Medan was formed from a rapidly growing tobacco plantation industry in the 1800s. The village area originally named Medan Putri slowly changed when Dutch investor Nienhuys saw the prospect of tobacco planta-tions in this region (Sinar, 2006). Medan Putri originally inhabited by the Malay community began to flourish and attract many people coming from various ethnic groups. The development of a plantation industry that requires a lot of manpower encourages investors to bring in more planter and laborers from outside Medan. Manpower imported comes from Java, Tamils and Chinese (Sinar, 2006). The transfer of the royal administrative center of the Deli Kingdom from Labuhan to Medan further increased the crowds of Medan at that time. This transfer of royal government center invites other ethnic groups from outside to come and open their business in this area. Other ethnicities that come next are ethnic Mandailing, Minang, Aceh and also Arabs.

Ethnic diversity in Medan City has been going on since the past and is now has been taken as consideration to find the character of the city of Medan itself. The research undertaken aims to reveal how this ethnic diversity will influence the formation of urban character now. The re-search will be conducted by conducting historical searches on related literature and doing direct observation in the field. This research will later reveal facts in the field related to the role of culture towards the formation of the character of a city.

Methods

This study was conducted using space-based approach based on historical analysis in the research area. Data collection techniques are done by doing some historical record that occurred in the area of research and then seen in sequence phenomena that occur in the region. Combined with direct observation techniques will be

* Corresponding author: hari.hajaruddin.siregar@gmail.com

seen the phenomenon that occurred in the current region refers to the history of the area that has been collected. This approach is considered very suitable because we can trace the phenomenon of culture and history that occurred in the area of research in order to obtain the character of the region or sense of place sought.

Each place or region has a local spirit or intelligence that makes the place come alive, this spirit is often referred to as a unique has an-other name "guardian spirit" (Genius Loci) that distinguishes one place from another (Norberg-Schulz, 1980). This research is explorative, where the data to be collected is obtained through direct observation techniques in the study area and collect information through interviews to the competent parties in the field such as cultural and history experts who under-stand the history of the region and cultural his-tory in the city of Medan.

This approach is chosen because it can explain the phenomenon that occurs in the research area so that it can find a sense of place or character of the region. In revealing the character or spirit of place of the design area, the same stages are done by Norberg-Schultz in identifying the locus genius in several case studies in his book. Diagrammatically, the concept used is in Figure 1.

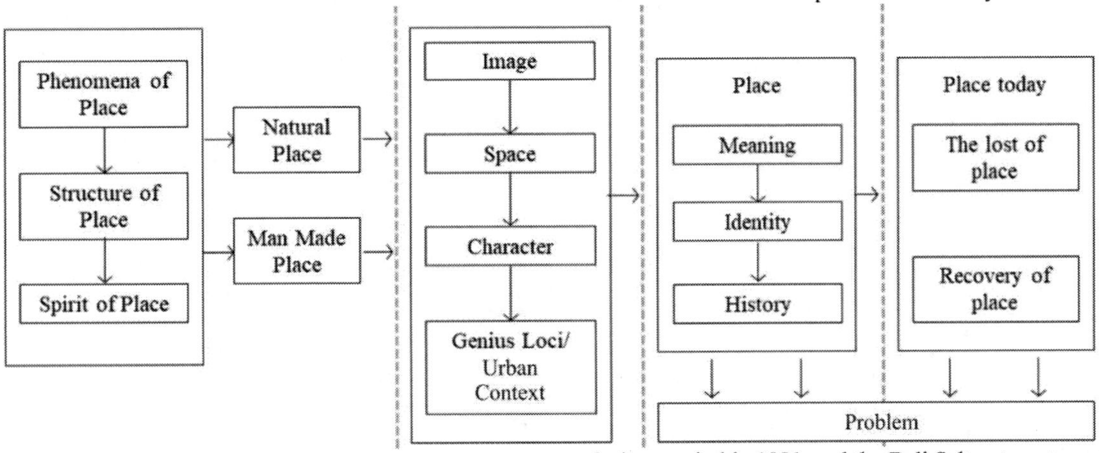

Fig. 1. The search process of finding Genius Loci.
Source: Norberg-Schultz, 1991.

The plot in figure 1 is a technique or the way Norberg-Schultz invented Genius Loci in his book Genius Loci: Towards a Phenomenology of Architecture. In this study, genius loci sought only limited to the understanding of the context of the region approaching the character of the study area. This is done to shorten the study period but still get the character of the area you want to find.

Results and Discussion

City Phenomenon

Traced from its history, the city of Medan is formed from a small village called Medan Putri. Based on a report written by Jhon Anderson, a British delegate, Medan Puteri was still a population of about 200 people only in 1923 (Sinar, 2006). Medan Putri Village is located close to the Deli River, one of the main rivers at that time which can still be used as a means of transportation (based on informant interview 2). In addition to the area of Medan Putri, other activity centers are in Labuhan in the north and Deli Tua in the south (Sinar, 2006).

In 1869 a Dutch investor came, Nienhuys, and started a plantation industry in Medan. Nienhuys set up his office and residence, Deli Mij, at the Deli River intersection because it was a strategic position for hauling the estate. At that time the road infrastructure is not good and still use the river as the most effective means of transportation. For transportation needs at that time then made several docks or harbor barge at Deli River intersection (Sinar, 2006). The first development ofplantation industry made by Nienhuys at that time, he brought some workers from outside Medan which is ethnic Javanese, Chinese and Tamil. Some of the existing buildings at that time were Hotel De Boer (Hotel Dharma Deli) as well as some Dutch resident buildings near the current House of Representatives (Sinar, 2006).

Since the tobacco plantation industry entered a glorious period in 1881, and the Deli Sultanate got great benefit from it, a plan to relocate the central government of Deli Sultanate from Labuhan to Medan had been started. Under the leadership of Sultan Deli IX, Maimoon Palace was built in 1888 and completed 3 years later. Since the establishment of Maimoon Palace followed by the construction of the Grand Mosque of Al Ma-Shun, Taman Sri Deli and also the Castle of Puri, the area around it be-came an exclusive and respected area at that time because it is the central area of Deli Malay Sultanate in Medan (Sinar, 2006). Regions, such as courtyards and Sri Deli Pool is a property belonging to the sultanate that not just anyone can enter it. Even the people traveling across the palace would descend from his bike and lead him to a certain distance due to his respect for the deli sultanate at the time. At this time more and more migrants who open businesses and live in the city of Medan among others Mandailing people, Minang, Aceh and Arab.

https://doi.org/10.1051/shsconf/20184104011

Since the social revolution that took place in East Sumatra in 1946, driven by the communists, there have been attempts to abolish the Malay kingdom system that exist in East Sumatra. During the year, many of the Malay sultanate assets were burned and destroyed (Kahin, 1952). After this revolution, the area began to open to the public and began crowded visited by the public because of the influence of the Sultanate of Deli at that time has begun to decrease.

leadership of the presidency. While the local population such as ethnic Malay, mandailing, Java, and Minang obedient and subject to the rules and orders of the sultanate.

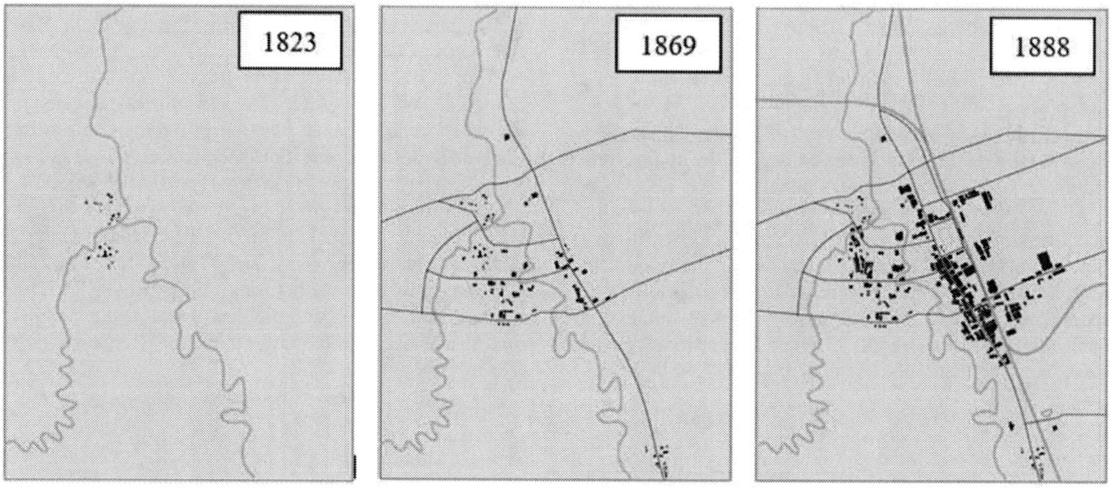

Fig. 2. Morphology of Medan City Center.

City Structure

Based on the morphology of the area in Figures 2 it can be seen that the area has evolved since the removal of the plantation industry office which was originally in Labuhan to Medan as well as the transfer of the Residency Center of East Sumatra from Bengkalis to Medan. Medan Putri initially only small village on the banks of the Deli River developed into a plantation area and commencement of construction of infrastructure and buildings sup-porting the plantation industry near the area. In the period 1879 - 1888 infrastructure develop-ment took place from the construction of roads, railways, sanitation networks, to tele-communications networks (Sinar, 2006).

Through the morphological picture, we can also see that there are two things that led to the development of the structure of Medan City area. The north of the developing region was driven by the need for the tobacco plantation industry and in the south of the developing region as it was driven by the displacement of the Deli Sul-tan Palace from Labuhan to Medan in 1891 after the completion of the construction of the palace which began in 1888 (Sinar, 2006).

The appointment of Medan as Gemeente in 1909 made the city divide in two areas of power, each of which was administered under the Sultanate of Deli and also the Dutch Residency. It aims to facilitate the regulation of the region as well as the laws prevailing in the region at that time. The people of Europe, ethnic Chi-nese, and tamils are obedient and governed under the

Image of The City

The image of the region is shaped by cultural and historical background that has existed since ancient times. The entry of Dutch investors resulted in the development of plantation industry so that emerging areas that are man-aged and inhabited by people who manage the plantation. In the center of Medan (which be-came the current Merdeka Square) the image of the area was formed by colonial architectural buildings as it was the center of the gathering of offices of the plantation industry as well as the seat of the Dutch residency government (figure 4). Moving to the south (the corridors of Kesawan and surrounding areas) the image of the area is shaped by the typical two-storey shophouse buildings that are managed and inhabited by ethnic Chinese who are granted business and residence permits by residency at the time (figure 4). The farther south (the palace and the surrounding area) the image of the region is influenced by Malay culture and seen in its influence on the important buildings sur-rounding it (figure 5).

The diversity of existing cultures strongly affects the image of the region formed. The existence of this diversity led to the image of different areas in every corner of Medan. Some of the cultural spreading that occurred in Medan City can be seen in figure 3.

The image of the city was shaped by the historical and cultural background that existed from the past to the present. Where this image gets influence from the culture of Europe, Asia and Archipelago. Medan as one of the largest cities in Indonesia has cultural diversity. Uniquely from the existing cultures there is no one culture that is dominant in terms of social life or culture. Medan became a melting point, a blend of cultures that until now exist and transformed into a new culture that is different from the culture of origin.

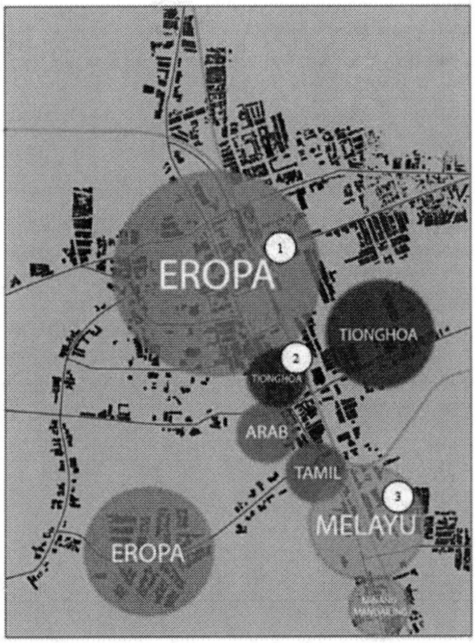

Fig. 3. Distribution of locations and settlements of various ethnic groups in Medan.

Fig. 4. Colonial-style buildings around Merdeka Square. Source: https://pussisunimed.files.wordpress.com accessed 23 Mei 2017.

Fig. 5. Old Kesawan area.

Source: https://szlovely.files.wordpress.com accessed 23 Mei 2017.

Fig. 6. Maimoon Palace and Puri Castle.
Source: https://www.wego.co.id accessed 23 Mei 2017.

Space

Space here is interpreted as an architectural container arising from the social interaction that occurred in it. This space can be both physical and non-physical in architectural terms (Ekomadyo, Zhara, & Najmi, 2012). The spaces created in the city of Medan are formed and caused by the activity of the resonator in the area. these spaces are unique and interesting to discuss, because the historical and cultural backgrounds that contribute to them are very diverse. In the city of Medan there is some space formed on the need for a place of inter-action between European societies such as Medan Club (figure 7). This place was founded in 1880 by the Societteite de Witte, a gathering of Dutch people. For more complex interactions, occurs in the spaces formed by the market. There are several markets that have different historical backgrounds in accordance with the cultural mix that plays a role in its formation. Medan Central Market (figure 8) is the biggest market and main market in Medan City. Founded in 1933 during the reign of the Dutch East Indies. This market is shaped by the space requirements of the growing economic demands of the plantation. Here there is an interaction between various ethnic cultures in the city of Medan at that time.

Next is Kesawan area, space formed due to the interaction between ethnic Chinese and also the people of europe. Aside from being a residential area of ethnic Chinese, in this region there are also some important buildings that became the headquarters of government and also the plantation industry in the past. There is also a Hindu market and the area around the palace which has spaces formed due to the interaction between certain ethnic groups.

Fig. 7. Current Condition of Medan Club at Jalan Kartini. Source: http://medan.m-heritage.org ac-cessed 24 Mei 2017.

Fig. 8. Old Medan Central Market.
Source: https://deliheritageclub.files.wordpress.com accsessed 24 Mei 2017.

Character

Medan City has a character that has been formed by a strong history in the field of plantation industry and has the uniqueness of various ethnic cultures that participate in contributing to the face of the city today. Indirectly, the role of this ethnic diversity has an equal position without any dominance of a particular ethnicity. In its history Medan has been known as a metropolis area because it has become a center of trade and plantation that has a population of 14,000 (in 1905) to 76,000 (1930) from different ethnic backgrounds, not just entries from the archipelago but Europe, Asia and Arabic. So in other words the city of Medan is currently known as a city with a cultural form that has undergone mixing so that no one ethnic that has dominance

Cultural differences that contained in the city of Medan indirectly provide a new face and character to the city. The combinations of these different cultures and ethnic groups give a new form of architecture that is different from its original form. These new forms, until now, be-come new characters and faces owned by the city of Medan itself, a city that has a diversity of cultures whose manifestations can be seen from its architectural character.

Genius Loci

From the phenomenon of the area that occurred can be concluded that the city of Medan is formed by the activity of a very advanced plantation in ancient times. This plantation activity raises the interest of the world community and local people to come and open their business in this area. From the explanation it can be said that the genius loci Medan City is the cultural diversity formed by the history of economic activity that occurred in the region.

Characters in the context of Medan can be either implicit or explicit. Both of these are often interpreted as the spirit of place or genius loci (Jiven and Larkam, 2003). Today, the implicit evidence can be social concepts applied by society in different ways and dependent on their respective cultural concepts. While the concept of exploit itself looks firmly from the various forms of architecture owned by Medan City and still exist until today. Both of these also provide an overview that the city of Medan is a city with the concept of a character that is formed from different cultural and social assemblage.

Conclusions

In his book, Norberg-Schultz, uses three key-words to summarize the results of his explora-tion of the "place" he studies. These three key-words are meaning, identity and historicity. The meaning of "place" in the city of Medan is formed due to a mixture of cultures that are based on the need for the same economic ac-tivity. With the combination of culture that oc-curred then the city of Medan has a different cultural background that can revive and move the city's economy until now.

Different cultural backgrounds create spaces for activities that have unique characteristics. This cultural diversity will indirectly bring about social interaction between communities in ac-cordance with culture and essentially require public space to perform such activities (Tan-jung, 2015). Cultural differences that exist one of them can be seen from the shape and mani-festation of architecture that until now still exist in the city of Medan. Where these buildings become a form of strong identity that can be used as a reflection of cultural diversity in this area physically.

The existence of the historic buildings in Medan city is a testament to the history of the strong development city in the past. Starting from the heyday of the plantation industry until the reign of Sultanate of Melayu Deli can be seen from the historical buildings that still exist today.

The city of Medan as one of the big cities in Indonesia, has uniqueness and distinctive char-acteristics of course different from other cities in Indonesia. Characteristics that characterize the city of Medan is the diversity of culture it has. This cultural diversity is formed due to the history of economic development driven by the plantation industry that existed since ancient times. This economic movement also continues to this day.

Acknowledgements

Thank God I pray to Allah SWT and all parties who have helped in the preparation of this re-search paper, from the beginning of data col-lection until the analysis process. To Mr. PetrusNatalivan and Mr. Agus Ekomadyo as thesis supervisors who have a lot contributed to the content of this research. Thanks to all those who have assisted in the data collection process in Medan, for informants who have taken the time to provide information related to history and culture of Medan city, such as Mr. Khairul (BWS), Mr. Edi Sumarno (Lecturer of History of FIB USU), Mrs. Rozanna (Chairman of the Faculty ofMalay Culture), Mr. Anto, Mr. Ridwan as the guards of the Great Mosque and Maimun Palace, Mrs. Isnen (Lecturer of Archi-tecture USU), Mr. Umar, Mr. Muzahar and Mr. Usman as experts in the field of antopology and history. As well as some local people who helped provide information related to this re-search area. Thanks to relatives and friends who have given their moral support and en-couragement until the research is done on time.

Ligula dolor vel ipsum posuere consequat grav-ida, mauris at, in suscipit magna libero enim mauris a. Sed ut imperdiet ridiculus.

References

Ekomadyo, A. S., Zhara, A., & Najmi, I. (2012). Public Market as Urban Social Nodes: An Architectural Phenomenology Approach. Arte-Polis 4 International Conference - Creative Connectivity and the Making of Place.

Jiven,G & Larkam, P.J. (2003). Sense of Place, Authenticity and Character: A Com-mentary. Journal of Urban Design, Vol. 8, No. 1, 67–81

Kahin, G. M. (1952). Nationalism and Revolution in Indonesia. Cornell University Press.

Leushuis, E. (2014). Panduan Jelajah Kota-Kota Pusaka di Indonesia. Yogyakarta: Penerbit Ombak.

Norberg-Schulz, C. (1980). Genius Loci: Towards A Phenomenology of Architecture. London: Academy Edition.

Sinar, T. L. (2006). Bangun dan Runtuhnya Kerajaan Melayu di Sumatera Timur. Medan: Yayasan Kesultanan Serdang.

Tanjung, F. (2015, September 30). Kompasiana. Retrieved from Kompasiana Rubrik: http://www.kompasiana.com/ftanjung/harapan-dan-tantangan-ruang-publik-kota-medan_560bdcbb0e97734008 1c61d7

Recharting The Philosophy of Technology in Contemporary Architecture

Revianto Budi Santosa[1,*]

[1] Department of Architecture, Universitas Islam Indonesia, Indonesia

Abstract. Technology is an indispensable aspect of architecture. In fact, it is being an essential part of the human effort in making architecture. Since the early modern era, technology that rapidly change has been seen as the sign of progress, not only pertaining to the technology itself, but also architecture and even civilization. Modern architectural theoreticians, from Sant'Elia to Le Corbusier, enthusiastically embraced the progressive side of technology and engineering. Philosophically, however, modern technology is regarded pessimistically. Heidegger and Jaspers considered technology as the source of alienation to the human being themselves and to the reality they face. To overcome this gap, Alan Drengson, proposed the four philosophy of technology to rechart the variety of tendency towards technology in Western society, consisting of (1) technological anarchy, (2) technophilia, (3) technophobia, and (4) technological appropriateness. In this explanation, he coined the terms "creative philosophy" to include many aspects and ways of thinking which might be incorporated in the creative activities like architectural design. This paper attempts to evaluate the appropriatenes of Drengson's philosophical scheme as a platform for architectural education in Indonesia in general, by relating his framework with the architectural theories and practices in Indonesia. The result of this effort is while the formulation of his scheme is the very inclusive and closely related with creative activity like architectural design, it contains bias of industrial technology appearing in the Asian scene brought by Western European colonials. Discussing philosophy underlying Gandhi's movement in India to reject oppressive technology, we may arrive at the conclusion that the philosopy of non-violence, truth and justice based on the principle of self restrained are relevant to figure out the ideal of appropriate technology in Asia.

Keywords: Philosophy of technology, appropriate technology, architectural design

Introduction

The advancement of technology in the last 100 years was fascinating for everyone. Flying high to reach the moon and far beyond, moving fast to compete with sound, to communicate face-to-face across the globe like in previously seen in science fictions, as well as to build taller and taller reaching the cloudy sky in the atmosphere. If Max Weber said that modern time was the era of disenchantment, that was true due to the progress of technology shifting the realm of ghosts, fieries and gods to the periphery since miracles could be performed by technology.

One of the most salients feature of technology is change, or in more anthusiastic tone is progress. Technology with all of its miracles is regarded as the saviour of human misery. By fully embracing the ever progressing technology, our houses, workplaces, and cities will be glorious. Writing in the first quarter of the twentieth century, Antonio Sant'Elia in his manifesto, proclaimed the arrival of Futurist Architecture and discarded the entire history of mankind in creating their built-environment:

"From an architecture conceived in this way no formal or linear habit can grow, since the fundamental characteristics of Futurist architecture will be its impermanence and transience. THINGS WILL ENDURE LESS THAN US. EVERY GENERATION MUST BUILD ITS OWN CITY. This constant renewal of the architectonic environment will contribute to the victory of Futurism which has already been affirmed by WORDS-IN-FREEDOM, PLASTIC DYNAMISM, MUSIC WITHOUT QUADRATURE, AND THE ART OF NOISES, and for which we fight without respite against traditionalist cowardice (Sant'Elia, 2007: p. 17, originally published in 1914)".

The speed of change was celebrated, written in all capital letters, "THINGS WILL ENDURE LESS THAN US. EVERY GENERATION MUST BUILD ITS OWN CITY". Cities which were previously seen as the most palpable signs of permanence, then were deemed disposable for the sake of embracing the ever changing technology.

In more a restrained manner but not less fundamental in promoting progress was Le Corbusier's pronounciation of "Architecture au Revolution". Social

* Corresponding author: revianto@uii.ac.id

unrest, for him, rooted in the unsatisfaction of life, which was mainly as the result of improper built-environment that people have created. With the power of technology to provide "space—daylight—greenery", the "essential joys of urban life" for all, then, revolution was unnecessary. Social justice, morality and progress could be achieved once we had a proper home and city to suit with modern life. Technology and rationality were the saviors of contemporary miseries.

Many architects, especially those from Western Europe during the interwar years, wholeheartedly embraced modern technology. For them the fate of humanity in the future depended on the way human able to foster and apply technology in their life. Engineers in other fields believe in such notion even more. The application of technology in the progressive spirit has made architecture into "Olympic situations" in which citius, altius and fortius can be measured.

Heidegger's Dilemma Of Modern Technology

Philosophy questions everything in life, including technology which greatly shapes the life of most people around the globe. While technology gives obvious axiological appearance on how it might be, it lacks ontological ground on what technology essentially is.

In his seminal lecture, "The Question Concerning Technology," initially published in 1954, the philosopers Martin Heidegger addressed the inquiry on the ontological foundation of technology. He went beyond the commonly accepted notions of technology being a machine or a set of technical procedures.

The main concern of Heidegger's philosophy is the question of being. Thus, when Heidegger claims that technology is a way of revealing, he means that it involves a specific kind of being. It rests upon a relation between human beings and reality, one involving a way of revealing in which human beings "set upon" and "challenge" what they bring forth as real. As a result, reality is understood in terms of what is available to and can be controlled by human beings.The way of revealing of the ancient Greeks provides a ready contrast. The word "technology," Heidegger points out, comes from the ancient Greek word "technikon", which refers to everything pertaining to technè. Technè comprised both craft and art alike. It was a form of "poièsis", which is generally translated loosely as "making." Human existence is emphasized in as creator in the application of technè.

In contrast from the technè, modern technology, in Heidegger's view develop a much different way of disclosure. Modern technology is created as processing device. Nature, therefore, presents as raw material or resources rather than reality to discover and disclose. Humans no longer encounter nature as entities that "emerge from conealment into unconealment". The surroundings simply serve as a storehouse of available raw materials.

In discussing the presence of tecnology in the form of construction, Heidegger (2001), in his essay, "Building Dwelling Thinking". In this writing, he questions the "thingness" of a building by being a "dwelling". Dwelling is the end, while building is its means. Not every building is a dwelling. Only buildings which provide meaningful habitation are qualified as dwellings.

Drengson's Inclusive Notion On The Philosophy Of Technology

Trying to redefine the word "philosophy", Drengson (1982, p. 27) proposes the term as "a way of life formed by attitudes and assumptions which, taken together, constitute a systematic way of conceptualizing actions and experiences by means of an implicit process of unquestioned judgments and conditioned emotional responses". This notion encompases three stages, from a certain patterns of thinking and practice, to the explicit formulation of assumption, axioms and other forms of verbal elaboration, and at the end to be a creative activity of conceptual inquiry in order to develop more appropriate cultural practices.

Observing the dynamics, variety and historical progress of Western industrial society, Drengson develop a framework to understand variety of fundamental attitudes toward technology.

1. technological anarchy,
2. technophilia,
3. technophobia, and
4. technological appropriateness.

Technological anarchy is the initial state of technological exploration and innovation. This situation is based on "a philosophy of exuberant, youthful curiosity and self-centeredness" being "an expression of optimistic self-assertion and individual opportunism" (Drengson, 1982, p. 29). It is a highly divergent period with few government regulation, direction or limitation. Market will determine which technology to support and to promote.

In the state of technophilia, certain technology is already in the advanced stage. Supported by the forces of the market, technology now becomes industry. It is the primary way to supply society, to gain financial benefit even later directed forces governing many aspects of life. It becomes the savior for the future of humankind and universe. Experiencing the state of technophilia or the love of technology, we are enamored with, and entrapped by technology. "Technology, which was originally pursued as an instrument to satisfy desires and needs, tends in such a context to become an end in itself" (Drengson 1982, p. 29). As technology permeates into, and direct our activities and existence, life becomes merely mechanism. The prominence of technology often causes the lost of control over technology, as human becomes mere operator or even appendage to the mighty industrial machine.

It is quite natural to realize in the technophilia state that technology threaten human existence and the natural world. Sensibility, sensivity and spontaneity of life are replaced by the ever growing hegemony of industrial machine manifested in myriad facets of life. The choice to develop and then to embrace technology is conceived as a big mistake. To redeem, some people attempt to make distance with the machine and even challenge themselves to do "de-technologization" as seeing technology becomes a dehumanizing force.

Drengson (1982, p. 30) calls this state as technophobia with underlying principles of rejecting technological autonomy and asserting human autonomy over it. Technophobia leads to two achievements by renewing commitment to human values and revitalizing arts, crafts, technics and skills. Personal development is more important than technological supremacy over human and nature. Maturity and wisdom are more meaningful than progress and advancement.

Technophilia and technophobia lead human engagement with nature in the opposing directions. One sees technology as the most certain promise of the glorious future; the other one considers technology is a monster threatening the very existence of life.

In this conflicting situation, one politically correct answer is notion of "appropriate technology" seen by Drengson (1982, 31) as the most mature state of attitude and philosophy toward technology. Diverse choice of thechnology and the ways to develop, appropriate, and exploit are availabe. He emphasizes on the role of appropriate technology in incorporating three other philosophies by considering the role of the mature selves in selecting technology and directing its progress without losing control over the devices.

With gigantic capitals supporting certain choice of technology with its massive greed, the idea of appropriate technology being benign technology able to revitalize the spirit of community sounds utopian. Drengson (1982, p. 33), however, attempts to propose some way to "tame" technology in order to suit with the spirit of appropriate technology in four ways, namely: 1) technological modification, 2) technological hybridization, 3) technological mutation, and 4) technological mastery and creation.

Asian Resistence

With its long involvement in the process of Industrial Revolution, Western European society see technology in its liberating forces freeing mankind from the burden of menial labor, albeit many discontents along its course. They are then engaged in the mission of civilizing the world with rationalism and technology.

Asia have ver much different history in their involvement with modern technology. This type of human advancement for them arrived in one pack with Colonialism. Entering the Asian land with the spirit of exploitation, technology and industrialization often conceived in its negative sides of dehumanization.

Mahatma Gandhi (1869-1948) is one figure prominently standing for Asian voice against Colonialism by liberating themselves from the bondage of techonology. He leads the movement to reject British industrial products as they are associated with the oppressive forces of Colonialism. He establishes nonviolent non-co-operation platform epitomized in the swadeshi policy—to boycott foreign-made goods, especially British industrial goods exported to India. He advocate to wear khadi (homespun cloth) instead of British-made textiles. Indian men and women, rich or poor, should spend their time spinning khadi to support independence movement.

Callicott and McRae (2014) investigate the philosophy underlying Gandhi movement. The centerpiece of Gandhi philosophy is "ahimsa" or non-injury. Based on personal non-injury commitment, this philosophy is extended to social, economic and political forces. The second prominent notion is "satyagraha" or or persevering in the truth without recourse to violence of any kind. The third underlying spirit of Gandhi is "asteya", means abstention from stealing or the misappropriation of the possessions of others.

Relying on these philosophical themes, Gandhi sees massive industrialization as practiced by the British is a betrayal of truth, realized with violent means with the spirit of greed stealing resources from the colonized land ad people.

Quoting Gandhi's words, Callicot and Mc Rae (2014, p. 20) considers that people fall from the pursuit of the ideal of plain living and high thinking at the moment they want to multiply his daily want. People's unability to distinguish between wants and needs multiply the unreasonable demands on the environment. Concerning with the exhaustion of natural resources, he says that "The earth provides enough to satisfy every man's need, but not every man's greed."

Appropriateness Of Technology And Architecture

Gandhi's philosophy of the self to restrain from the over exploitation of the people and environment to discover the inner truth and higher spirit inspires the world in many respects. In the field of architecture, a number of figures can best understood in Gandhi's spirit even though none of them is an avowed follower of Gandhi, especially those receiving the Aga Khan Award of Architecture such as Hassan Fathy in Egypt, Barefoot Architects Group in India and Mangunwijaya of Indonesia. Fathy's insistence on the application of adobe to use abundant resources with people's own hand, Barefoot Architects with their social organization, and Manguwijaya with his effort to ennoble marginalized people, might be understood in revealing the truth of humanity in just ways.

The aforementioned philosophy of self restrained manifested in the notions of non-violence, truth and justice may enhance and direct the Drengson's philosophy of appropriate technology to better serve humanity and human's role as cutodians—rather than possessors—of the earth; including in the field of architecture.

References

Callicott, J.B. and J. McRae (2014) *Environmental Philosophy in Asian Traditions of Thought*. Albany: State University of New York Press.

Drengson, A. R. (1982) "Four Philosophies of Technology". *Philosophy Today,* pp 103-17.

Heidegger, M. (1977) "The Question Concerning Technology" in *The Question Concerning Technology and other Essays*. London: Garland Publishing.

Heidegger, M. (2001) "Building Dwelling Thinking" in *Poetry, Language and Thought*, New York, NY: Harper Collins, pp. 141-160.

Verbeek, P. A. (2005) *What Things Do: Philosophical Reflections on Technology, Agency and Design*. Pennsylvania State University Press.

Learning to Know and Think: Computing for Architecture Course

Aswin Indraprastha[1,*]

[1]Architecture Program, Institute Technology Bandung, Indonesia

Abstract. Computational technologies for solving design problems become increasingly important in architecturalpractice. In responding, architectural education has encouraged the use of this tool and method in the curricula. As the technology and environment are ever changing, the curricula should be evaluated and updated to adapt and to teach method and skills necessary to the students. Over the last four years, Architecture Program of ITB has inserted new computational courses into undergraduate levels. These courses are the mix of a skill focused computational workshop that is compulsory for the second-year students and introduction of computational design as an elective course that can be enrolled both by third and fourth year students. This paper delivers a report of our methods and findings from our continuingstudy of the courses including analysis of student outcomes, student evaluation of the course structures, assignments, and feedback as well as computational abilities after completion of the courses. The aim of the study is to have a grounded validation of computational courses in architecture curricula and to improve courses goal based on the evaluation. The result of our study reveals challenging issue to teach computational thinking undergraduate level rather than only providing them a set of computer skills for production and presentation techniques of the design.

Keywords:computational thinking, computational course, architectural design

Introduction

Computational technologies for solving design problems become increasingly important in architectural practice. In responds, architectural education has encouraged the use of this tool and method in the curricula. The needs and expectancy for innovative architectural design that is aesthetically pleasing and responsive to the environmental loads and energy (i.e. green design, energy-efficient design) meets with new methods and tools to front-loading[1]design processes in order to achieve higher performance design through means of a digitally-assisted system (Kalay, 2004).

Computing as the act of calculating which has root in mathematics plays significant roles in shaping
today's design methodology to make sense the design output that potentially bring back architect to be master builder as once was.

Among the first notable critical works of mathematics in architecture occurs in the Notes of Synthesis of Form(Alexander, 1964) and The Geometry of Environment (March & Steadman, 1971). Those works made important contributions to the design method, particularly on what we know as the relational theory, one of the basic principles of architectural computing. In 1996, William Mitchell wrote ideas to develop architectural formal language by dissecting and developing inter-relation among building functions and elements in The Logic of Architecture (Mitchell, 1996).

Since then, the discourse over computation involvement with the architectural design process has been an interplay among scholars, designers,architects, engineers, even programmers, biologist, among others (Burry, 2010), (Terzidis, 2006), (Kolarevic, 2005), (Aish, 2013), (Menges, 2017),(Meredith 2007). By some, architectural design became a testbed for playing with formal yet liberating methodology called programming. John Maeda pioneering this approach to explore flexibility and platform-free of programming in design. Later, he involved in the development of *Processing*, one of the most popular programming platform today for non-programmers (Maeda, 2001).

This issue and trend left a challenge for architectural education both in pedagogy, or method of teaching and subject that involve technology that is always growing and changes constantly. There are some findings pointed that computation and architecture education (and architecture students as well) are hard to blend (Varinlioglu, 2004), (Colakoglu, 2007), (Abell, 2013). Abstracting and thinking in terms of formal, explicit and procedural way is hard to learn, especially when by nature, in architectural education creativity as an act to discover, is learned through many methods and less in a formal way. This opposition also contributes to the fact that many architectural students are hesitant to learn computation

[1]A method in the workflow of design and construction phase where some processes which previously carried out in linear and sequential fashion now can be completed in parallel and move forward to schematic design phase (e.g, design analysis, schematic, quantification and scheduling).

* Corresponding author: aswin@itb.ac.id

because it is hard and they cannot envision how it will help to become a better designer (McCollough, 2006 in Senske, 2013).

On the other hand, computation as a *lingua franca* for all digital apparatus that assist design-construction process nowadays transforms itself into software and hardware that constantly developed and changes. Programming then, become an essential skill to master these new systems. McCollough in 2006 suggested that programming has been rediscovered in architectural design mainly by two reasons: first, there are more incentive to express design based on machining process. Second, this form expression is increasingly informed by other domain of knowledge that already familiar with the notion of the beauty of generative algoritmic: namely Biology, where harmonies, proportion, recursion can be codified into algorithmic and parametric formula (McCollough, 2006).

By this fact, we believe there is a potential to include the new way of thinking and working into architecture curricula. This paper delivers report of our development incorporating computation-based courses and the result in form of feedbacks to assess the course outcomes and the effectiveness of its execution.

Background

As part of First Common Year program, in 2008 ITB has introduced an introductory course on programming that is mandatory for all students. This course presumably planned to introduce basic computational logic through series of lectures and workshops emphasizing on basic programming skills using C++ programming language.

Although the course is based on university evaluation to strengthen basic STEM (Science, Technology, Engineering, Mathematics), it is found that the majority students from SAPPD (School of Architecture, Planning and Policy Development) had difficulties to grasp the most important goals intended for this course namely, to understand how to think computationally through procedures (algorithm) and to understand how to execute program through systematically written syntaxes in form of codes. This condition has apparent significant reason and confusion among students as some students asked why they learn to code and how it will be relevant in their future career.

Nonetheless, at the second year in architecture program, the freshmen will be taught computational design studio (3 credits), which is designed to provide basic skills and knowledge of CAD, 3D modeling and BIM. Production of drawing, transformation and creative elaboration of 3D objects, and component-based modeling are the goals in which spread into three modules that conducted in sequence. This course in principle was designed to be a supporting course for their architecture design studio project in subsequent years.

The computational design studio however, did not adequately equipped students with computational thinking or computing abilities. They know how computational processes work through series of exercises using various sofware and could produce solution over certain geometrical problems, but they cannot explore beyond was taught in the lab session. This formal method usually limits student's creative thinking down to only the function and features that already been taught.

The computational design studio has another issue that fails to recall knowledge and understanding of computational logic they learned in the first year. From the lecturer's perspective, this condition is undertandable given the fact that learning to use multiple software requires time and in this case, there was no room to teach computational logic.

Furthermore, in third and fourth year, we offer two elective courses with topic of algorithmic design and BIM respectively. These elective courses named Algorithmic Approach to Architectural Design (2 credits) and BIM Approach to Architectural Design (2 credits) are offered for the third and fourth-year students. Both courses designed to be basic introduction courses to architectural computing using two distinct approaches.

Workshop I: Computational Design Studio

Computational Design Studio (AR2250) is a mandatory course for the fourth semester of second- year architecture students that was conducted through three subsequent modules:

1. Production of technical drawing using 2D CAD
2. 3D geometry exploration and modeling
3. Component-based modeling of virtual building using BIM

The challenge was to provide the same level of abilities to all students of CAD, 3D modeling and BIM, having no prior experience in CAD, nor 3D modeling and BIM. We advised that at the second year, the appropriate learning method is through two stages:

1. Learning to draft: drawing and modeling bu imitating a reference. Limited flexibility to expand skills through design. The computer is being used as the tool for drafting.
2. Learning to expand skills and abilities: further step after imitating is expanding abilities by modeling details, adding meta-information of elements, or producing documents.The computer is being used as a tool for modeling.

This workshop meets once a week, first with a lecture session and then followed by two hours lab session. We still conducted the scheme of the lecture-lab sequence as oppose to lab-lecture i.e. to work with concrete form first then learn the abstract later (Bransford et al., 2000).

 https://doi.org/10.1051/shsconf/20184105001

The partial result showed in this Figure 1.

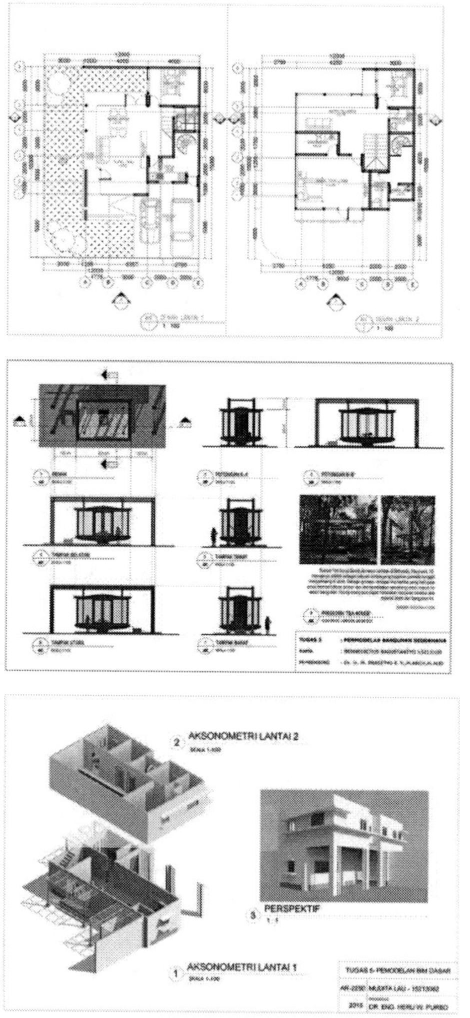

Fig. 1. (Top) CAD Drawing; (Middle) 3D modeling; (Down) BIM Modeling.

In the lab session, our method of teaching is following our hands-on training material and facilitate an active process where each student works at their own pace, the tutors are available to answer any questions or any technical difficulties.

The assessments for this workshop were designated on par with the Architectural Design Studio II where modeled objects are a single and small house. In 3D modules particularly, we advised students to expand their ability to model object as detailed as possible.

Serving as the basic introductory studio to learn to use software, most students felt that the course was important to support their study and their future career. Among other feedbacks gathered through a post-class survey, we discussed our course evaluation as it is presented in the later section of this paper.

Workshop II: Algorithmic Approach to Architectural Design

The Algorithmic Approach in Design (AR4121) is an elective course (2 credits) designated for the third and fourth year students and conducted in the fifth semester. This course was planned and designed to be an introductory course for architectural computation that includes computational design thinking and the methods of formal languages. The goal is set specifically towards computational methods of formal design exploration and analysis through algorithms.

Since an elective course, the syllabus, assessment and course evaluation was tailored and evaluated each year based on student feedbacks and their performance.

The course was conducted using learning by doing approach, starting with lecture and followed by programming session, in two hours in a week.

On early implemention of this course (2013-2014), overall topics of the syllabus for 15 sessions is as follows:

A. Fundamental
- Introduction to computational design, parametric design, computational thinking.
- Parametric functions, mathematics, and logic, set, operators, variables using Rhinoceros & Grasshopper as visual algorithm editor.
- Data structure, tree and manipulation
- Vector, Curve and Surface parameterization and manipulation

B. Application
- Parametric patterns
- Parametric analysis: structure, tensile, optimization and environmental analysis.

Assignments were given in two stages: the first project is to be carried out individually. Student required to design, develop or explore scripts within GH VAE (Grasshopper Visual Algorithm Editor) with an architectural application such as a pattern of the wall, roof shelter, or industrial products such as lampstand. The output is a poster depicting algorithmic process and its output. The second project is a group project to design and construct a scale model of a shelter, bench, or art installation, based on parametric and algorithmic process.

The teaching pedagogy we experimented during the past four years is combining a theoretical and a top-down style lectures on the logic and principles of computation, followed by hands-on exercise on the logic of the workflow of the GH VAE, its structures and syntaxes, data management and manipulation, and geometric parameterization and manipulation.

After self-evaluation, the topics of this course during 2015-2017 are arranged as follows:

A. Basic Understanding
1. Introduction to Rhinoceros and Grashopper: workflows, principles, navigation, commands.
2. Program logic and workflow, syntaxes/components, network.
3. Data structure and manipulation.
4. Point, vector and curve.
5. Surface and polysurfaces.

B. Control and Manipulation
1. Folded, grid and porosity logic
2. Force field and image mapping

3. Dynamic relaxation
4. Optimization methods

During last three years of the course, the weekly course-lab subjects and assignments have been continuously re-evaluated and refined. Part of the result of assignments is depicted in Figure 2.

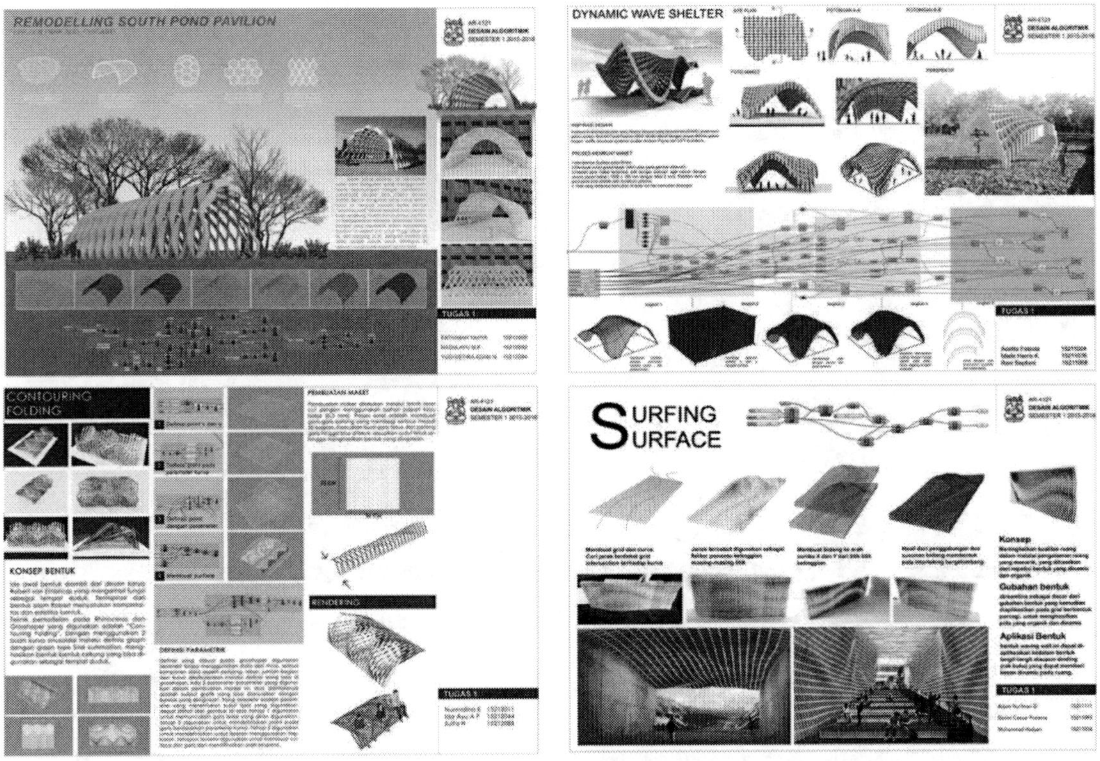

Fig. 2. AR4121 Student Works on AlgorithmicApproach.

We faced multiple issues mostly from pedagogical aspects and learning methodologies. These issues can be summarized as follows:

1. It is a mandatory to provide a ground and fundamental lecture combined with hand-on exercises on computer programming and logic of computing during short period of the course. It is found that majority of Architecture Students had high expectation on the parametric design side without knowing that they had to learn how to program.
2. In consequence, students were trying hard to grasp basic prinsiples and understanding some fundamental concepts in computing: functions, operator, set, etc. it is found that the programing principles they had in their first year, has a little trace in their third and fourth year.
3. Among other aspects, it is found that some students failed to expand or to explore, computing principles beyond precooked program or definition that has taught. They know the workflow, they un derstand the principles and methods but they feel the hardest part is to transform real life problem i.e. concrete problem into computational problem.

Workshop III: BIM Approach to Architectural Design

The BIM Approach in Design (AR4222) is another elective course (2 credits) intended to equip students the advanced knowledge, skills related with Building Information Modeling. This course is divided into two modules:

1. Architectural BIM: advanced parametric modeling, component-based prototyping: hosted and hosted parametric component, schedule and quantification.
2. Collaborative BIM: introducing coordination methods, clash detection, introduction to scheduling and technical documentation.

As with Algorithmic Course, during the past four years of its implementation, we evaluated and refined the course subjects, assignments based on student performance and feedback.

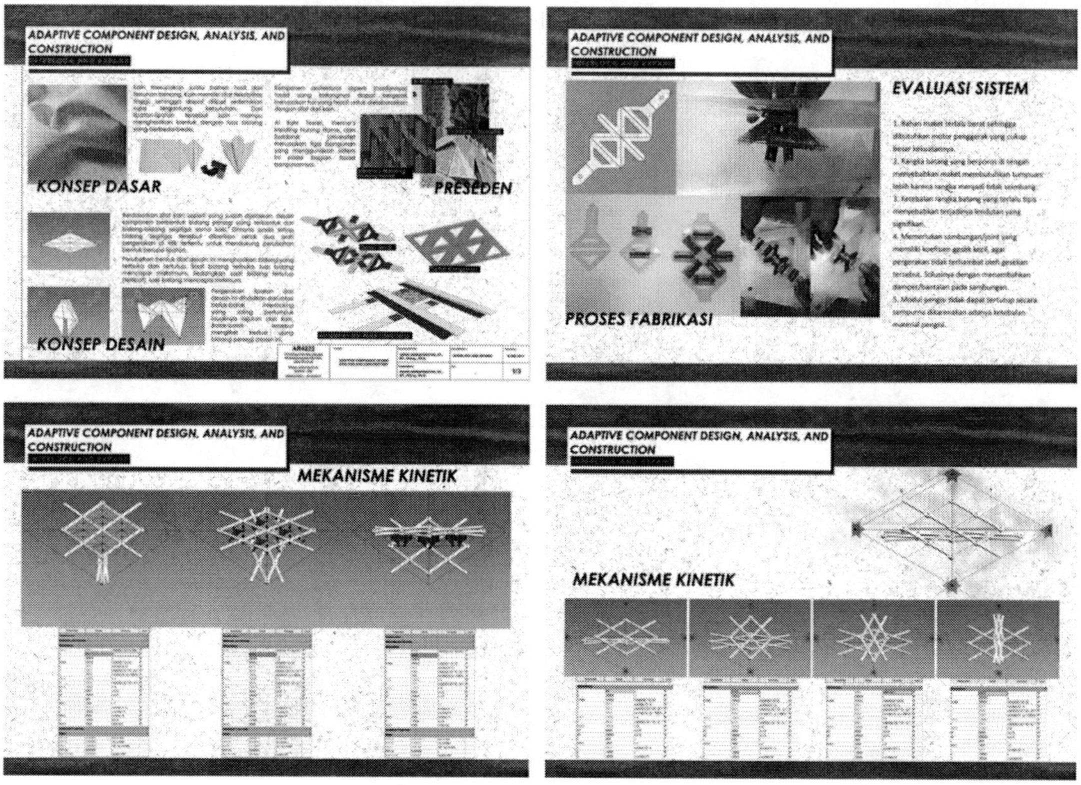

Fig. 3. Student Works on Adaptive Component using BIM Application.

The BIM course is supposedly providing student method and focused on developing fluency BIM software. However, we believe that in addition to skills and rather using software as tool, computational thinking is the most significant aspect that should change their mindset to use computation well, regardless its role in design process.

During the first module, architectural BIM for instance, there is a potential to develop computational thinking using parametric family (component) modeling in family. By BIM software, in this case we use Autodesk Revit, the platform and pipeline are already in place so that any parametric object can be used to generate drawings, quantified, and analyzed.

However, our experience suggested that was not the case. Two hours per week sessions in the total 15 weeks is inadequate to cover learning curve of 3D BIM. Although students starting to rely on BIM software for their Architectural Design studio projects, the necessity to think that BIM is a platform for architectural computation and a tool for form and design exploration is a far fetch. They still used BIM modeling for pragmatic purpose.

This condition may be forgiving if we consider for undergraduate students the knowledge and skills of architectural computation are for practical purposes. The goal of the undergraduate program of architecture is to prepare knowledge, skills, and values required for professional practice in architecture.

In 2017, we advised another approach to develop computational skill using BIM by conducting a collaborative project with Universitas Pelita Harapan (UPH). In this joint workshop, our students were required to design a dynamic and adaptive component while students of UPH developing and constructing kinetic system using either digital or analog mechanism. Each team composed of ITB and UPH students then colaboratively construct and present their kinetic installation.

This workshop was effective to teach computational processes from design inception, parametrization of a component of a system, fabrication, and prototyping in addition to work as a team in different places. Students on both universities engaged in works process where they can learn from each other (Figure 4).

Fig. 4. Collaborative Workshop on BIM and Kinetic Installation.

Although the vehicle in this context is BIM, the pre-workshop lectures and exercises on parametric modeling gave students fundamental principles of Computation.

Some of the result of student projects are depicted in Figure 3 that reflect their abstraction of real/concrete problem, to be solved using computational method.

Feedbacks

We conducted a post-class survey and collecting feedbacks for all courses. This survey is mandatory by the university for all students and included multiple coice with Likert Scale (1-5; 1=unlikely, 5=likely) and short written answers. The result of this survey s is presented in the table in Figure 4. The statement of questionnaires mainly focused on course's outcome and operation/process as shown below:

1. *Outcomes or Goals are achieveable.*
2. *I obtain all knowledge and information needed to accomplish the outcome*
3. *The lecture was designed and conducted so that students can accomplish goals and assessments.*
4. *I can master determined goals for this course.*
5. *The workshop is well prepared.*
6. *The workshop was well-organized.*
7. *Lecturer and tutor have good communication*
8. *Lecturer and tutor care about students being able to grasp knowledge and skills to accomplish the goals or outcomes.*
9. *Lecturer was being fair to the students.*
10. *The credit of this course is questionable compare with the assessments and lab activities.*
11. *Lab and infrastucture is adequate for conducting workshop.*
12. *There are other additional resources to support this course.*

On AR2250 as we may inspect at the chart in Figure 5, students felt that the credit (3) is too small for such workshops with assessments, lab sessions and lectures while in average, this course was well-received. More than 90% of the students agreed that outcome of this course was achievable (average 3.42 in 2013-2017).

On AR4121, our concern was the course credit (question #10) and goals (question #4). The operational or process of the course has average 3.13/5.00 while outcome has average 3.44/5.00. Judging by this result, we have reason to believe that the course syllabus is approachable but need improvement to engage students so that they are motivated to keep learning and expand their skills after completion of the course. Another finding based on this result is that students agreed that they had sufficient knowledge and skills to achieve course's goal (average 3.49/5.00) but they felt they cannot reach to the level

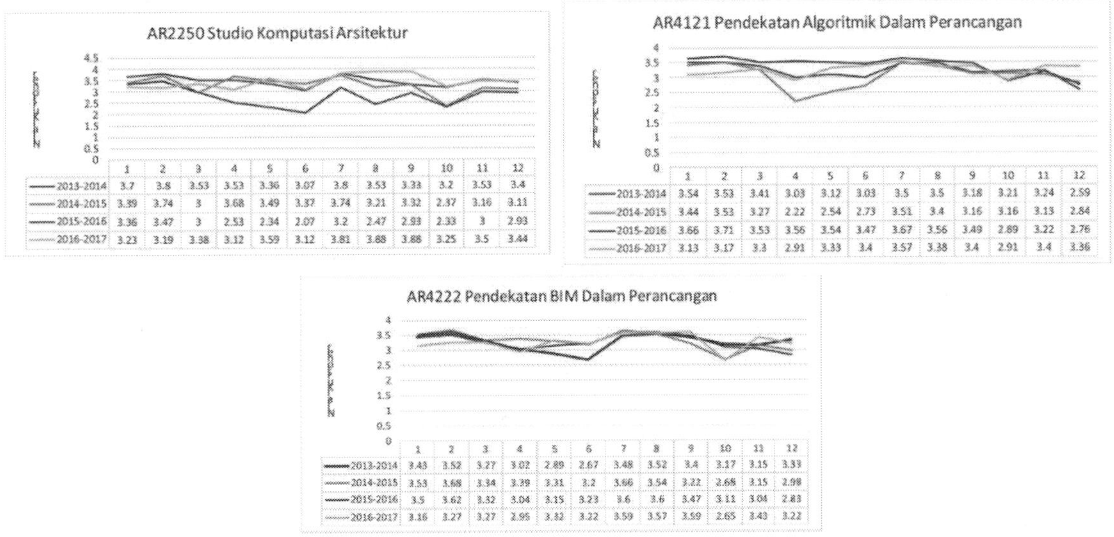

Fig. 5. Student Feedbacks.

of mastering this course. Unfortunately, we cannot determine how they understand computational thinking through their projects and network of components in GH VAE.

On AR4222, our finding suggested that course credit (question #10) and course preparation and organization (question #6) had low average among others (average 3.08/5.00 and 2.90/5.00 respectively). The BIM course over year from 2013 to 2016 always invited practitioner as external tutors for the module of Collaborative BIM. In this module, we simulate workflow of collaboration among users to modify, inspect and create development of a project. This requires better technical preparation of IT infrastructure and additional resources that our university need to plan it out in the future.

Aside from those results, it saves to note that the computation-based courses appear to be successful at teaching and provide knowledge and skills to many students (average participant for AR2250 is 100 students, while AR4121 and AR4222 usually 40-45 students). However, we arestill curious on the course's goals whether we can improve on the outcome to increase student's performance in the computational thinking that it will lead to evaluation of our teaching and exercise methodologies.

Reflection and Discussion

This paper presents our self assessment and ongoing evaluation for the computational-based courses.

Over four years of implementation, there are several lessons that may provide insights improvements. We still believe that in our case, the computational-based course for undergraduate program focuses on the know-how aspect of pedagogy.

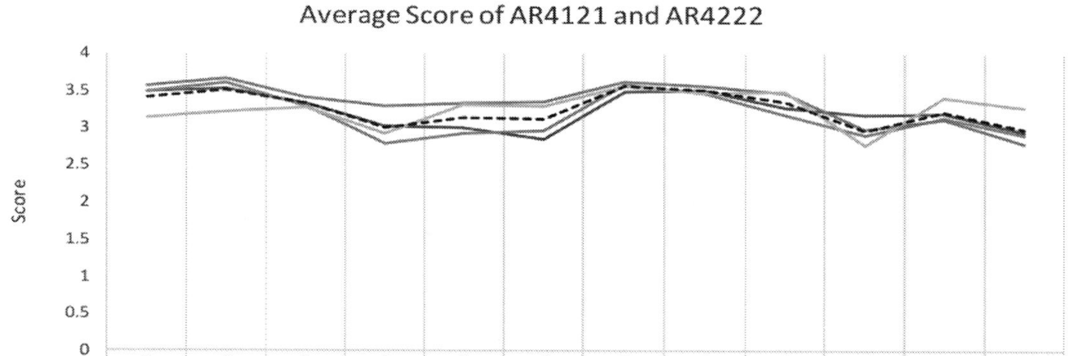

	1	2	3	4	5	6	7	8	9	10	11	12
2013-2014	3.485	3.525	3.34	3.025	3.005	2.85	3.49	3.51	3.29	3.19	3.195	2.96
2014-2015	3.485	3.605	3.305	2.805	2.925	2.965	3.585	3.47	3.19	2.92	3.14	2.91
2015-2016	3.58	3.665	3.425	3.3	3.345	3.35	3.635	3.58	3.48	3	3.13	2.795
2016-2017	3.145	3.22	3.285	2.93	3.325	3.31	3.58	3.475	3.495	2.78	3.415	3.29
Combined	3.42375	3.50375	3.33875	3.015	3.15	3.11875	3.5725	3.50875	3.36375	2.9725	3.22	2.98875

Fig. 6. Summary of Student Feedbacks on AR4121 and AR4222.

The goal is to know how to use, how to solve a certain problem using some set of tools or algorithms. Therefore, the basic principles of computation still relevant to be taught in the first year to provide a fundamental understanding of how computation works.

Given its significant impact on computational skill,the Computational Design Studio (AR2250) remain to befocused on the know-how aspect of learning various software, priority is given to the pragmatic solution of given drawing and modeling assignments. Another additional goal is to leverage computational skills among the students.

Algorithmic and BIM elective courses (AR4121 and AR4222) however, can be further sharpened into specific goals and lay a foundation for further exploration in the subjects of parametric design and analysis, and BIM applications. However, it needs further innovations and other supporting improvements.

As we may inspect in Figure 6 where both Algorithmic and BIM courses suffer in the aspects ofthe method of teaching, workshop execution and process, as well as additional resources.

a. Topic

Topics or object of learning both in Algorithmic and BIM courses focus on computational method. In Algorithmic Approach course the main topics are mathematics logic and functions, data structures, elements of geometry, and parametric applications. While in BIM Approach course, the main topics are parametric modeling and collaborative BIM.

It is should be taken into consideration that the topics being refined, is based on experiences and evaluation of student's performance in the period of 2013-2017. Students characteristic of each batch may vary and have an impact on the overall class performance as well.

b. Activity

Student activities during course should be carefully planned in order to give them both challenge and time to enjoy programming both in GH VAE or BIM platform. The additional joint workshop is preferable but should be planned and arranged well to maximise efficiency. One aspect important related with activity is it should be fun and enjoyable to do. Student assignments is a tricky part since it should balance between encouraging the student to push to their limit of creative exploration through programming and time required to complete them.

c. Additional Resources

Additional resources to support both courses is inevitable and should be improved. From lecturer's stand point, some topics both in Algorithmic and BIM courses can be enriched by digital apparatus such as CNC milling, 3D printers in order to give experiences of prototyping and hand-on with real material properties, and constructing real elements.

Another important improvement should be relying on the methodology of teaching computation for architecture students. We suggest that a visual-based algorithm editor such as GH is a good tool to teach such principles. The ability to understand and explore computational thinking is something that need constant learning through programming. We should encourage students to explore beyond class hours and provide insight on the computational works that they will face in their lines of work in the future.

In our brief conversation with WassimJabi[2] about teaching methodology of computation for architecture student, one of the approach is the *Deep Pool* approach, meaning that in other to gain more understanding of computing, students must do scripting in which they not only learn workflow of the program, they also learn how to think as a computer program works, learning to manage resources, learning to discover and to think rather than to know. This insight should be a great idea and already implemented in some architectural schools. In UNC Charlotte, North Carolina for example, learning programming in the early years of architectural education is mandatory to expose the student to computational thinking, which may help them learn software and other tools, motivates them to continue learning in the future (Senske, 2013).

Both cases are adequate and fit for their context and academic cultures. We are aware and fully understand that computational thinking and methods will extend beyond these courses. The potential, accessibility, and advancement of computation will democratize work of design, prototyping, and construction. It is our responsibility as academia to respond and to adapt to this inevitable phenomenon in order to stay relevant and innovative.

As Mike Silver pointed out in the editorial notes in Architectural Design, that that what computation now serve are the founding concepts, intuitions, and desires that can only emerge from a varied and creative practice. Computing should be regarded as an endeavor of creativity and arena of play, beyond any restrictions. That will be achieved through the development of computation culture in academia. It will require time, paradigm shift both for faculty members and students regarding computational thinking and further study, but we are preparing towards that direction.

References

Books

Abell, John, H. (2013), Hand Crafting Computational Design Thinking in Basic Design Studio, Proceedings of the AAE Conference.

Aish, Robert (2013), First Build Your Tools, in Architeural Design: Inside Smart Geometry- Expanding the Architectural Possibiities of Computational Design, Wiley, London.

Alexander, Christopher (1964). Notes on the Synthesis of Form. Harvard University Press. New York.

Burry, Mark and Jane Burry (2010), The New Mathematics of Architecture, Thames and Hudson, London.

Colakoglu, Birgul and Tugrul Yazar (2007), An Innovative Design Education Approach: Computational Design Teaching for Architecture, Proceedings of the METU JFA Conference.

Jabi, Wassim (2013), Parametric Design for Architecture, Laurence King, London.

Kalay, Yehuda, E. (2004), Architecture's New Media: Principles, Theories and Methods of Computer-Aided Design, the MIT Press, Massachusetts.

Kolarevic, Branco (2005), Architecture in the Digital Age: Design and Manufacturing, Taylor and Francis, New York.

Maeda, John (2001), Design by Numbers, the MIT Press, Massachusetts.

March, Lionel & Phillip Steadman (1971), The Geometry of Environment, the MIT Press, Massachusetts.

McCollough, Malcolm (2006), 20 Years of Scripted Space, in Programming Cultures: Art and Architecture in the Age of Software), Architectural Design, July-August 200 6Wiley-Academy, London

Menges, Achim, et al. (2017), Advancing Wood Architecture, a Computational Approach, Routledge, London.

Meredith, Michael, et al. (2007), From Control to Design: Parametric/Algorithmic Architecture, Ingoprint SL.

Mitchell, William J. (1996), The Logic of Architecture: Design, Computation and Cognition, the MIT Press, Massachusetts.

Senske, Nick (2013), Building a Computational Culture: A Pedagogical Study of a Computer Programming Requirement, Proceedings of ARCC 2013 Conference: Pedagogy: New Visions and Revisions in Architectural Education.

Silver, Mike (2006), Programming Cultures: Art and Architecture in the Age of Software (Editor), Architectural Design, July-August 2006Wiley-Academy, London.

Terzidis, Kostas (2006), Algorithmic Architecture, Architectural Press, Massachusetts.

Varienlioglu, Guzden, et al. (2004), Computational Thinking and Architectural Curriculum: Simple to Complex or Complex to Simple, Proceedings of the 34th eCAADe Symposium: Design Tools.

[2]AssociateProfessorat the WelshSchool of Architecture, Cardiff University. Chair, Digital StrategyGroup.The conversation took place during ASCAAD Conference in London, November 2016.

Composing the Layer of Knowledge of Digital Technology in Architecture

Mikhael Johanes[1,*], *Yandi* Andri Yatmo[1]

[1]Department of Architecture, Universitas Indonesia, Indonesia

Abstract. The use of digital tools in architectural practice has been evolving significantly. In following such developments, architectural practice has been incorporating digital technology not only to meet the current demand but also to pursue the vast amount of possibilities ahead. However, the integration of digital technology in architectural knowledge has been reasonably operative that produces uncritical understanding, and it tends to put architects as a passive user of technology. This paper argues that there are layers of knowledge that nees to be acknowledged and nourished accordingly in embracing the use of computation tools yet avoiding the overly simplistic.understanding. It attempts to explore the methods of digital technology in archietctural design practices as well as dicussions that follow to create a critical evaluation of its roles and potentials. The review is conducted theoretically in which the use of digital in the design process is explored in such a way to reveal its importance in architectural design methods. The review also crosses beyond the disciplines of architecture to construct more comprehensive understanding that bridges the logic of digital technology and architecture. The resulted map of methods of the digital thus can be used to develop a framework for digital discourse that bridge the operative knowledge of technology to the more critical perspectives.

Keywords: digital architecture, information, design method, architectural education

Introduction: From Representation to Information

Architects commonly use analogue representation such as drawings and models as a medium to materialise their ideas. It requires skill to relate the visual presentations of the medium as well as the tactile of hand involved with the real experience of the building for which takes years to develop during the training of an architect (Scheer, 2014). The materiality of the building and ideas is believed to relate directly to the representation that used by the architects. Various stages of architectural production processes determined the mode of representation that depends on its intention of creation (Robbins and Cullinan, 1994). For instance, architects often use a series of vogue sketches in the early stages of the design process where the ideas are synthesised from different consideration. Meanwhile, the abstract yet clear conventional construction drawing is used in the final phase of design in which the precise delivery of information to construction stage is necessary. The practice of architecture thus shapes its internal system of representation that enables the communication and production of knowledge that is nourished in architectural education systems.

Digital technologies "abstractly schematize" the analogic quality of representation into bits of information, structured and processed into its medium (Allen, 1997). The use of digital technology in architecture transforms its representational medium into an information system. This transformation inevitably changes the way architects work with their ideas in architecture. The traditional representational system enables architects to articulate their thoughts by cultivating the immediacy between hand and paper (Picon, 2003). On the other hand, the information of architecture within the digital medium enables particular operations in a specific way according to its mode of information and underlying system (Christenson, 2016) which both allow new possibility yet causes difficulties. This change is somewhat inevitable in the current architectural practice as well as our everyday life. The ubiquity of information enabled by Internet shapes our primary perception thus shifts our physical reality as well. The changing of the architectural working medium from meaning-seeking representation platform into performance-based information system perhaps become the only options that we have now in responding to such development.

This paper argued that the development of information based architecture profoundly challenges the notion of materiality in architecture. Picon (2003) argues that digital development redefines our relation to materiality by changing our perception and behaviour through its medium. The immediacy of human gesture in sketches and drawings which allows intuitive coupling

* Corresponding author: mikhael.johanes@eng.ui.ac.id

between ideas in its representation provide convenience medium for architects to work. In contrast, the structured and systematic nature of digital medium creates a thick layer of complexity that breaks the convenience. The following sections thereby attempt to break down these thick conceptual layers of digital medium in which the materiality of architecture is transformed into information to put it into different "reality".

The materiality of Architectural Information

Mastering the use of computers does not mean that we need to learn all program available in the market. In contrast, the ability to work with computers is measured by the capability to extend its functionality beyond its designated utilisation. We can attain this capacity if we can understand how the information is processed and manipulated in its digital medium. In other words, it is necessary to understand the materiality of the digital information analogous to building elements on the building construction. Computers create the new "perceptual entities" and objects that enable architects to manipulate different kind of transformation that is unavailable to traditional means of architectural representation (Picon, 2003). While the perception of architecture in the conventional representational medium often oscillates between concrete experience and abstract ideas of materiality, digital tools fundamentally alter this relation by creating a different realm of materiality.

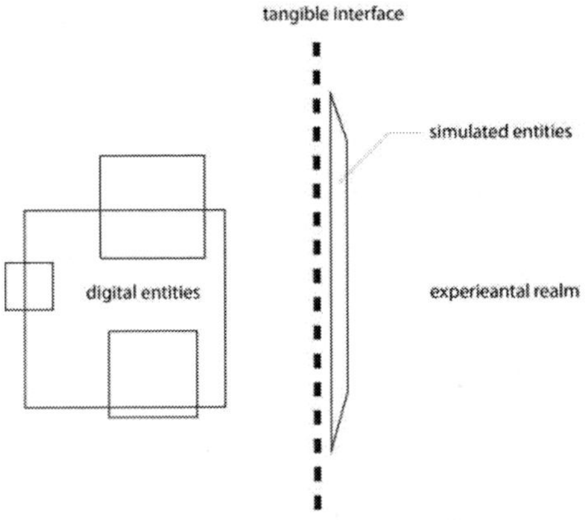

Fig. 1. Two Essential Elements of The Digital Medium.

We can conceive the digital medium such as computers according to its two essential elements: the surface demarcated by the interface on which people can interact with the medium and its interior that is the underlying system that works behind our vision (Figure 1). In its surface, the visible interface of digital medium creates a simulated experience in which its entities are perceived as reality, or even becomes the new reality itself (Scheer, 2014). The typical manifestation of this kind of digital surface is found in a smartphone, where the abstract information is experienced as reality using its intuitive touchscreen. The ever increasing development of virtual reality marks our fascination toward this simulated environment. This progression of highly simulated digital environment inevitably affects the practice of architecture as well as our lives in general. This ubiquity of simulated materiality thereby creates limited understanding because it mostly perceived as simulated experience and left its underlying abstract entities behind.

The interior of digital medium has less fascinating existence for people but perhaps holds more critical roles in understanding the digital materiality. The invisible, abstract mathematical entities (Emmer, 2005) inside the computer are the building blocks in digital architecture that are located within this interior as information. This information is somehow abstract, but its computability promises a wide range of opportunity if one can understand its operations. However, the current digital culture is seemingly flocking into the same direction in which a lot of things is done by similar blueprints with lack of inventiveness (Axel Killian in Burry, 2014). This condition reflects the shallow understanding of digital tools which only scratches the surface of digital without really digging deep by into its interior state. We argue that the knowledge of the digital should go beyond this surface, cultivating the space of abstract information within its medium that can sustain this development.

Christenson (2016) creates an analogy between grain in the physical objects to the materiality of digital information in architectural practice. He reveals that each mode of information has a different way of interpreting "reality" into digital environment thus affecting the kind of "force" that can be applied to it. Unlike the real physical objects that respond to natural forces such as gravity, weathering, and resistance in unity; the digital entities fragments and structures the reality into the various modes of information which can only be affected by specific operation enabled by its underlying systems. The more profound understanding of the way of information operation in digital architecture becomes crucial both to expose its limit and reveals its wide range of possibilities.

The Operational Layer of Architectural Information

The use of digital technologies is inevitably operative. However, the passive use of technologies puts its real operational capacity still intact, in which the nourished knowledge is more often only touching its surface of the interface. We propose that there are layers of operation by which the information of architecture potentially cultivated in an active manner, allows the tactical use of technologies instead of surrender to the given strategies (Figure 2). Those layers are the generation, translation, materialisation and simulation. Generation involves the operation of creation of digital entities which argued critical in the ideation of architectural possibility yet often limited by the existence of digital interface. Translation enables the fragmented nature of digital entities into the realm of conversations, to complement each other thus develops a richer architectural information system. Materialization and digitalization enable a specific mode of translation from and to digital realm into our physical realities. Those specific translations are argued to have important roles in bridging the two realms of architectural materiality. Simulation enables digital entities to reflects reality by performing it according to particular laws. While the generated data has significant value in anticipating the real performance of architecture, more critical evaluation of its law is arguably importance thus open architectural practice to a broader field.

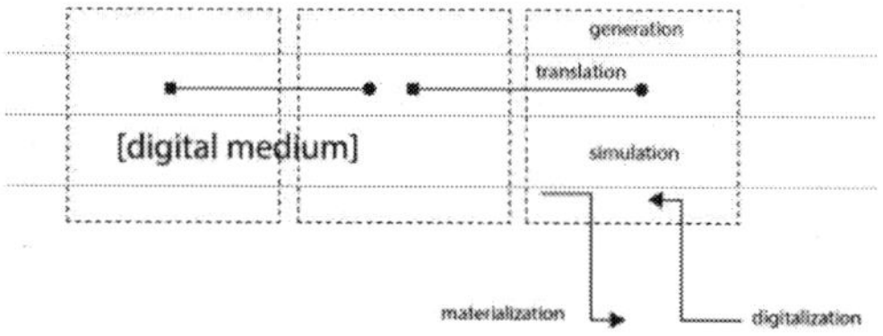

Fig. 2. Layers of Operation by The Information of Architecture.

Generation

Using digital medium requires the ability to generate desirable entities in which those objects convey particular information. There are three generation methods that dominate the current digital architectural process: directly creating using specific designated application, parametric systems and scripting (Dunn, 2012). Generating information using the assigned computer program is a pretty straightforward approach. In this method, the process is enabled but constrained by the interface of the program. For example, creating free-flowing continuous surface is more feasible using NURBS based modelling application rather that polygon based modelling application that is more appropriate to develop arbitrary surfaces from a collection of faces, edges and vertices (Corazza, 2012). This method requires the simulated environment to present the created abstract information directly to application's interface. Interfaces that both enable and limit the interactions between digital medium and its users define the very action that can be performed thus tend to hinder us for cultivating its latent possibilities.

Parametric systems enable architects to explore different possible outcomes from one abstract formal system by changing the finite value set of parameters. Parametric models are built from a set of components that embody specific operation, assembled as such to perform a somewhat complicated process and dealing with its abstract data structure (Woodbury, 2010). Parametric systems can extend the process of information generation beyond the designated interfaces by deliberately composing them. The tricky part of creating parametric system lies on the configurations of operations and the generated data to establish a desirable semantical relationship (Christenson, 2009). Fundamentally, creating a parametric system is a construction of a mathematical function that transforms a set of finite parameters into visible geometry. Mathematics thus holds an essential position in the establishment of such systems (Legendre, 2011) due to its very abstract nature behind the instrumentality of computation. Parametric systems bring mathematics into life by enabling operation that allows people to get results from its conceptual process in the form of geometry representation. However, people tend to avoid the difficulties in tackling this kind of abstracts problem and constraint itself using easy but limited operation in a well-designated application.

While the creation of architecture geometrical information using a specific application as well as parametric systems is performed through a well-designated interface, scripting allows us to bypass the limitation and interface to perform the particular operation defined by specific syntactical computer language called codes (Burry, 2014). Script exploits the procedural process of computation to generate architectural information from its logical cores. It enables people to extend the operation of an application that is not defined in its interface yet possible. In other words, scripting opens a wide array of possibilities of action enabled by its medium rather than specific application. Understanding computer codes require the abstraction of procedures in which the process is composed which is not easy for everyone. The dichotomy between user and programmer creates a paradigmatic gulf that hinders the pollination between digital medium and architecture (Derix and Izaki, 2014). However, Kay (2012) argues that the digital language perhaps holds important roles in the future of learning analogous to literacy system that enables the forming of past civilizations. Once the difficulties of reading and writing of digital entities are surpassed, it amplifies our ability to cover different ground thus enable us to think differently.

Translation and Interoperability

Christenson (2016) describes translation as the act of interpreting information into another form such as interpreting a text as a visual image. In translation, the loss is inevitable, because each mode of data is designated as such according to its intended meaning and delivery purpose. Translation only allows particular information to be passed. In the case of conversion from text to an image, it neglects the textual data and maintains its graphical appearance. The loss also implies the translation is sometimes irreversible, which means we can never recover the missing information once it performed.

Each mode of information suggests a different operation according to its particular logic. Translation is not supposed to be seen as a form of loss. Instead, it enables various procedure that extends the meaning of initial information. Translation allows the abstract information to be processed "outside" its original form. In architectural processes in which the multiplicities of information are inevitable, translation enables transformative forces to be applied in the different situation thus provide a vibrant area of experimentation where new connections and associations are made

(Christenson, 2016). Translation is argued as necessary in digital architecture to form a more diverse architectural design process.
Interoperability between systems in digital medium enables a kind of translation that has a slightly different aim from the former. Interoperability is defined as the capability of the various system to exchange its data (Jeong, Eastman, Sacks and Kaner, 2009) and prevent data loss as much as possible. While the translation between different modes of information aims to changes its method of interpretation and operation, interoperability seeks to create a shared understanding of separate systems to minimise the loss of information then it can be processed in the entirely different system. The more complex information in one system, the more laborious exchanges to be performed because every system has its way in structuring its information. Standardisation of such complex data is thus necessary to face those difficulties. For instance, the use of universal exchange format in BIM industry enables the different platform to work with each other (Jeong et al., 2009). However, total standardisation potentially will restrict the possibility of a different system to have various features which limit the possibility of extending the capability of the digital medium. The differences are necessary to maintain the possibility of diverse extensions. Both translation and interoperability bridge those gaps and thus allows a healthy diverse digital ecosystem to grow and work together.

Materialisation and Digitalisation

Digital fabrication fundamentally is a form of translation that transforms geometry information from the digital medium to physical materials through machinic operation controlled by computers (Ox-man and Oxman, 2014). Digital fabrication requires the ability to transform digital abstract information into process-based information to be executed by specific machines. This stage involves the understanding of available material process such as cutting, subtractive, additive and formation fabrication (Kolarevic, 2003) as well as the strategies that available for specific materials (Beorkrem, 2013). The capability to differentiate the process of materialisation enables building construction industry to shift from mass production that is based on standardisation into mass customisation from which differentiated building elements can be produced economically. Digital fabrication in building industry extends the possibility of construction thus giving architects more space in realising their ideas.

Digital fabrication also enables architects to quickly materialise her ideas during the process using rapid prototyping (Oxman and Oxman, 2014). It is a form of translation between abstract digital entities that enables iteration between the abstract materiality of the digital and physical materiality within design process which can bridge the gap between these two domains. The bridging between digital and physical reality allows architects to reconfigure its architectural information systems according to the feedback from physical realisation thus giving the digital-enabled process for having a "reality check" during the process. Digital fabrication thereby makes architecture as both spatial and material practice, practically blurring the boundary between design and construction.

On the other end of translations between digital and physical, digitalization denotes the reverse vector of translation which enables realities to be denoted as information (Allen, 1997). The most of loss is tend to happen in this kind of translation in which idealisation and structuration are inevitable. However, the ability of the digital medium to manage a large sum of information enables it to be a better medium of working rather than writing and drawing. While writing frees us from the necessity of memorizing and enable us to think otherwise (Kay, 2012), digitalization extends this prosthetic capability even forward. What is more, digitalization extends our capability of thinking through computation.

Simulation and Performativity

Computation enables the digital medium to perform digital simulations. Simulation fundamentally is a replication of reality that provides the possibility to study real-world phenomenon without direct intervention to its real context (Groat and Wang, 2013). Digital simulation holds an instrumental value that conventional medium cannot offer. The importance of simulation lies in its ability to generate information and forming iterative process between design and analysis. Simulation enables architects to anticipate building performance (Scheer, 2014) before being built into reality. High-performance architecture becomes possible due to the available computation technology in optimising different aspects of the building that sometimes have competing values which is harder to solve using conventional means of calculation.

There is no universal simulation yet and perhaps there never will be. Different simulation is needed to analyse various aspects of performance and requires a different set of information. For example, simulating structural integrity of the building requires information on the structural system that is not available in the 3D modelling application that only gives geometrical information. Building Information Modeling (BIM) system was thus invented to overcome this limitation by fundamentally providing better information management system and enable interoperability between different systems. BIM systems create "mirrored" reality of a building by forming a structured information that covers almost every aspects of the building including spatial, material, structural, operational, cost, schedules and even life-cycle factors over long periods of time (Groat and Wang, 2013). The development of digital simulation has come to the point that almost every performative aspects can be managed, evaluated and finally anticipated.

However, there is a limit. No matter how sophisticated and complex the simulation created in replicating real-world phenomenon, it is only a mode of analysis that should not be confused with the reality itself. The knowledge that can be gathered is highly operative from which the only observable effects is already anticipated by the "law" that underlies behind the simulation (Scheer, 2014). While the "law" aim to explain and predict a particular real phenomenon, simulation is created to perform according to this law. The way to expand the understanding the of reality through simulation should begin from its underlying theoretical basis. The multidiscipline effort is important in which different disciplines can work together to develop collective understanding in performing a simulation. Simulation can be seen as a door to open dialogue outside the discipline of architecture where the assessment and evaluation of performance requires a fair amount of theoretical understanding. Simulation can ground architecture to its reality only if it is constantly evaluated and criticised. Otherwise, the detachment of architecture from the "reality" is perhaps inevitable.

The Layers of Digital Knowledge: Beyond Rationale of Vision

The development of digital has come to an era when digitalisation of architectural practice is inevitable. However, the current development mostly puts architectural materiality as a simulated experience of information using its interface. The information based architecture requires further understanding of how the actual logic of the digital works in the medium. Digital entities embody different materiality (Picon, 2003) in which architectural ideas can be generated. Generating architecture in a digital medium both enable and constraint architects in using its interface. Bypassing the constraint of the interface thus becomes crucial in opening a wide range of possibility that lies in the digital interior system. Derix and Izaki (2014) suggest that mathematical thinking and computer programming is necessary to create intuitive coupling between digital media and its user. Intuitiveness in using digital media does not come from a seamless interface between user and media, but from familiarising oneself to the logic of how information is interpreted and operated digitally.

The transformation of architectural communication from representation into information enables procedural mode of presentation of architecture. Parametric system and algorithmic architecture show that design can be presented as a set of procedures containing various values and rules. The resulted architectural forms thus are only consequences from such processes. Reviewing such architecture thereby cannot be judged solely based on its final result. Instead one has to evaluate the very logic of values and rules that operate behind. It is not impossible if the future of architectural design publications will contain sets of procedures showing the logic behind rather than a set of drawings that solely represent its formal consequences. This transition is perhaps the best bet for architecture to escape from the rationale of vision that already dominates the architectural practice from the beginning of its development (Eisenman, 1992).

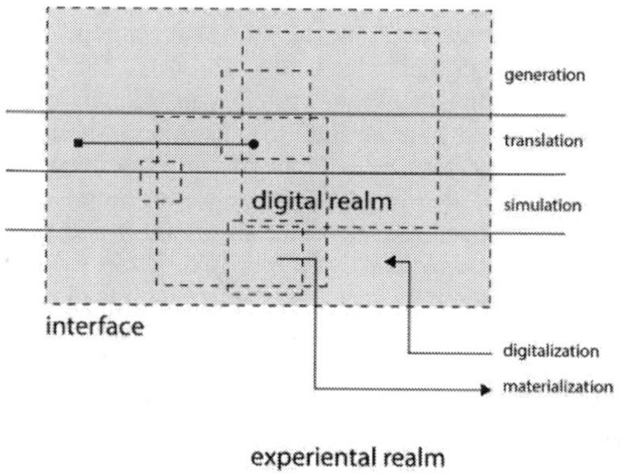

Fig. 3. Whole Layers of Digital Knowledge of Architecture.

The emergence of the digital process also enables architecture to reconsider its relation to materiality. The development of digital is often seen as a movement away from architectural materiality (Picon, 2003). It is true that the virtualization of architecture in the digital medium allows the material reality to escape from the grasp of our perception, but the dialogue of the digital and physical in digital medium goes beyond our perceptual experience. The fabrication process of digital lessens the gap of digital and physical where the material aspects inevitably should be taken into account. On the other hand, simulation puts the materiality of architecture in the medium of the digital domain in its abstract form based on particular theory or law. The theorization of materials enables a wide range of hypothesis to be tested before being realized into built forms. However, simulations and their underlying principles must be continuously checked to its real-world context empirically. Both simulations and fabrication oscillate the materiality of architecture within the realm of digital and physical, creating entanglement between both domains and forming the iterative architectural material practice.

Figure 3 shows the whole layers of digital knowledge of architecture. It can be conceived in two form which perhaps holds an essential role in the development of the architectural education curricula: the layers of the medium and the layers of methods. The layers of the medium denote the interface and interior in which the abstract entities inside should be understood, allow us to work in surpass the interface and work more tactically by understand its operating language. Computer language and its abstract foundational knowledge such as mathematics are arguably held important roles in digital practice as notation and convention also form the language of architectural representation. The layers of method denote the operational layers in which the entities of architectural information are generated and translated from and to the digital medium. Although those methods of operation can be performed in every layer of the digital medium from interface level to its deeper interior level, the deeper the layer of the medium used, the more possibilities can be founded thus provide a richer area of cultivation. On the other hand, these layers of methods also hold important roles in bridging the fundamental knowledge of digital medium with the more practical architectural design processes.

Concluding Remark: Missing Information

The transformation of architectural representation into information, however, has a limit in itself. The abstraction into digital realm forces architecture to be translated into entities and sets ontological equalities between them in which the status of relations between entities must be clarified (Loo, 2013). This translation implies that there will always be missing information in the digital architecture. The very abstracts nature of mathematical and algorithmic creates a stubborn reality of quantities has been limited itself in grasping the bodily feeling of spatiotemporal experiences (Parisi, 2013). Although digital transformation enables the materiality of architecture to shifts to new realm thus allowing it to evolve as such, the immaterial aspects of architectural materiality are yet still missing and perhaps can never be incorporated in the digital medium.

References

Allen, S. (1997). Object to Field. In M. Toy (Ed.), Architecture After Geometry (Vol. 67). Academy Press.

Beorkrem, C. (2013). Material strategies in digital fabrication. New York: Routledge, Taylor & Francis Group.

Christenson, M. (2009). Testing the Relevance of Parameterization to Architectural Epistemology. Architectural Science Review, 52(2), 135. https://doi.org/10.3763/asre.2009.0022

Christenson, M. (2016). Beginning design technology. New York: Routledge, Taylor & Francis Group.

Corazza, S. (2012). Modelling Techniques-Differences between NURBS and Polygon. Retrieved October 3, 2017, from https://seraphinacorazza.wordpress.com/2012/12/28/modeling-techniques-differences-between-nurbs-and-polygon-modelling-new-york-city-rooftop-3d-architecture-project-research/

Derix, C., & Izaki, A. (2014). New Curricula: Syntonic Systems. Architectural Design, 84(5), 122–129.

Eisenman, P. (2015). Architecture After the Age of Printing (1992). In M. Carpo (Ed.), The Digital Turn in Architecture 1992-2012 (pp. 15–27). Hoboken, NJ, USA: John Wiley & Sons, Inc. https://doi.org/10.1002/9781118795811.ch1

Emmer, M. (2005). Mathland: The Role of Mathematics in Virtual Architecture. Nexus Network Journal, 7(2), 73–88. https://doi.org/10.1007/s00004-005-0023-1

Groat, L. N., & Wang, D. (2013). Architectural research methods (2. ed. [updated and expanded]). Hoboken, NJ: Wiley.

Jeong, Y.-S., Eastman, C. M., Sacks, R., & Kaner, I. (2009). Benchmark tests for BIM data exchanges of precast concrete. Automation in Construction, 18(4), 469–484. https://doi.org/10.1016/j.autcon.2008.11.001

Kay, A. (2012). The Future of Reading Depends on the Future of Learning Difficult-to-Learn Things. In B. Junge, C. Zwick, & Kunsthochschule Berlin-Weißensee (Eds.), The digital turn: design in the era of interactive technologies. Zürich: Park books.

Kolarevic, B. (Ed.). (2003). Architecture in the digital age: design and manufacturing. New York, NY: Spon Press.

Legendre, G. L. (Ed.). (2011). Mathematics of space. London: Wiley.

Loo, S. (2013). Abstract Care. In H. Frichot & S. Loo (Eds.), Deleuze and architecture (pp. 245–261). Edinburgh University Press.

Parisi, L. (2013). Contagious architecture: computation, aesthetics, and space. Cambridge, Massachusetts l London, England: The MIT Press.

Picon, A. (2003). Architecture and the virtual: towards a new materiality? Retrieved from https://e-pub.uni-weimar.de/opus4/files/1236/picon.pdf

Robbins, E., & Cullinan, E. (1994). Why architects draw. MIT press.

Scheer, D. R. (2014). The death of drawing: architecture in the age of simulation. London ; New York: Routledge.

Woodbury, R. (2010). Elements of parametric design. London ; New York, NY: Routledge.

Generative Algorithms in Alternative Design Exploration

M. Galieh Gunagama[1,*]

[1]Department of Architecture, Universitas Islam Indonesia, Indonesia

Abstract. The improvement of design capabilities with the help of computers has been very commonly discussed recently. Many studies have presented the topic of evaluating a condition on a design proposing a form by applying an algorithmic approach. However, this is still a discourse when it comes to design exploration at the preliminary design stage, which in general, design initiation is still dependent on humans. This research seeks to bridge the existing gap by devising an algorithm that can help create alternative designs at preliminary design stage. The case scenario presented in this study is for 1 story building on sloped land. Design limitations and site conditions are the primary considerations. The discussion focused on the dynamics of the algorithm process that was arranged for the acceleration of alternative design creations.

Keywords: Generative Algorithm, Design Exploration, Preliminary Design Stage, Architecture

Research Background

Optimization of computer capability in the design has been a dream and hope of the designer since the past four decades. Currently, design with the help of computer has become commonplace. However, the use of computers as a tool to rapidly arrange alternative designs began to be developed and not fully utilized. Previous studies have not been able to meet the expectations of a computing process that enables optimum design automation. This condition is influenced by many things, including the many variables that should be considered in the design by the architect (Lobos & Donath, 2010) and the extent of the disciplines that must be integrated into a design (Turrin, Buelow, & Stouffs, 2011)

Utilization of computer capabilities in the preliminary design stage has been done before. However, the design that using a computer in the preliminary design stage for some researchers is only seen as a mere replacement tool, which is to move the manual media into digital (Aliakseyeu, Martens, & Rauterberg, 2006) (Pranovich, 2004). Besides, simulating the design related to environmental conditions both inside and outside the building has been done before (Attia, 2011) (Attia, Gratia, Herde, & Hensen, 2012). In general, the simulation is intended to evaluate the condition of the building by predetermined standards and criteria or to find a form of building that is considered optimal in a particular domain

The development of computational design allows the design to be created freely and without limit. (Humppi, 2015) (Khabazi, 2012) So aside from being used as a tool to evaluate, should the use of computer capability does not stop at that point only. But it should be used to encourage the design process towards creating a large number of alternatives that can speed up the preliminary design process. With the help of computer skills, the limited human cognitive power to produce alternatives should be expanded, especially as a catalyst for the creation of new possibilities (Humppi, 2015)

Although considered capable of improving human capabilities, computers are not always aware of the environment. Terzidis (2006) argues that the mental processes of design are conceived, imagined, and processed in the human mind. Thus, computers are only seen as a means of calculation, organization, productivity, and presentation. Because of that, the algorithmic results generated by computational design must refer to the human mind, i.e., programmers or designers, so that the process of criticizing the algorithm results is always directed to the designer using the algorithm. In other words, the one responsible for the final result is the designer, not the algorithm that helps generate it (Terzidis, 2006).

The discussion in this study attempts to open up the possibility of accelerating the exploration of alternative designs at preliminary design stage assisted by computational design. The absence of algorithms in computational design that refers to land with slopes and sites with vegetation encourages the idea of building algorithms that seek to solve the problems in the scenario.

This research tries to achieve that idea using Rhinoceros and Grasshopper, especially the Galapagos component. Galapagos simulates a generative process that mimics evolution to find the most optimal composition based on predetermined settings (Rutten, 2010). Although not the best method, the evolutionary solver in the Galapagos is considered capable of generating an alternative answer

* Corresponding author: galieh.gunagama@uii.ac.id

to the given problem, which in general, each new answer will be better than the previous answer (Rutten, 2010).

With these capabilities, to rapidly generate large numbers of design alternatives in the preliminary design stage, generative algorithms are considered capable to simulate the answers for the model to be built. The built model will focus on the design on land with a certain slope, with the size of the space, the existing vegetation conditions, the width of the border of the land, and the predetermined distance between building masses. This research proposes an algorithm that can be used to obtain design alternatives that have optimal value in response to site conditions

Parametric in Preliminary Design

The design process in Architecture can be divided into two stages, namely the initial design stage and final design stage. In the initial design phase, the problem has not been well defined, the thing that is done is still within the scope of exploration which includes the process of describing the sketch, schematic, and building the model. Activities undertaken are explorations in the effort to formulate design issues. Its characteristics include generating alternatives, shifting from one idea to another because of a lack of commitment to one idea, general information input, and rough visualization to reinterpret the design at different times (Pranovich, 2004).

The final design stage has the characteristics that the problem is well defined. The resulting product has been elaborated with working drawings, details, and threedimensional models. The activities that occur are elaborating solutions to the problem of space, which includes trying to formulate solutions to existing problems, sharpen ideas in more detail, including more specific information, and define ideas more precisely (Pranovich, 2004).

In other words, the initial design or preliminary design stage is the beginning when the design is conceived. This section is the starting point before the design is sharpened in the next process. In the preliminary stage, the designer will focus on how the design will be constructed both regarding partial and holistic compositions, and also the concept of beauty will be defined.

The demand to propose many alternatives to clients in this phase becomes very important and significant. Designs prepared from the best alternative will ease the next design stage for the designer. Although it is commonly known, the number of alternatives created is very limited to human capabilities. At this point the role of design with the help of computers is necessary.

This research seeks to facilitate the composition of the mass of the building at the preliminary design stage. In addition to the human's cognitive limitations, sketching, schematics, and physical models for 3dimensional studies are considered less practical and require a relatively long time compared to compiling a computer simulation. Therefore, this study seeks to enhance the design capability of the initial design stage by proposing a model constructed using an algorithm and will represent a 3dimensional form of simulation with the ultimate goal of accelerating the creation of alternative designs at the initial design stage.

Algorithms and Design Restriction

Generative algorithms are algorithmic and parametric ways that handle geometry in design problems. With this algorithm method, designers can utilize all the capabilities and possibilities regarding geometric computing, as well as in managing large amounts of data, numbers, and calculations better than conventional geometry methods. Discussions to be conducted with this method are not limited to predetermined experiments, but to explore the unlimited potential. There are many ways to set up a design algorithm, although there is an issue of the limitations on the commands provided in the software. However, it still allows the preparation of alternative solutions that can be offered and gives unlimited possibilities (Khabazi, 2012).

Simplistically, the algorithm can be likened to a cookbook containing a step-by-step guide to making food. Maldonado (2014) argues that, as in cookbooks, recipes can be articulated as a series of interconnected steps or actions that govern the interactions between the various components. The variation of the parameters that inform the interaction potentially alters the process's final results significantly. To produce alternative dishes, substantial modifications in the process need to be done after the fundamental interactions of the given recipes are understood (Maldonado, 2014).

Therefore, if the algorithm is a system, then the variables and parameters needed to run the system. Variables and parameters can be analogous to the limitations in the design. Variables are entities that change in a particular system, whereas parameters are entities that are used to connect or unify two or more variables of an equation. Parameters may or may not be able to have the same dimensions as the variable (Difference Between Variable and Parameter, 2012).

To create a system that works well, data inputs and the boundaries need to be appropriately defined in the beginning. In architectural design, things that can be categorized as data such as land properties, including land size, land contour, existing vegetation, and climatic conditions. The design boundaries include building regulations, structural constraints, and specific values in designs that defined by the designer, such as space programming and relationships between spaces.

There are five conditions to be considered for the model made in this study. The conditions are, (1) the building is placed on sloped land, (2) the building should not be out of the boundary of the border rules of a certain size, (3) the building should not violate the preexisting vegetation growing on the site, (4) the room masses is expected not to touch each other, and (5) the ideal distance between adjacent spaces should not exceed a predetermined number.

https://doi.org/10.1051/shsconf/20184105003

The scenario used is a land area with a size of 35 x 20 square meters, for 1floor building, and the building will consist of 10 rooms. All sizes in buildings and spatial information are arranged in syntactic plugins to facilitate organization and space management (Nourian, Rezvani, & Sariyildiz, 2013). In this study, Syntactic is used as a tool to define space and size only, while the built model will be evaluated using the Galapagos. The genome input to be used is the coordinate of the room located on the site and the rotation angle of the room that is modeled. While the target fitness to be addressed is to get an alternative setting for the room on the site with the optimal value of the conditions required above.

Modeling and Simulation

Modeling

The model used for the simulation is built in several stages. The first stage is the creation of the site for the simulation; the second is the preparation of room modules in the building to be simulated; the third is the modeling for the test limits; the fourth is the preparation of tests to be used.

All models are built in Grasshopper and Rhinoceros are used as visual viewers only. The Use of Grasshopper from modeling stage is done to make the produced model can be more easily modified, so it also can be applied to similar cases with different settings.

The land that will be used for simulation is a rectangular and simulated slopeshaped area. The selection of the land arrangement in such a way is to test the model in the 2dimensional basic shape, i.e., square on the X and Y axis, as well as at the point that has transformed to the height of the slope, i.e., on the Z axis. The selection of site simulation with this kind of form also allows the study to see which part of the room module that will sink into the ground with that setting. The model for the site shown as a surface with a contour line as a reference for its slope.

The preparation of building room module will be based on the points on the X and Y axes that refer to the length and width of the land. The reference point is arranged independently, i.e., the value of each X and Y for each point can be determined by itself. These points are then projected on the ground surface to obtain the coordinates of the Z axis that bound to the land height.
The independently arranged reference point allows the plotting of room modules to be more flexible. It can also be seen as a simulation of freedom in plotting the building for the designer. The arrangement in such a way allows the occurrence of two points that have the same X and Y values. This condition is not expected to happen, so different tests are needed.

The points with the coordinates X, Y, and Z are used as the midpoint on the square which is the basic form of the room modules. The room name data and the size of the room for each module have been determined in advance. Space organizing is done with syntactic components. Part of this plugin allows space to be traced by giving names to spaces and color sign on the display corresponding to specific room information.
Then each square for the room module is rotated with the rotation axis at the center point of reference created beforehand. The next step is to provide a height measurement for 2dimensional room module so that the form achieved is in the 3dimensional box. The height of the module can be adjusted with the prearranged room name data.

The rotation angle for each room module is set to be freely changeable. This allows the buildings arrangement to be not rigid on facing one side only. This condition can also be analogized as the freedom of the designer in determining the orientation of building masses.

In this simulation, 10 rooms are modeled. Variable point X, point Y, and rotation angle will be used as a genome in Galapagos. The variable for this simulation uses real numbers according to the site length and width with 1 digit after the decimal point to determine the points X and Y, and the integer value is equivalent to 180 degrees for the rotation angle.

The next stage is modeling for test. It has been mentioned that there are some limitations, i.e. (1) the building must not exceed the site boundary limit; (2) the building does not violate the preexisting vegetation on the site; (3) between room modules do not touch each other; and (4) the spacing between the room modules does not exceeds the set limit.

The limit of the border for the site is obtained by offsetting inwards for the surface of the site based on the size that has been set. In the simulations performed, the offset is the same size, so there is no difference in size between the road border and the border between the buildings. The boundary of the border that has been obtained is then given height on the Zaxis and formed into a box.

This border box will be a test for the room module box. In the test, the border box will be the outermost boundary for the room module box is allowed to be on site. So, the number that is expected to appear is the maximum value for 10 room module boxes is declared within 1 border box.

The preexisting vegetation on the site is made by using a random population that produces 10 dots in the field. It is used to simulate the location of existing trees that are also random. The existing dots are then transformed into cylinders to symbolize tree trunks.

The cylinder is the closest boundary for the room module box that can be placed on the site. That is, if the room module box exceeds this limit, then the room module can touch the tree or crash into the existing tree. This is a condition that should be avoided, so the test is made to find the maximum number for 10 room module boxes to be declared on the outside of 10 cylinders of existing vegetation.

In this simulation, the room module boxes are required not to collide with each other, which means there can not be the same X and Y value. So the test is built is to find the maximum value which states that every room module boxes that exist are not colliding with another room module box. Although not colliding with each other, it does not mean that all room module boxes must be entirely separate and unrelated. So the distance between the room module box should be set for the room that is considered interconnected. This is to simulate the proximity of distance that usually arises between interlinked spaces. Because the distance criteria between room modules should not exceed the specified limits, the modeled simulations need to add some adjustments.

The volume centroid for each room module box needs to be known. Then, the centroids for the proximity spaces are connected by a straight line. The length of the line is then entered into the test to determine whether the line length has met the specified distance limit. Thus, the test is to find the maximum value stating that each interconnecting line is within the specified range.

The distance adjustment is greatly influenced by the 3dimensional box boundary distance to the centroid, and this occurs on both tested objects. Although the gap between the boxes has entered the specified criteria, this allows a problem to happen, which is a smaller box located in the larger box. To minimize the effects of this issue, one test needs to be added to reinforce the final result.

Adjustments need to be done by including a total volume test. If the distance between the centroids is too close, the two boxes will intersect. This makes the total volume of the boxes will also be reduced compared to if the boxes stand alone. Therefore, it is necessary to limit the extent to which the total volume of the final interconnected box may be reduced compared to the original state when there has been no contact. The defined test is to find the appropriate volumetric configuration and composition, which must be greater than the minimum permitted percentage limit.

The maximum value of each test that has been done will be formulated into a mathematical equation that will be the basis of value research in fitness in the Galapagos. The value will be a benchmark whether an alternative mass building configuration is considered to meet the requirements that have been determined.

In this research, the algorithm will be tested 4 times with evolutionary solver in Galapagos to get a different alternative. Since the initial condition of the input to the genome affects the final result of the algorithm calculation, then 4 initial conditions need to be created so that the simulation can produce different alternatives. Four initial conditions are prepared by inserting the X and Y coordinate values for each building mass by configuration (1) when the X and Y values are minimum, (2) when X = maximum and Y = minimum, (3) when X = minimum and Y = maximum, and (4) when the value of X and Y is maximum. The initial condition for the building mass rotation on the simulation will be set to the minimum value of the range of provided angles.

Simulation

To know the maximum expected value of the algorithm, it is necessary to see the highest value for each test variable. The expected value for each variable is then compiled in the equation that will cover the entire variable. The calculation result of the equation will be used as a fitness input in the Galapagos.

The simulation in Galapagos begins by defining the relation between object A and B and converting it into numerical form. The space mass box (object A) will be tested whether inside, intersect, or outside the boundary box of the site (object B). The result of a relation is encoded into 3 numeric categories, i.e., 0 if object A is inside object B, is 1 if object A intersects with object B, and is 2 if object A is outside object B. Based on that, it can be seen if all the room mass boxes are in the boundary box of the site, meaning the maximum value that must be obtained is 0.

Next is a test to see the relation of the room module box (object A) to the preexisting vegetation cylinder (object B). The value that needs to be defined is the statement "true" or "wrong" in the Boolean Operation. Since the required data is numerical, each statement needs to be converted into numbers. In Grasshopper, each false is 0, whereas each true is 1. This means that if there is an object A declared "true" collided with object B, then the value that appears is 1.
For the logic in the simulation algorithm to be in line with the predetermined requirements, the statement needs to be adjusted. The required condition is that there should be no room module box that can be tangent to the preexisting vegetation cylinder. So the corresponding statement is if there is an object A declared "not true" collided with object B, then the value that appears is 1. Based on that, if there are 10 boxes that do not collide with 10 cylinders, then the maximum value should be gained 100.

Then to test whether each box of the room collides with another box of the room, the same way with box and vegetation cases can be used. However, there is a difference in understanding the logic of the test results, especially since the input data comes from the same object. The treatment for this test result needs to be scrutinized because it will produce an output that is logically considered valid, but the value will be regarded as unsuitable.

If there is one group of objects tested whether it collides with the same group, then there will be a result for 1 piece that will always be considered "true" to collide. This is because if object A is tested to the same object, then the test value will always be "true" to collide with each other. Logically, this is considered valid because it means that object A resides within object A. Similar results will also occur when object B is tested with object B, object C with object C, and so on. But it will not happen if the object is different like when object A is tested with object B, C, and so on. Thus, for minimum value search, this needs to be an additional calculation, i.e., to eliminate the test result value between the same object as the exclusion criterion. Otherwise, it can capture the maximum value for tests on different objects as inclusion criteria.

Based on that, it can be seen if 10 objects are tested with the same 10 objects, then the result will always be 10 test results that will always "true" collide with itself and 90 other test results that have not been determined as "true" or "false." Therefore, in the calculation to find the maximum value, the value of 10 for the "true" test result will be eliminated, and the expected maximum value is 90 for the "incorrect" statement of the building module box colliding with each other.

To evaluate the distance between the boxes, the test used is different from the other parts. Instead of using a 10 point building module, in this test using 11 points is to add 1 line that determines the entrance area into the site. This one point will simulate the closest distance from the site entrance to the earliest building module encountered with circulation, i.e., carport or garage. This is a consideration because the site alone without any circulation is considered unlikely to happen. Therefore, if the line length is tested for a specific range and if it is entirely "true" then the maximum value is 11.

As for the evaluation of the total mass volume of the building, the test will be made to compare the sum of the original volume of the entire mass of the building and the volume when it has come into contact. In Boolean Operations, merging or unification (union) of multiple solid objects may occur. This allows the changes in both surface area and overall volume.

The changed volume is used as the tested variable, while 90% of the original volume is the reference variable. Determining the percentage of the initial volume is an attempt to simulate the degree of freedom in creating the composition of the building mass, and to what extent the percentage of space can change compared to the original size. The use of this system in addition to aims to separate between one object with another also provides an opportunity on the existing object in order to be tangent to simulate freedom in designing. The calculation result in this test will only take 1 "right" condition that is when the volume of the object after tangent is higher than 90% of the original volume.

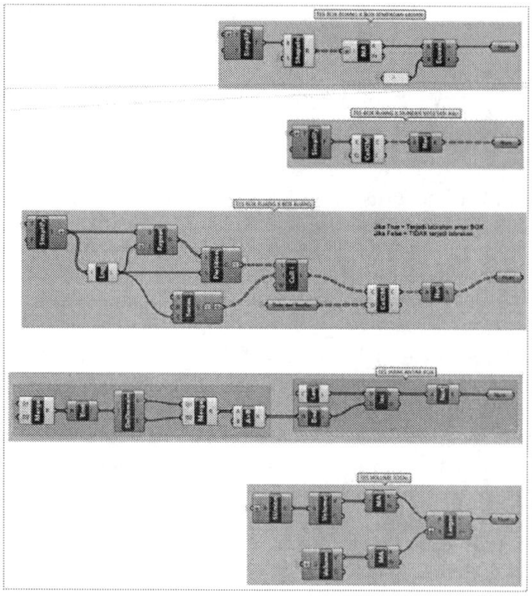

Fitness

Khabazi (2012) argues that the idea of finding solutions or optimizing products based on design problems that have been set is fascinating for various reasons. It may encourage the designer to think through the logic of each designed product, to formalize this logic with the language of mathematics and geometry, and to encourage constant movement toward better solutions by adding more criteria to evaluate a product (Khabazi, 2012).

The process of the algorithm making in this study also experienced a similar thing. For the final aim to be achieved, additional criteria and updates are done each time the algorithm is tested on the model. The results in the initial stage simulation were analyzed to find the weakness and improved the test in the next opportunity. The process continues to be repeated until the algorithm gets closer to the expected condition.

Based on the process mentioned earlier, the results of each test will be symbolized in the variable and arranged in an equation that will be used to assess fitness. Therefore, the symbol for the test results of the room module box for the bounding box = α; test result of room module box to cylinder of preexisting vegetation = b; test results of room module box toward the same room module box = c; test result of evaluation distance between boxes = d; and test result for total volume of the boxes = e.

In the compiled algorithm, the evolutionary solver is set to find the maximum value so that some values of the variables need to be adjusted in their fitness equations. Because variable a will be minimized and is a prerequisite that must be fulfilled, so it is placed as a multiplier. The test results of variable α must be = 0, which means that none of the space module boxes violate the border box. If this can be met, then it can be defined as true with value = 1. Thus, as long as the value of α is still not reached or false, the fitness value will be ignored because of the final result of calculation = 0. In such conditions, the solver will repeat the simulation and automatically search a better alternative fitness value.

The variable values b and c are the values that will be maximized, but still, allow the existence of the intersection between the space module boxes with the vegetation cylinders and the module boxes with itself. So both are used as the search for the optimal value to be divided. However, the value of these two variables will only be calculated if the value of variable e as multiplier get the maximum value, which is true. Same as variable α, as long as the value of e is still false, then the equation will be locked at 0, and the fitness result is deemed ineligible. This will force the solver to continue the search for better fitness value.

The variable d will act as a divider on this equation. Although the required value d is the maximum, that is all the distance will be true if it is at a predetermined number, but still allowed for any value that does not match the requirements. This value is adjusted to get the opposite value described above, then squared. This is done so that if there is only 1 value that false in the variable d, it will decrease the final result. That is, the true value is still maximized by minimizing the false value.

The number 1 is the constant required for the equation to be calculated. If the value of d = 0, which means the calculation is successfully maximized, and that value becomes the only divider, then the whole fitness will be an error because all that is divided by 0 yields undefined value. To avoid this, the number 1 as constant will be the only divider in that scenario, which allows fitness to be calculated.

Based on the formula, the ideal value obtained is 190 with the ideal value for variable b = 100 and variable c = 90. While the other variables because it works on Boolean Operation as the multiplier (minimum variables a = 1 and e = 1), or work to find the minimum value (d = 0), so the resulting value does not appear in the final result, but it will determine fitness success. The simulation will be stopped after it is considered close to the ideal number and before passing through the 140th generation. This is to shorten the time as the primary purpose of this research, which is to accelerate the creation of alternative designs.

In the first simulation with initial conditions both X and Y minimum, up to 3rd generation has obtained good condition with fitness score 41.0. After gaining a score of 164.0 in the 19th generation, the scores for the given alternatives rose gradually until the 102nd generation began to stagnate at a score of 185.0. In the 139th generation, the simulation was stopped because the fitness value was considered stagnant and no longer produced a new alternative.

In the second simulation with initial condition X = maximum and Y = minimum, what happens is up to 10th generation simulation has not found suitable condition with fitness score only 6.96. However, after a score of 171.0 in the 11th generation, the scores for the given alternatives improved considerably until the 76th generation began to stagnate at a score of 186.0. In the 105th generation, the simulation was stopped because the fitness score had stagnated and no new alternatives were completely different.

The third simulation starts with initial condition X = minimum and Y = maximum. In the 5th generation, the algorithm has been able to produce a soaring fitness score that is from 39.5 to 167.0. The score gradually increased until the 73rd generation got 186.0 and after that, the scores for the given alternatives began to stagnate. The simulation was stopped in the 111th generation when completely different new alternatives were not found.

The initial conditions for the fourth simulation are both X and Y maximum. The simulations show that up to the 32nd generation fitness scores are only at 6.96 and just 42.0 in the 33rd generation. This score goes up gradually until the 71st generation to 45.0, which then increased sharply to 178.0 in the 72nd generation. This score is improved but not significantly until the simulation is stopped because the algorithm does not provide a different alternative, i.e., in the 121st generation with a fitness score of 184.0.

In the four simulations, with different initial conditions, the working process also seems different. Some simulations seem quite quick to provide solutions with high scores in the early generations, some of which need up to tens of generations to score high. It occurs in addition to being affected by the initial conditions, also influenced by the way the evolutionary solver works.

As mentioned in the previous section, the evolutionary solver is an imitation of the evolutionary process (Rutten, 2010), i.e., the probability of a generation resulting from the parent gene pair. Simulations that have a low score up to several generations occur because the solver seems to be fascinated with a gene alternative with low value and does not seem to be able to switch to another possible gene that is unusual, although it may be better. The composition of the gene with the low value appears to be the best, even though it is not ideal yet. This kind of simulation condition will last until found better value.

The opposite condition also occurs when the early generations have produced high fitness scores. The solver will automatically eliminate genes that produce low value generations. Because high values are preferred, so in the next generation solver will work only on selected combinations of genes only. For the long term, the same score for several generations will be considered redundant, and fitness will appear stagnant.
This is a weakness of the evolutionary solver as Rutten (2010) has disclosed. Nevertheless, the ability of this solver in the creation of design alternatives quickly is the intended point in this study. The results obtained in this study may be different from other reviews because of the initial conditions and the nature of the solver that indeed allows differentiation to occur

Discussion on Alternative Design

The design alternatives generated from the al-gorithm are quite varied. From 4 initial condi-tions, 4 alternatives were selected which had the highest score and were in the last genera-tion before the simulation was stopped. The four alternatives obtained have their ad-vantages and disadvantages that can stimulate a better idea at a later stage.

Like an alternative is offered from the first sim-ulation. The pool area is on the lowest ground and directly adjacent to the back patio area which may encourage discussion to maximize visual to the lower side of the land. However, in this simulation plotting of the front porch and lounge areas looks unusual and separate from other buildings.

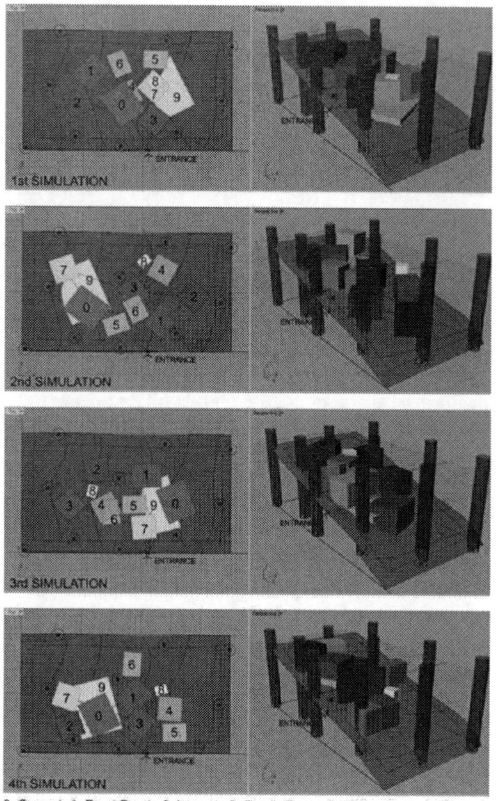

0: Carport; 1: Front Porch; 2: Lounge; 3: Family Room; 4: Dining Room; 5:Kitchen; 6: Main Bedroom; 7: Backside Porch; 8: Bathroom; 9: Swimming Pool

While on the second alternative, the location of the kitchen that prevents direct circulation from the entrance area to the carport can be consid-ered less suitable. Although the mass of the building looks more separated than other alter-natives, the location of the family room in the middle of building masses can be seen as a binder between several building masses.

The third alternative design allows discussion of design ideas by lowering the carport area and putting the pool in a higher position beside the carport. So get additional open area above the carport that can be utilized for the activity. Meanwhile, the kitchen area is the link between the swimming pool, the two terraces and the dining area. However, the location of the lounge and living room are quite separate from other buildings, and the combined dining and bedroom rooms may not be so desirable that they can be fixed again at the next design stage.

Meanwhile, the fourth alternative design allows the location of the bedroom that can maximize the atmosphere of a quiet room. The location of the unobstructed bedroom with another building enables visual optimization towards the lower site. However, placing the carport piled with swimming pools and adjacent to the two terraces can be considered unusual and less expected.

At a glance, design alternatives offered can be called meaningless because of various things. For example, through a more conservative per-spective, the space that should be able to con-nect directly, but on the alternatives offered is not interconnected, can be regarded as a bad design. Besides, discussions about spaces that are displayed separately in their respective boxes without visual attachment as a building may complicate security control or in organiz-ing utilities. Also, the simulated display in the form of a box can give rise to a very rigid de-sign impression and cannot accommodate oth-er geometric shapes.

In fact, alternative designs at the initial design stage can also be translated as a guide for de-veloping the model at a later stage. The dis-course is primarily about the ability of the algo-rithm to explore the ideas of arranging the composition at the initial design stage. So the alternative can be used as a discussion and consideration. Then, from this point on, the designer can enter into further design discus-sions to interpret and make adjustments so that the design results can be accepted.

Conclusions and Recommendations

Based on the explanation in the previous sec-tion, generative algorithms are considered ca-pable of producing various alternatives. How-ever, due to the breadth of the optimal design definition and the lack of ability to translate verbal ideas to mathematical, the resulting al-ternatives are still limited to the minimal. The addition of complementary variables and the deepening of the effort to convert the ideas into mathematical notation will be a good start-ing point for subsequent research.

Because of the lack of variables mentioned above, the design aesthetics produced are also very likely to be discussed and further criti-cized. The proposed alternative can be consid-ered reasonable and rational for some design-ers, and otherwise can even be regarded as fanciful and unusual for some others. Although opening an experimental discussion may open up the possibility of receiving the alternative, the issue of aesthetic perceptions gained from this simulation remains a significant concern that needs to be followed up.

The algorithm that is built is still on a simple level. This can be seen from the building used as a case study, i.e., a standard 1-storey house on sloped land. Because this research tries to generate design alternatives with an algorithmic approach in such cases, the discussion limits do not include the expansion of more complex alternative possibilities, such as the multi-story buildings or buildings with more mass. Hence, discussions to improve the ability of algorithms with higher levels of complexity can be done on other occasions.
The high number of genomes that became in-put on the simulation in this study may allow infinite design alternative. However, fitness de-signed to be strict and have a clear direction becomes important in finding the expected de-sign alternative. Nevertheless, both of these results in slower computer calculating perfor-mance.

Preparation of models and simulations in Grasshopper is considered more flexible re-garding modification. Of course, the discus-sion of the role of a designer in computational design can be more referring to two things, namely the designer as a human who builds algorithms and designers as humans who interpret alternative design results. This condition is to emphasize that computational design is the result of the human designer who in the pro-cess is assisted by the machine. This process aims to accelerate the creation of alternatives given primarily at the initial design stage.

References

Aliakseyeu, D., Martens, J., & Rauterberg, M. (2006). A computer support tool for the early stages of architectural design. Interacting with Computers 18, 4, 528-55.

Attia, S. (2011). State of the Art of Existing Early Design Simulation Tools for Net Zero Energy Buildings : A Comparison of Ten Tools. Louvain La Neuve, Belgium: Architecture et climat, Université catholique de Louvain.

Attia, S., Gratia, E., Herde, A. D., & Hensen, J. L. (2012). Simulation-Based Decision Support Tool for Early Stages of Zero-Energy Building Design. Energy and Buildings, 49. doi:10.1016/j.enbuild.2012.01.028

Difference Between Variable and Parameter. (2012, Mei 31). (DifferenceBetween.Com) Retrieved September 29, 2017, from http://www.differencebetween.com/difference-between-variable-and-vs-parameter/

Humppi, H. (2015). Algorithm-Aided Building Information Modelling; Connecting Alorithm-Aided Design and Object-Oriented Design. Tampere University of Technology.

Khabazi, Z. (2012). Generative Algorithms. Morphogenesism.

Lobos, D., & Donath, D. (2010). The problem of space layout in architecture: A survey and reflections. Arquitetura Revista, 6, 136–161.

Maldonado, M. P. (2014). Digital Recipes; A diagrammatic approach to digital design methodologies in undergraduate architecture studios. eCAADe. 1. Fusion-Data Integration at its best.

Nourian, P., Rezvani, S., & Sariyildiz, S. (2013). A SYNTACTIC ARCHITECTURAL DESIGN METHODOLOGY : Integrating real-time space syntax analysis in a configurative architectural design process. Ninth International Space Syntax Symposium.

Pranovich, S. (2004). Structural sketcher : a tool for supporting architects in early design. doi:10.6100/IR576067

Rutten, D. (2010). Evolutionary principles applied to problem solving. Retrieved Januari 17, 2017, from Grasshopper3D: http://www.grasshopper3d.com/profiles/blogs/evolutionary-principle

Terzidis, K. (2006). Algorithmic Architecture. Burlington, MA.: Architectural Press.

Turrin, M., Buelow, P. V., & Stouffs, R. (2011). Design explorations of performance driven geometry in architectural design using parametric modeling and genetic algorithms. Advanced Engineering Informatics, 25(4), 656–675.

Parametric Design in Timber Gridshell Tectonics

Syarifah Ismailiyah Al Athas[1,*]

[1]Universitas Islam Indonesia, Indonesia, Indonesia

Abstract. This paper begins with a simple proposition: rather than mimicking the geometric structures found in nature, perhaps the most effective modes of sustainable fabrication can be found throughunderstanding the nature of materials themselves. The material becomes a design parameter through the constraints of fabrication tools, limitations of material size, and most importantly the productivecapacity of material resistance a given material's capacity and tendencies to take shape, rather than cutting shape out of material. Gridshell structures provide an intriguing case study to pursue this proposition. Not only is there clear precedent in the form finding experiments of frei Otto and the institute for lightweight structures, but also the very nurbs based tools of current design practices developed from the ability of wood to bend. Taking the bent wood spline quite literally, gridshells provide a means that is at once formally expressive, structurally optimized, materially efficient, and quite simply a delight to experience. The the larger motivation of this work anticipates a parametric system linking the intrinsic material values of the gridshell tectonic with extrinsic criteria such as programmatic needs and environmental response. Through an applied case study of gridshells, the play between form and material is tested out through the author's own experimentation with gridshells and the pedagogical results of two gridshell studios.The goal of this research is to establish a give and take relationship between top down formal emphasis and a bottom-up material influence.

Keywords: parametric design, timber construction, gridshell construction

Introduction

The tension between form and material is neither a new one, nor is it necessarily a digital matter. As early as 1934, henri focillon suggests that matter imposes its own form upon form (focillon 1934). With today's digital fabrication tools, material is all too often seen as a homogenous substrate with the outputs of digital fabrication tools essentially creating physical representations of digitally derived form. Useful and at times inspiring, yet clearly this matter does not impose its own form upon form. Said another way, there is not an in- herent material feedback loop in the digital design process.

Material morphogenesis, as the name would suggest, is more than simply a material feedback loop in the design process, but suggests a material system that derives form. Contemporary fabrication tools are not simply about material output, but are a means to extend the material capacity, complexity, and variation of a material system. Research practices such as ocean north are a noticeable proponent of this approach. They note that "natural morphogenesis" develops its complex form from the systematic interaction between internal material capacities and external environmental influences and forces. In- spired from this natural morphogenesis, they situate their "computational morphogenesis" on the complex interrelationships between material capacities, manufacturing constraints and logics of assembly, and external micro-climatic conditions. In achim menges' teaching research at the architectural association's design research lab (aadrl), focus is placed on micro systems on material assemblies and their aggregate effect on larger structures. They emphasize a bottom-up approach, critiquing the "indulgence" of 1990's formalism (menges 2007).

Literature Review

The potentials for a bottom-up, material centric design approach are far more than sim- ply pragmatic issues, but have strong philosophical, conceptual, and perhaps even ethical values. Working backwards through a chain of philosophical influences from gilles deleuze and felix guattari to henri bergson, manual delanda, like focillon, is working against the philosophical indifference to material. This platonic indifference suggests that ideas are simply actualized by the addition of matter—a view that matter is an inert receptacle of ex- ternal ideas. Rather, the view delanda takes up is that material is an active participant in the genesis of form. He focuses on two central aspects of material qualities: their capacity and tendencies (delanda 2004). To oversimplify this case in the wood spline, its capaci- ties are its bending strength and its tendencies develop from its axis in bending (bi-axial vs. Uni-axial). These material capacities and tendencies are material constraints. From a material morphogenetic point of view, constraints are not simply limitations, but become pro- ductive by shaping design parameters. Quite literally, material constraints ground design. Viewing constraints as productive can have a powerful affect on design concepts. After all, attentiveness to materials directs ones attention beyond the symbolic and representational nature of design and instead prioritizes the

* Corresponding author: syarifah.alathas@uii.ac.id

environmental, kinesthetic and haptic experi- ence of architecture. Consequently, the primary visual interface of digital media may be balanced by the material influence through taking the wood spline quite literally. Through this, there is a connection between material values and digital parameters. All too often the "parameters" of digitally derived work are abstract points—if abstracted at all—from which form is instrumentalized. These 'datascapes,' as reiser + umemoto contend, are "an un- fortunate consequence of design in the semantic mode" (reiser 2006: 217). Suggesting that values and parameters are not, in the end, synonymous, provides an opportunity to productively connect the two rather than conflate them as one and the same thing. While many may argue that the free-form fascination in contemporary architecture is a consequence of nurbs based software, an overlooked historical fact is that material con- straints and parametric flexibility were the foundations of the system. While pierre bezier, among many other mathematician/engineers, is the father of today's nurbs systems, the motivation to develop the system was based on a coupling between a simple parametric structure in response to the physical constraints of computer-aided manufacture. One year before his death, pierre bezier recounts: "to sum up the basic ideas of the system, it can be said that it came from the ability to work, think, and react in the rigid cartesian world of machine tools and, at the same time, in the more flexible, n-dimensional parametric world" (bezier 1998). Of course, the very physical spline that bezier abstracted into what we now know of as the bezier curve, contained both the material resistance of the wood spline and the geometric constraints of the weighted ducks. Ironically, in abstracting the geometry of the constraints, the materials capacities and tendencies were abstracted out of the sys- tem. Simply put, material constraints are not built into this digital system. In exchange for these lack of material constraints, and the restriction to plane curves (2d curves) and fixed spline lengths, we get an incredibly adaptable flexible system built upon space curves (3d curves) and infinitely extensible curves. In architecture this has given a new formal renais- sance, and the blob as a new typology. However, the accessibility of nurbs based soft- ware long before the accessibility of today's digital fabrication tools has only extended the philosophical ideas that material is a receptacle of externally driven forms. Consequently, taming this wild geometry is so frequently done through slicing and egg crating that these techniques have become synonymous with digital fabrication.

This research takes the material spline quite literally by looking at its capacity to take shape, rather than cutting shape out of material, as well as the tendencies of the cross-sec- tional area of the spline to bend in a very constrained manner. While clearly a conservative and very constrained approach, the result is a give-and-take relationship between material and geometry that is easily constructible, materially efficient, and structurally expressive architecture. In the end, the goal is not to suggest gridshells are the answer, but rather, that the precedent of gridshells may be a pedagogical tool developing material values in a digi- tal design culture.

Gridshell Precedent

The first examples of gridshells were developed through the partnership of architect frei otto and a young ted happold, then at ove arup, exemplified by the timber gridshell at mannheim in germany built in 1975. Although very few gridshells have been built, there has been a resurgence of these structures in the last eight years through shigeru ban's japanese pavilion and jacques herzog's expodach for expo 2000 in hanover, helsinki university of technology wood studio's timber bubble at the helsinki zoo (2003), edward cullinan's downland gridshell (2002), and most recently glenhowells architects' savill building (2006). Typically these structures are fabricated from a flat mat of straight laths, or paper tubes in ban's pavillion, and then raised and/or lowered into shape. The expodach is a prefabricated lamella approach and the timber bubble is constructed piece by piece of pre- formed laminated laths.

Structurally, gridshells have the properties of a structural shell, which gains its strength and stiffness through curvature, with a shell formed from double curvature as the most ef- ficient in terms of minimum use of material. Gridshells are typically constructed through a criss-crossed pattern of straight laths. Synclastic surfaces are curved only in tension or only in compression. For example, an arch is formed using only compression, and a dome is a rotated arch. Geodesic domes are examples of synclastic shapes. Anticlastic surfaces, or saddle shapes, have tension forces in one direction and compressive forces in the other stabilized by the tension forces. The significance of anticlastic surfaces is that they are more flexible in their formal morphology and structurally more efficient from the balance between tension and compression forces.

Gridshells Applied

This applied gridshell research developed over the course of about one year and included teaching two gridshell studios. Presented first are the techniques and challenges of devel- oping gridshells with conventional digital tools, and then briefly presented are the varied approaches taken in studio work. Despite the curving surfaces of gridshells, they are typi- cally constrained to straight barrel vault warehouse like spaces as a result of the flat matt technique of raising or lower lattice gridshells. To challenge this plan-oriented constraint, the base surface for this experiment developed from a 90-degree turn through four bound- ary curves and one center curve.

Lath Pattern: Projected, Applied, Geodesic

A pattern can be projected onto a surface, however this is limited as the pattern can only be projected in one direction. Another method is to simply "apply curves" to a nurbs sur- face. While at first it appears this is an easy method to depart from the opposing uv logic of nurbs surfaces, in fact, the pattern is stretched proportionally across the surface in because each node is located at the surface normal, the node connection is square to the flattened unrolled lathe. Consequently, when the laths are put in place, the node holes

https://doi.org/10.1051/shsconf/20184105004

will align only when the laths take their correct position as a kind of pin registration. Through a trial and error process, the applied grid pattern was developed such that an unrolled lath could fit on a 2x8 board of clear oregon pine. A section model was built as a proto- type testing the accuracy of these developed surfaces and the pin registration at the nodal points (figures six and seven). While the model was successful and easily assembled, cut- ting crescent shapes from solid boards of clear oregon pine, only utilizing about 60% of the material, was a contradiction in the goals of the project to do more with less.

Surface Relaxation: Form and Geodesic Pattern Revisited

In the form-finding procedures of frei otto, the hanging chain models find a minimum en- ergy form—their relaxed state. In contemporary examples such as the downland gridshell, a software based dynamic relaxation technique was employed. Connecting the minimum energy geodesic curve with a minimum energy (relaxed) surface seamed a plausible ap- proach. Through a surface relaxation plug-in (www.reconstructivism.net), the base surface was relaxed with the four boundary curves constrained. The grid pattern was applied, and from these same endpoints a geodesic net was developed.

In the relaxed surface, a geodesic net is evenly distributed, and was therefore able to be developed from straight laths. This simple example proves a signifi- cant point: a give-and-take relationship between top-down formal emphasis and a bottom- up material influence is necessary.

Gridshell Studios

The gridshell studios were given as a context to teach a bottom-up materials first approach in developing complex form. While the relevance of this studio was conceived at a school that was digital savvy, the studios were actually taught at two different schools with upper

Division students who had very marginal, if any, digital modeling skills. The studios began through graphic case studies of built gridshells, as well as building physical prototypes of these case studies to feel the forces at play in these structures. Physically testing material to its limits was of interest to these students, yet their digital skills lacked the geometric attention that the above approaches necessitated. From these case studies, the students were welcome to critique the gridshell's appropriateness for this project, as well as to de- velop new approaches to gridshell inspired designs. Four approaches are illustrated here.

Gridshell Mold

The most conservative though straightforward approach was to develop a 3d mold and then apply laths individually over the mold. This was the first time for this graduate student to use a cnc router, and he was noticeably excited to see this surface emerge from a block of foam. After seeing and touching this surface for flat spots, he quickly located areas for revision, though regretfully he felt it was too late in the term to alter and re-mill his mold. He then meticulously glued a four-layer lath over the mold using t-pins to hold the laths in place.

An alternate mold approach, inspired by the pre-fabricated shells of the expodach, em- ployed a modular mold using overlapping rings instead of straight laths. The ring pattern was applied to the surface and unrolled as developable surfaces. Each ring was laminated from two paper strips to hold their circular shape, and then connected at the quadrant points through overlapping bridal joints.

Standardized Unit / Variable Shim Approach

One team greatly resisted that each piece would be unique and pursued an approach that used a standardized unit with the space between the units providing the variation. A series of laser cut boxes with variable shims between the boxes was used to physically model the various configurations this approach could take. Ironically this straight- forward approach is exceedingly difficult to digitally model with conventional tools. While their digital representations were approximations of this approach, they also built a full- scale mock-up to test this modular approach.

Periodic Laminated Approach

Perhaps the most sophisticated approach was developed through a periodic surface and a hexagonal tessellated pattern. Lacking sophisticated skills and tools to track a series of unique pieces, the periodic surface and pattern configuration created a unique, two directional, waveform from only 12 unique pieces. Like the prefabricated lamella structure of the expodach, prefabricating a modular lamella gridshell surface over these larger spans could attain further enclosure.

Conclusion

The intent of this applied research is to propose an economy of means through under- standing the nature of materials rather than mimicking the aesthetics of natural systems. While digital fabrication tools are a welcome and significant addition to the architects' tool- set, the 'digital' aspect of these tools typically suggest form first only much later to con- sider material, if at all. Gridshells inverse this relationship suggesting a bottom-up mate- rials first approach to form finding. However, the matt technique of developing gridshells yields a very constrained formal morphology, and while elegant, has limited application. The gridshell tectonic developed here intends to balance the geometric constraint based on a physical laths capacity and tendencies to take shape with a more globally flexible form suggesting a give-and-take relationship between a bottom- up and top-down formal orga- nization. It also serves to critique the proportional distribution of surface sub-division and

 https://doi.org/10.1051/shsconf/20184105004

applied patterns in parametric tools which are still tied to the non-material nurbs surface, in favor of approaches that use geodesic or minimal energy curves derived from a materi- als capacities and tendencies to balance form and material.

References

Achim, Menges. (2007). "Computational Morphogenesis: Integral Form Generation And Materialization Processes," Proceedings Of The 3rd International Ascaad Conference On Em'body'ing Virtual Architecture.

Bezier, Pierre. (1998). "A View Of The Cad/Cam Development Period," Ieee Annals Of The History Of Computing 20 (2): 37-40.

Delanda, Manuel. (2004). "Material Complexity," In Neil Leach, David Turnbull, And Chris Williams, Digital Tectonics. London: Wiley-Academy Press.

Focillon, Henri. (1934/1992). The Life Of Forms In Art. New York: Zone Books.

Harris, Richard, John Romer, Oliver Kelly, And Stephen Johnson. "Design And Construction Of The Downland Gridshell," Building Research And Information 6 (31): 427-454.

Reiser, Jesse. (2006). Atlas Of Novel Tectonics. New York: Princeton Architectural Press.

Virtual Reality As A Spatial Experience For Architecture Design: A Study of Effectiveness for Architecture Students

Luhur Sapto Pamungkas[1], *Cinthyaningtyas* Meytasari[1], *Hendro* Trieddiantoro[1,*]

[1]Technology University of Yogyakarta, Indonesia

Abstract. Studios. This ability gained through visual design thinking. The spatial experience honed by three dimensional thinking from the medium diversity. The spatial experience learned through a room layout, proportion, and composition. This research used an experimental method and the primary data obtained by a "Likert" scale questionnaire. The Respondents are 50 students of the Architectural Design Studio. Moreover, the analysis focuses on the VR for spatial experience. The result was a descriptive explanation of the effectiveness of Virtual Reality for a spatial experience of architecture students at Technology University of Yogyakarta.

Keywords: Design Cognition, Spatial Experience, Virtual Reality

Background

There was an agreement that the design was one of a product from problem-solving, which consist of a cognitive framework of analysis, synthesis, and evaluation that carried out by the designer. Hubers (2007) describes the goal of a designer was making a good design, such as a space with the good quality of function, aesthetics, structure, and cost calculation.

The architects had a responsibility to provide a good design, just as the architecture students were required to understand the design as well as how to design. Moreover, the architecture students guided by visual design thinking with several cases of designs. For instance, this was a learning activity that honed through me-dium diversity such as scale models as three-dimensional medium and floor plans as a two-dimensional medium. Technology provides a variety of mediums in architectural design, such as Virtual Reality that began to be known as simulation tools. Afterwards, the development of Virtual Reality already used in many fields such as surgery simulation, military strategy, and engineering. According to the background, our main objective is to evaluate the effective-ness of Virtual Reality as a spatial experience for architecture.

Architecture Design

Design is one of a product of logical solution that in Zande's (2006) view could be formed as crafted model or drawing also layout space in Hubers's (2007) that obtained from the design process, which consist of a cognitive frame-work of analysis, synthesis, and evaluation. Both Abdelhameed (2004) and Hubers (2007) added designing is problem-solving in a crea-tive way that involving complex cognitive tasks with a cyclical process. Moreover, the design-ers will get comments on their designs or ideas when those are presented or discussed. The new decision appeared as a reaction of evalua-tion, in other words, the architect proposed a tentative analysis and explore the possible consequences as synthesis, this argument similarly described in the book by Gross (1985, p. 136) that the decision selected on the rea-sonable cause, but it can be changed later.

Spatial Experiences And Simulation

The spatial experience is a subjective assess-ment that emerged through visual perception. According to Mihelj et al (2014), some of the elements that make up the human visual per-ception are the perception of light, the percep-tion of color, and the perception of depth.

Bill Scott (1988) said that medium was an in-termediate that used in the delivery of infor-mation from the speaker to the audience. The purpose is the efficiency of information or message. The architectural presentations tend to use a media in the form of drawings and models. The delivery of information not only through verbal language, but also non-verbal. He added in his book "The Skills of Communi-cating" that the obstacles in communicating caused by the listener may not hear and may not understand what is the speaker means, also listener may not approve the information from the speaker that cause lack of feedback in communication.

Banks et al (2005) said that simulation is the imitation of the operation of a real-world pro-cess or system over time. Many researchers agree that Virtual Reality (VR) is the term, which can be referred to as immersive multimedia or computer-simulated life, replicates an environ-ment that simulates the physical presence of the real world or imagined worlds. Banks et al (2005) added that by changing variables in the simulation, the predictions may be made about the behavior of the system.

* Corresponding author: hendro.trieddiantoro@gmail.com

Methodology

Respondents are 50 people from Architecture Design Studio. All of the respondents are archi-tecture students in their 2nd year of study. Likert scale questionnaires were given to the respondent to make an assessment of Virtual Reality.

VR Simulation Design

The VR simulation design is an office building from the Architecture Design Studio project. The design was done by the student with Sketchup software. Moreover, the VR images were rendered using Unity3D and Kubity. The VR tools in this research are Oculus Rift, a portable Head Mounted Display for PC build and VR BOX with a smartphone as a screen.

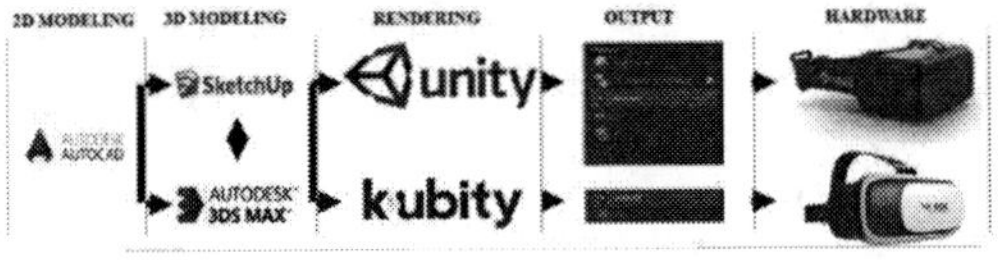

Fig. 1. VR Simulation Design Process.

The introduction of VR simulation was carried out in the computer laboratory of the Faculty of Science and Engineering at the Technology University of Yogyakarta. VR simulation training activities were attended by 15 students of architecture students. This 1.5-hour activity resulted in several VR simulations.

Fig. 2. VR Simulation Training.

The VR simulation design is the development of 2nd-year student in Architectural Design Studio. The design is an 8-story office building that has been created by students was developed during VR simulation training.

The VR simulation includes exterior and interior visualizations. Interior simulation such as office area, workspace layout, and food court. VR simulation that made by students are then assessed by respondents using a Likert scale questionnaire.

Questionnaire Design

The instruments for measuring the spatial expe-rience is a 7 range of Likert scale questionnaire. The questionnaire contains of 8 positive state-ments regarding to the ability of the media to explain the spatial experience. Each media rep-resented from Strongly-Agree to Strongly-Disagree.

Table 1. Likert Scale Questionnaire Design.

Questionnaire								
No.	Pernyataan	Answers						
		SA	A	AS	N	DS	D	SD
A1.	I'm sure to know the size of space in the room through Virtual Reality simulation							
A2.	I can feel the human scale in the room through Virtual Reality simulation							
A3.	I'm sure to know the type of material in the room through Virtual Reality simulation							
A4.	I can clearly see the lighting in the room through Virtual Reality simulation							
		VG	G	SG	Un	SB	B	VB
B5.	Representation or quality of room size through Virtual Reality simulation							
B6.	Representation or quality of human scale through Virtual Reality Simulation							
B7.	Representation or quality of material through Virtual Reality simulation							
B8.	Representation or quality of lighting through Virtual Reality simulation							
Virtual Reality Write down your feedback for Virtual Reality technology								
SA : Strongly Agree A : Agree AS : Agree Somewhat N : Netral DS : Disagree Somewhat D : Disagree SD : Strongly Disagree		VG : Very Good G : Good SG : Somewhat Good Un : Undecided SB : Somewhat Bad B : Bad VB : Very Bad						

Analysis And Result

Questionnaires are divided into statement A and B. Statement B is a confirmation of Statement A. The end of the questionnaire is a feedback or comment. The conversion of 7 range Likert, from strongly agree to strongly disagree were range from -3 to 3. The results displayed into graphical diagrams and describe in qualitative form.

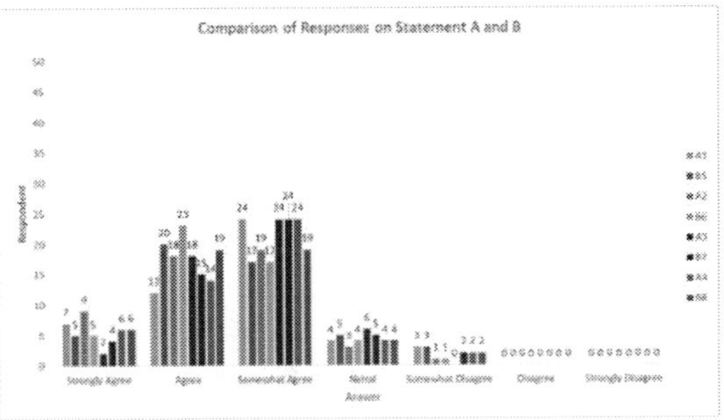

Fig. 3. Comparison of Responses on Statement A and B.

Figure 3 illustrates the comparison of responses on statement A and B. The change of voter amount in the answers was insignificantly different. The insignificant changing of voter value between statement A and B indicates the validity of the answer.

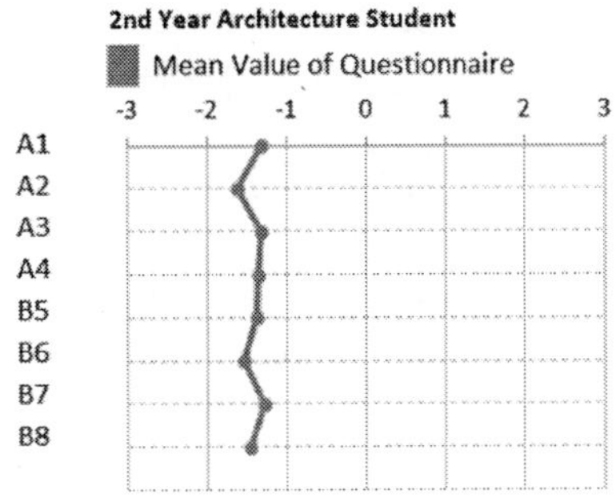

Fig. 4. Mean Value of Questionnaire.

Figure 4 shows the average results of the as-sessment of the questionnaire. Diagram reports that the average results are in the range of -1 and -2. This means that the results of the as-sessment of the VR simulation as a learning experience are in a good range.

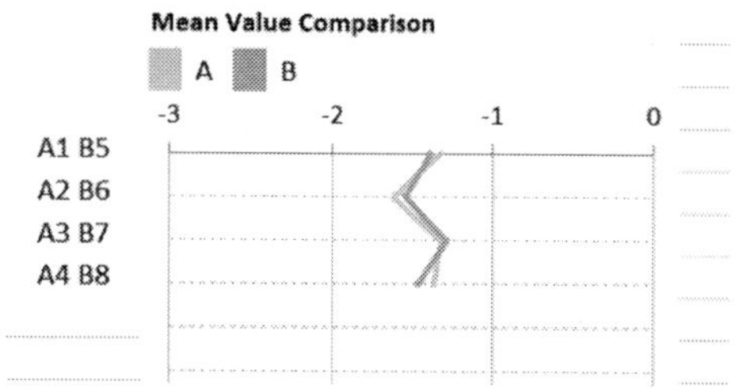

Fig. 5. Mean Value Comparison.

The diagram reports the changing mean value of statement A and B. The mean value of statement A number 1 or called A1 is -1.32 and statement B number 1 or called B1 is -1.38. The mean value of statement A2 is -1.62 and B2 is -1.54. Mean value of A3 is -1.32 and B3 -1.28. Mean value A4 is -1.36 and B4 is -1.46. The insignificant changing of mean value between statement A and B indicates the validity of the answer. So that this agrees the VR simulation has a good impact on a spatial experience of architecture design for 2nd year architecture students.

Feedback analysis was done by classifying the comments into several categories. Feedback categorized into 3, such as hardware, software, and personal perception.

Table 2. Feedback Response.

(hardware)		(personal)		(software)	
(+) Feeling Comfortable	(-) Uncomforthale, dizzy after using VR tool	(+) Bring up the ability to feel the scale of space, interested in VR	(-) not interested in VR, not understanding VR	(+) Good Quality of visual and lighting	(-) Poor Quality of visual and lighting
2	9	18	0	21	3

The hardware category is the assessment of the VR simulation tool, the software category is the assessment of the quality of VR simulation, and the personal category is the personal perception or response to VR technology for an architecture student. The feedbacks are de-scriptive and then transform into the chart form.

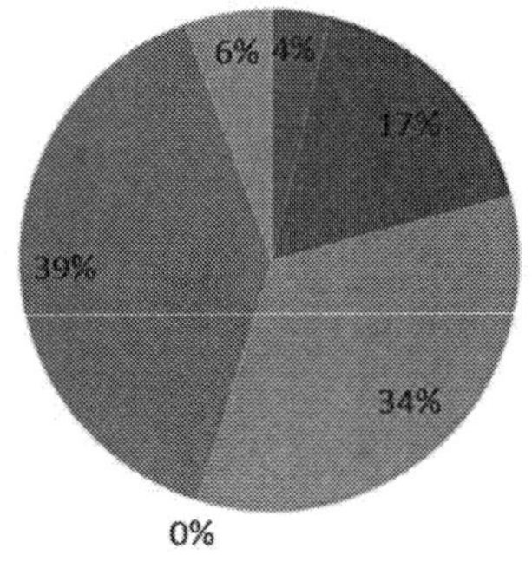

Fig. 6. Percentage Of VR Feedback.

Figure 6 shows the percentage of feedback. The highest percentage is 39% of feedback on the VR software, which response to the good quality of visual, and lighting from the VR simu-lation. The next second highest percentage is 34% at the personal perception, which VR simulation bring up the ability to feel the scale of space and they interested to learn more about VR. But as many as 17% of respondents said that they are feeling uncomfortable and dizzy after using VR tools simulation.

Conclusion

The questionnaire analysis showed that the changing of the mean value and voter amount from statement A and B were insignificantly changed. So that this agrees the VR simulation has a good impact on a spatial experience of office building design for 2nd-year architecture students.
Furthermore, the effectiveness of VR as a spatial experience for architecture design was shown from 39% of the 2nd year architecture students agreed with good quality of visual and lighting from The Kubity application as VR software. Followed with 34% of 2nd-year architecture students states that VR simulation brings up the ability to feel the scale of space and they are interested to learn more about VR. However, 17% of respondents felt uncomforta-ble and dizzy after using VR tools.

Suggestion for the next research and for other fellow researcher is in the preparation phase of VR simulation. Similarly with hardware and software selection tools for simulation. Virtual Reality simulation needs an optimization for hardware and software in architecture design.

Acknowledgements

Acknowledgment is addressed to Ristekdikti and Department of Architecture, Faculty of Science and Technology, Technology Universi-ty of Yogyakarta.

References

Abdelhameed, W. (2004). Visual Design Thinking in the Design Process as Impacted by Digital. Architecture in the Network Society [22nd eCAADe Conference Proceedings], , 90-94.

Burden, E. (1992). Design Presentation: Creating Marketing and Project Proposal Second Edition. New York: Me Graw Hill Inc.

Campbell, D. A., & Wells, M. (2003). A Critique of Virtual Reality in the Architectural. USA: University of Washington.

Ching, F. D. (1996). Interior Design Illustrated. New York: Van Nostrad Reinhold.

Djaali. (2008). Psikologi Pendidikan. Jakarta: PT. Bumi Aksara.

Gross, M. (1985). Design as exploring constraints. Cambridge: MIT.

Hubers, J. (2007). COLlaborative Architectural Design. PhD Thesis, Delft University of Technology, Faculty of Architecture, Netherlands.

https://doi.org/10.1051/shsconf/20184105005

Mihelj, M., Novak, D., & Beguš, S. (2014). Virtual Reality Technology and Applications (Vol. 68). (S. G. Tzafestas, Ed.) New York, London: Springer.

Paranandi, M., & Sarawgi, T. (2002). Virtual Reality In Architecture: Enabling Possibilities. CAADRIA 2002 [Proceedings of the 7th International Conference on Computer Aided Architectural Design Research in Asia / ISBN 983-2473-42-X] Cyberjaya (Malaysia) 18–20 April 2002, 309-316.

Pop, D. (2013). Space Perception and Its Implication in Architectural Design. Acta Technica Napocensis: Civil Engineering & Architecture Vol. 56, No. 2, 211-221.

Scott, B. (1988). The Skill of Communicating. England: Wildwood House.

Sekuler, R., & Blake, R. (2002). Perception. New York: McGraw Hill Book.

Steuer, J. (1993, October). Defining Virtual Reality: Dimensions Determining Telepresence. Social Reponses to Communication Technologies, 73-93.

Sugiyono. (2011). Metode Penelitian Kuantitatif, Kualitatif, dan R&D. Bandung: Alfabeta.

Zeinmeister, A. (2015). Spatial Turn – Perception in Architecture. In C. Perren , & M. Mlecek, PERCEPTION in Architecture: HERE and NOW (p. 11). UK: Cambridge Scholars Publishing.

Comparative Study of Daylighting Calculation Methods

Ariani Mandala[1,*] and *Amirani* Ritva Santoso[1]

[1]Lecturer at Department of Architecture, Universitas Katolik Parahyangan, Indonesia.

Abstract. The aim of this study is to assess five daylighting calculation method commonly used in architectural study. The methods used include hand calculation methods (SNI/DPMB method and BRE Daylighting Protractors), scale models studied in an artificial sky simulator and computer programs using Dialux and Velux lighting software. The test room is conditioned by the uniform sky conditions, simple room geometry with variations of the room reflectance (black, grey, and white color). The analyses compared the result (including daylight factor, illumination, and coefficient of uniformity value) and examines the similarity and contrast the result different. The color variations trial is used to analyses the internally reflection factor contribution to the result.

Keywords: daylighting methods, daylight factor, room reflection

Introduction

Daylighting has a great potential to do energy saving in buildings. Daylighting calculation is becoming the valuable need for the designer to optimize its potential in architectural design. Hopkinson (1966) cited by Acosta (2011) in his book Daylighting reveals the complexity of the calculation procedure in daylighting. Currently, a designer has many options in using the daylighting calculation methods. Until now, hand-calculation method with graphics and diagrams still used because of its versatility in use. A scale model simulation still remains to become one of the choice because it is a familiar method used in architectural design and also provides a representative object for presentation media. The emergence of a variety of specialized software for lighting (daylight and artificial light) allows designers to have a more complete and complex analysis of the lighting condition through a digital simulation. A wide choice of these calculation methods helps designers to choose the method that suits the design needs.

Different calculation methods also give a difference in the result. This is caused by many factors, each methods have their capable of accommodating the complexity daylighting calculation factor and the difference of calculation algorithm. This research aimed to evaluate several methods of calculation of dayighting that are commonly used in architectural education. The methods used include hand method calculation with SNI/DPMB method, calculation with a BRE protractor (Building Research Establishment), the scale model simulation on artificial sky and simulation software using Dialux and Velux program.

This study is limiting the daylighting calculation parameter to avoid the overly variation result and produce preciser calculation. A room model used in this research is a simple geometry with one side opening (window). Simulation is done in the overcast sky conditions to avoid the complexity of direct sunlight effect to the result. A room reflectance trial also simulates in this research using a different color in the room surface (black, gray, and white). Analysis of the comparative form of daylighting conditions includes the Daylight Factor value, illumination, and light distribution (pattern and uniformity). The room surface color trial will be analyzed to reveal its role in daylighting performance.

This research was conducted with several aims. First is to assess and compare the various methods of calculation that mentioned above with analyzing the similarities and differences of daylighting calculation result using daylight factor (DF) value and coefficient of uniformity (CU) value of those methods. Second is to study the room reflectance contribution to daylighting performance. This research is expected to provide the following benefits. The findings can be a learning material to optimizing daylighting design in architectural study and also becomes a valuable input for updating the national lighting standard in Indonesia. The analyses of the reflection factor contribution can be developed to be a practical guide to optimize the daylighting potential in the building design.

Daylighting Calculation Methods

Factors that affect the appearance in visual activities are types of activity, lighting conditions, and the observer's condition (Lechner, 2007:386). The level of illumination is one of the factors in lighting conditions that most frequently observed and easily measured. With the daylighting design, illumination is often associated with the concept of daylight factor. DF is the ratio of the illumination in the inside (room) and the outside counted in percentage. The measurement

* Corresponding author: ariani.mandala@unpar.ac.id

conditioned in a cloudy/overcast sky (uniform illumination) to minimize lighting level fluctuations caused by direct sunlight in different dates and times. In daylighting design, uniformity of lighting distribution is difficult to achieve. Artificial lighting source can easily be organized and modified. Daylighting source (an opening / window that brings the sunlight) has limitations so that various strategies need to be applied to obtain the uniform light distribution. Coefficient of Uniformity/CU is used to rate the light distribution. CU is the ratio of minimum and maximum illumination in a room.

There are three main components in the daylighting calculation, the Sky Component (SC) the Externally Reflection Component (ERC) and the Internally Reflection Component (IRC). This study calculates SC and IRC components into the calculation. ERC is not included into this trial.

The hand method calculation used in the study is SNI and BRE methods. The SNI calculation method refers to the national standard that applies in Indonesia. SNI (Standar Nasional Indonesia) is issued by the national standardization institution in the form of SNI 03-2396-2001 about the daylighting design system in a building. SNI method only accommodates sky component without internally and externally component. BRE (Building Research Establishment), which was formerly called BRS developed series protractor to read the percentage value of sky component. The BRE calculation method can calculate the SC and the ERC. BRE also have a Nomogram that was able to complete the calculation with components of the IRC. This research using the uniform sky protractor series to calculate SC and Nomogram to calculate the components of the IRC.

A scale model measurement is done in an artificial sky simulator. Testing on an artificial sky allows to get the more consistent results compared to testing in the real sky. A study by Acosta was done in 2011 when he compared the daylighting calculation methods in artificial sky and lighting software (Lightscape, Radiance, Lumen Micro, Ecotec, and Dialux). As a result, the value of DF from measurements using artificial sky is relatively larger than the software. While the value of the CU relative approach. In this study, testing done in the artificial sky simulator in laboratory of building physics belonging to Parahyangan Catholic University. The room has hexagonal-shaped in white paint, roof dome shape and uses mirrors around the room in. 18 pcs of white fluorescent lamps used as a representation of the skylight. Three models were made in 1:20 scale. The inside of each model painted with black, gray, and white for testing with the room reflection trial. Two digital lightmeter was used to measure the illumination inside and outside the building at the same time.

Fig. 1. Artificial sky trial simulator.

Software used in this research are Dialux and Velux. These programs used with consideration of ease of use. Iversen (2013) revealed that Dialux and Velux is a program with an easy to learn compared to other programs such as Radiance, Desktop Radiance, and Daysim. Another consideration is that both programs do not require a license and can be easily downloaded via the official website. The fundamental difference of these two software is the calculation methods / rendering type. Velux using photon mapping (bi-directional ray tracing) while Dialux using the radiosity rendering to calculate the result. Photon mapping calculation is able to optimize the simulation so that it can be used for more complex lighting scenarios, such as light pipes, and faster than the previous way backward ray tracing. While radiosity renderation is excellent for the application of translucent material with a spreads light source and also efficient to simulate the light distribution in space. But it has limitations in simulating natural lighting and can only be used for evaluation of space with simple shapes (Iversen, 2013).

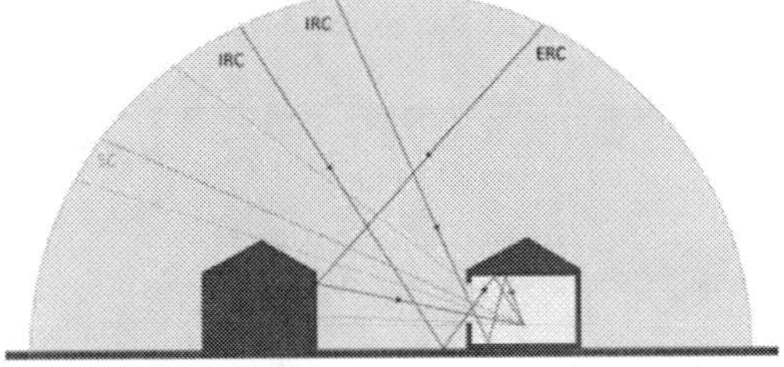

Fig. 2. Split flux formula's components.
Source: Iversen, 2013.

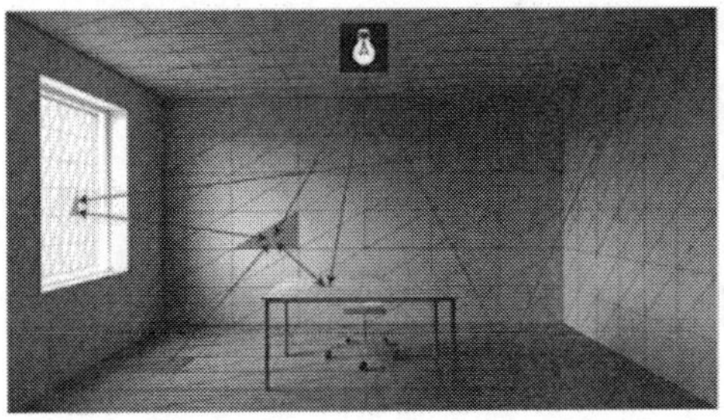

Fig. 3. Radiosity calculation method.
Source: Iversen, 201.

 https://doi.org/10.1051/shsconf/20184106001

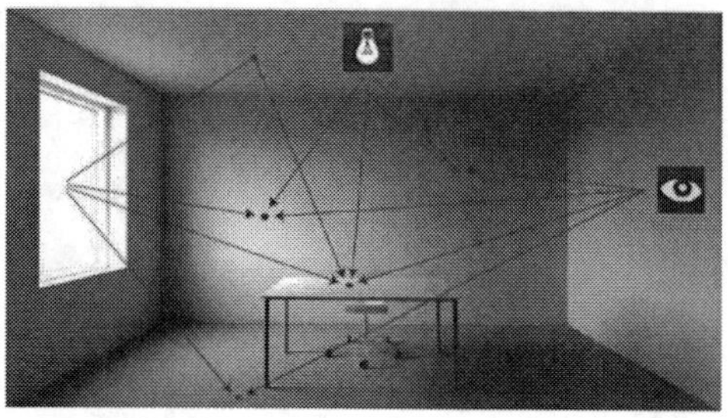

Fig. 4. Photton mapping calculation method.
Source: Iversen, 2013.

Research Methods

This research will compare calculations result of DF and CU by five daylighting calculation methods. Each method has limitations in the accommodating the daylight parameter. This study will limit the parameters required so that it can be accommodated by all the calculation methods. We used a square-shaped room model of 8 x 8 x 4.5 meters with an opening on one of its sides (Figure 4). Three models were made with different color surface for room reflection trial. We used white color (reflection factor 70%), gray (reflection factor 30%) and black (reflection factor 4%). The opening of 3,5 x 6,30 meters is centered on one side and has no filter so the light can enter the room completely.

There were 12 calculation points (CP) with 80 cm height, spread in the room in a grid pattern. The maximum illumination level is on CP 2, and the minimum illumination level is on CP 10 or CP 12 (because of its symmetrical shape). The measurement points position can be seen in Figure 4.

Table 1. Daylighting Calculation Parameters.

Factors			Varian
Sky Condition	Time		June, 21st 2017
	Geographical position		Bandung City (6^0 S)
	Sky component		Overcast
Building	Room geometry		Room orientation: North opening Room shape: square Room dimensions: 8 m x 8 m
	Window	Window geometry	Window orientation: facing north Position: center Qty: 1 pcs Shape: rectangle Dimension: 6.3 x 3.15 m
		Material	Transparency: 100% Reflection & texture: none
	Reflection factor	Internally reflection	Ceiling, wall, floor surface color: black (ρ =4%), gray (ρ =30%) and white (ρ =70%)
		Externally reflection	None
	Artificial lighting contribution		None
	Shading	Internal	None
		External	Wall thickness: 15 cm
	Maintenance		None
	Pollution		None
	Furniture		None

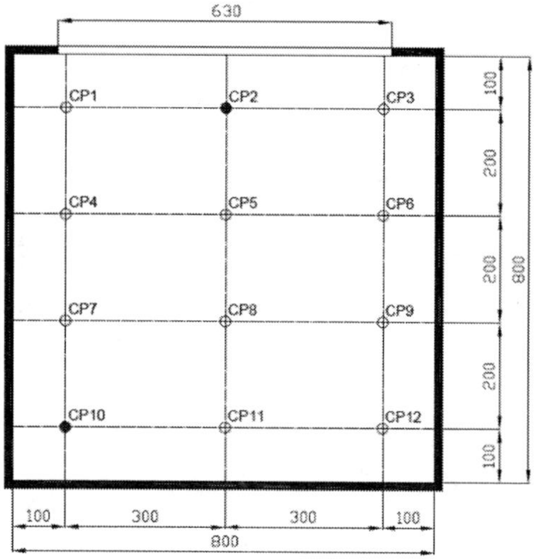

Fig. 4. Calculation points plan.

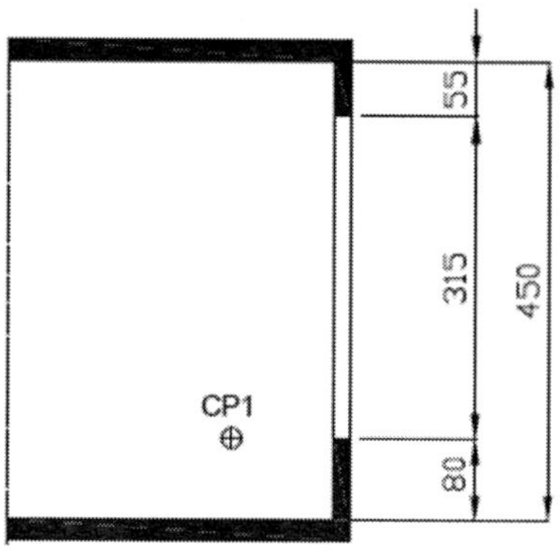

Fig. 5. Models' section.

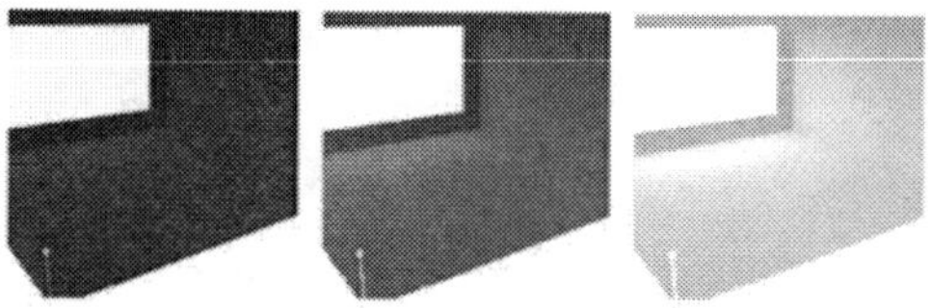

Fig. 6. Model simulations with color variations of the surface material.

Table 2. Calculation parameter of each method.

Hand calculation: SNI		
Calculation type		SC & E
Sky condition		Uniform (10.000 Lux)
Room material	Internally reflection	Black (ϱ =0%)
Window		No filter
Hand calculation: BRE		
Calculation type		DF & E
Sky condition		Uniform (10.000 Lux)
Protractor		Uniform sky-vertical glazing
Room material	Internally reflection	ϱ = 4% , 30%, 70%
Artificial sky simulator		
Calculation type		DF & E
Sky condition		Uniform sky
Room material	Internally reflection	Models are made with different surface color (Black, gray, White)
Window		No filter
Scale model		1 : 20
Model material		Gray cardboard (+paint)
Calculating Instrument		Light meter digital
Software: Dialux		
Calculation type		DF & E
Sky model		Overcast sky
Calculation options		Very accurate
Room material	Internally reflection	ϱ = 4%, 30%, 70%
Window	Transmission factor	100%
	Transparency factor	100%
	Reflection factor	0
	Pollution factor	None
	Framing factor	1.00
Maintenance		1.00
Software: Velux		
Calculation type		DF & E
Sky condition		Overcast sky
Render resolution		High
Room material	Internally reflection	ϱ = 4%, 30%, 70%
Window	Transmittance	1.00

Daylighting Calculation Result

DF and CU Comparison

Table 3. Light Pattern Comparison

Hand calculation: SNI

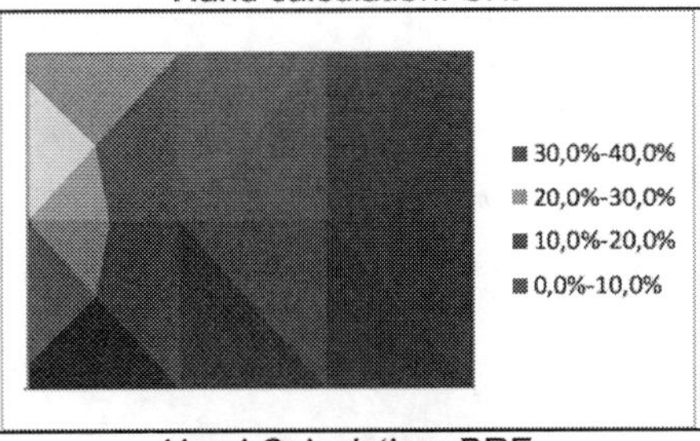

Hand Calculation: BRE

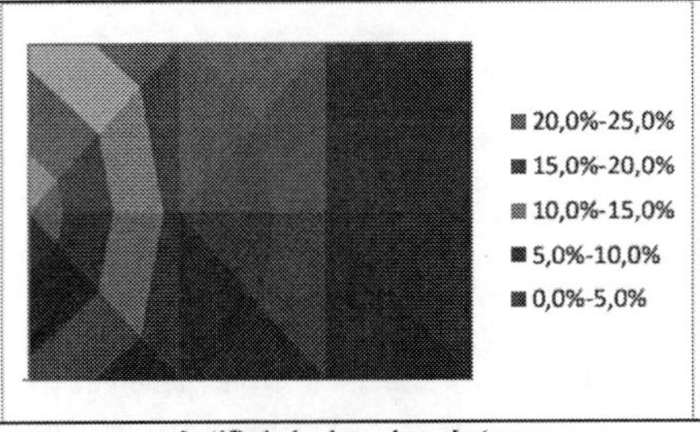

Artificial sky simulator

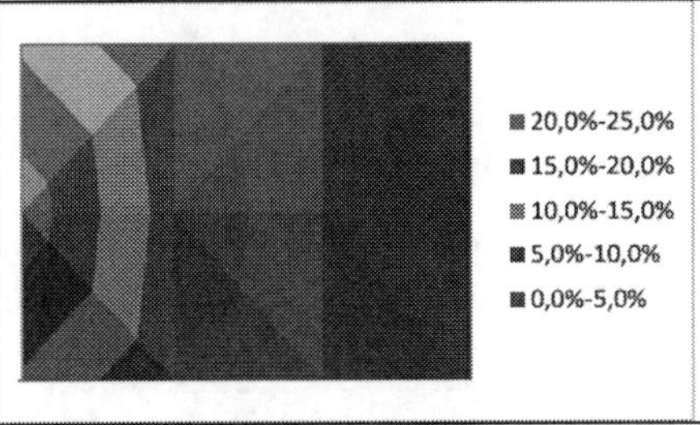

Software: Dialux

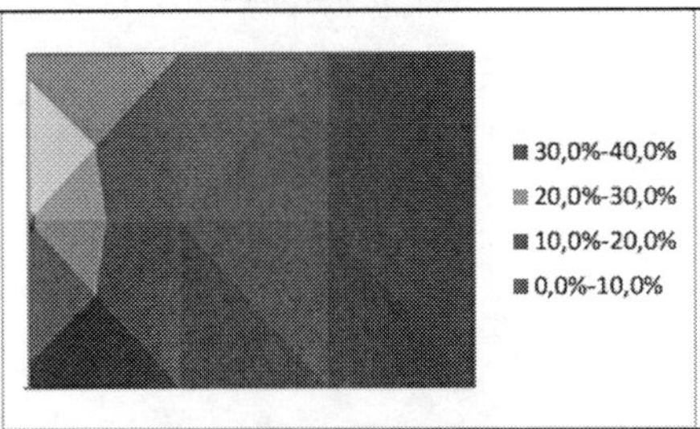

Software: Velux

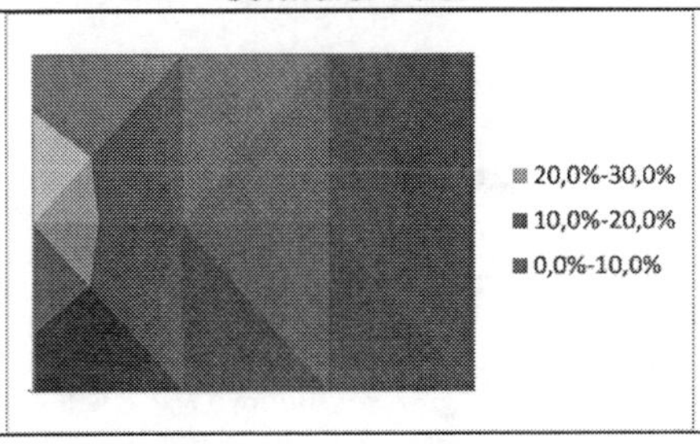

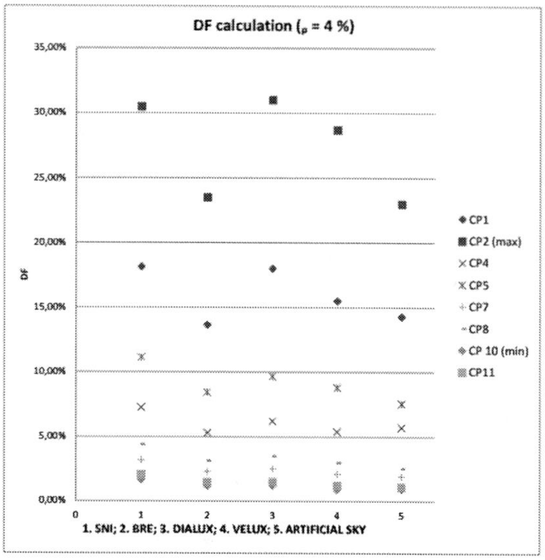

Fig. 7. DF Calculation Pattern ($_\varrho$ = 4%).

The results obtained in DF calculation on black room surface show the similarity of the light pattern on five calculation methods (Table 3). The higher luminance level is reached near the opening's area. The DF range is from 20%-30% on the highest measurement point value (CP2) and the lowest value is around 1% (CP 10 or 12). As is observed, SC value of the SNI method is relatively higher in all measurement points compared to the DF value from other methods. This is contradictory, since SC scores only light from the sky factor and does not count the internally reflection factor from room surface. In this trial, room surfaces are painted in Black with reflection contribution 4% based on Lechner (2015:407). Although the value is small, the DF value obtained is supposed to be higher than SC value. A study about logarithmic calculations used in the SNI / DPMB sky factor table is required for further analysis. BRE also has a sky factor table with different SC values with SNI sky factor table. BRE sky factor values are lower than the SNI sky factor table value. This result shows the importance of national standard evaluation that still referring to SNI calculation.

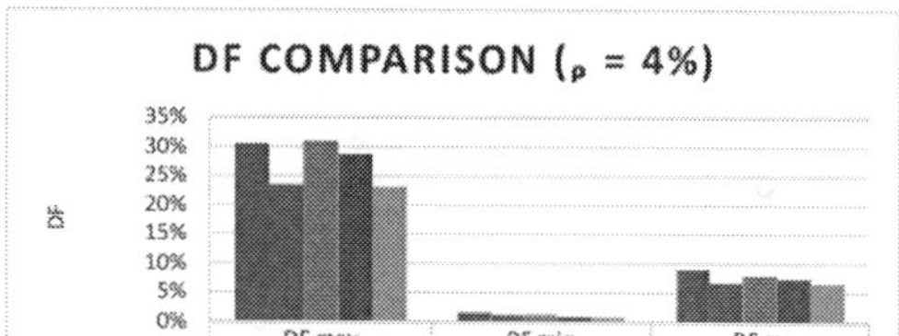

	DF max	DF min	DF av
SNI	30,48%	1,65%	9,04%
BRE	23,48%	1,18%	6,78%
Dialux	31,00%	1,23%	8,00%
Velux	28,70%	0,90%	7,39%
Artificial Sky	22,98%	0,94%	6,66%

Fig. 8. DF max, DF min, DF average graphic ($_\varrho$ = 4%).

It can clearly be observed in figure 8 that DF values are variable. Overall, the maximum SC / DF levels and average DF level of SNI method are practically close to the software result. While the measurement with the artificial sky simulator are close to BRE result. Minimum DF levels of each method are relatively close, but SNI result is still has the highest value.

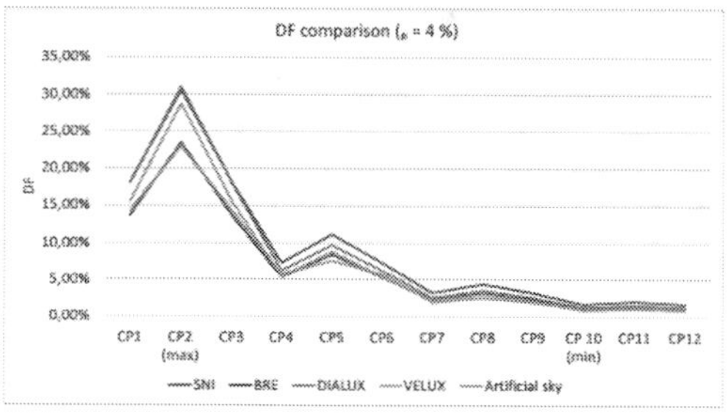

Fig. 9. DF at Each Calculation Points ($_\varrho$ = 4%).

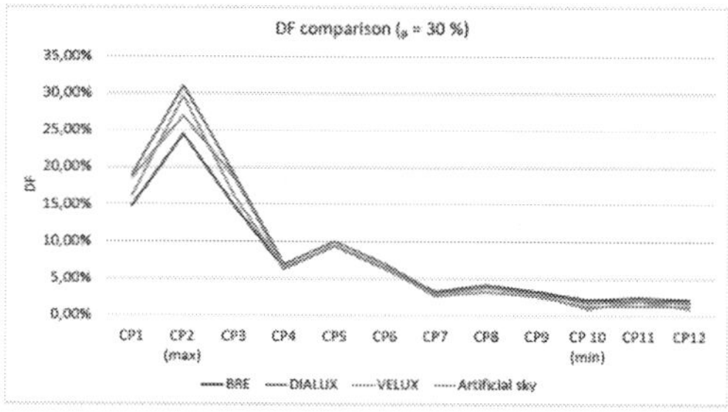

Fig. 10. DF at Each Calculation Points ($_\varrho$ = 30%).

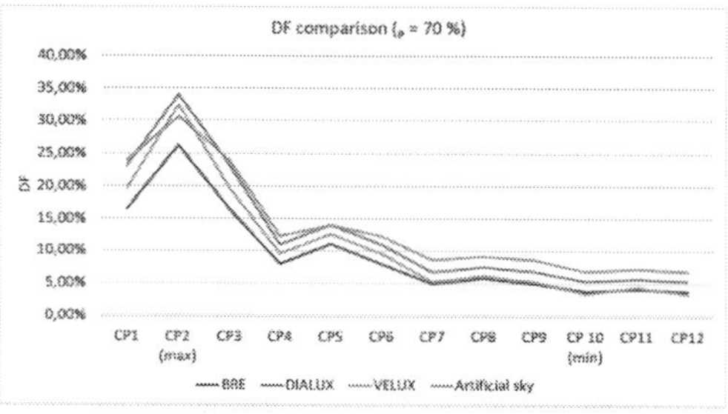

Fig. 11. DF at Each Calculation Points ($_\varrho$ = 70%).

DF values of software simulation on CP 1, CP2, and CP 3 (measuring points position are close to the window) have higher than the DF values of BRE and artificial sky result. However, DF values on other points are relatively having a constant result. These findings need further research to explore and observe.

Another finding is that the DF value of the model with the higher room reflection (painted in white) of the artificial sky result is much higher than the other methods result. Otherwise in the model with medium room reflection factor (painted in gray), DF value tends to be lower than other calculations. This significant difference occurs in the area away from the openings. A Research by Acosta (2011) show that the DF value of the artificial sky calculation is relatively higher than obtained by other methods (he compared with various computer simulation programs). As is observed, this condition may be caused by imprecise model making from the difference between color selected for a room model in artificial sky and others method. On BRE, Velux, and Dialux method, the use of reflection factor coefficients can be appropriately taken into account. BRE method has table coefficient value of 30% and 70%. The software program can set the glass material transmission value manually according to desired coefficient. The paint color chooses in artificial sky model paint has many variants and its difficult to get the reflection precise coefficient that suitable to the needs. The luminance meter instrument should be used in future research to occur more accurate result.

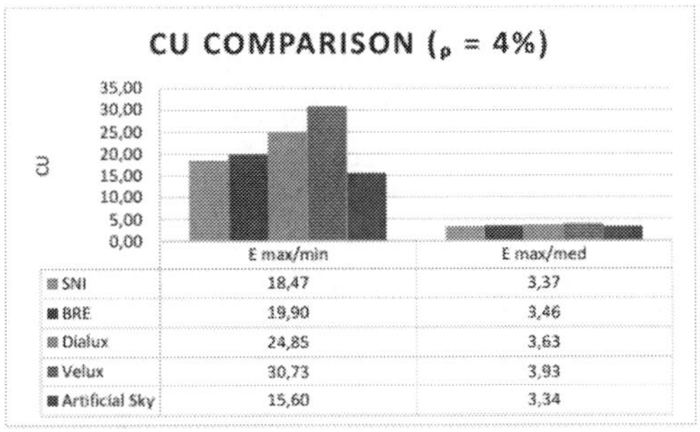

Fig. 12. CU Comparison (ρ = 4%).

This study also analyzed the coefficient of uniformity value in each method. The result shows the max/min illuminance level are variable between five methods while max/med illuminance level are relatively similar. Velux program has a higher CU than other methods, while those obtained with the artificial sky method are the lowest.

Room Reflection Trial

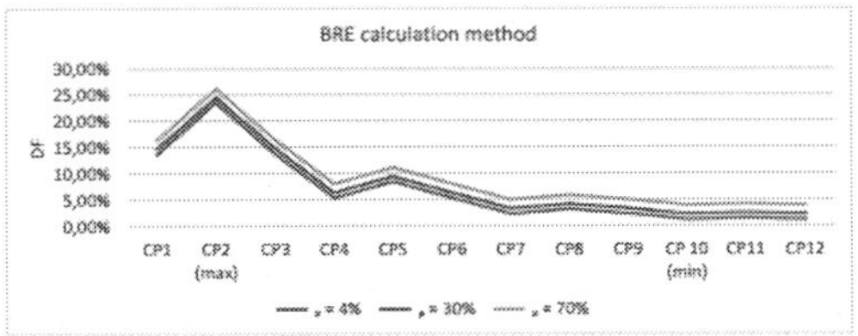

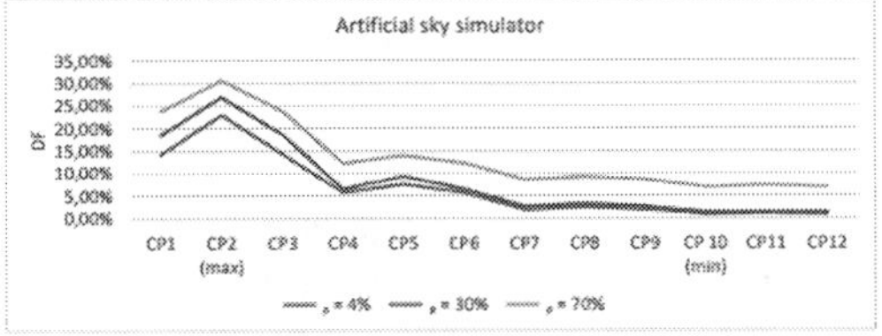

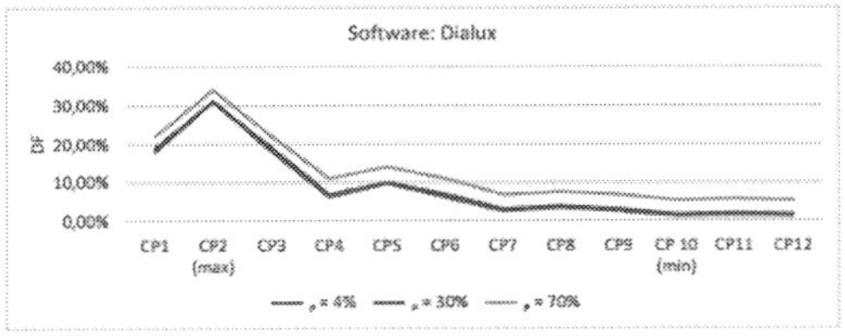

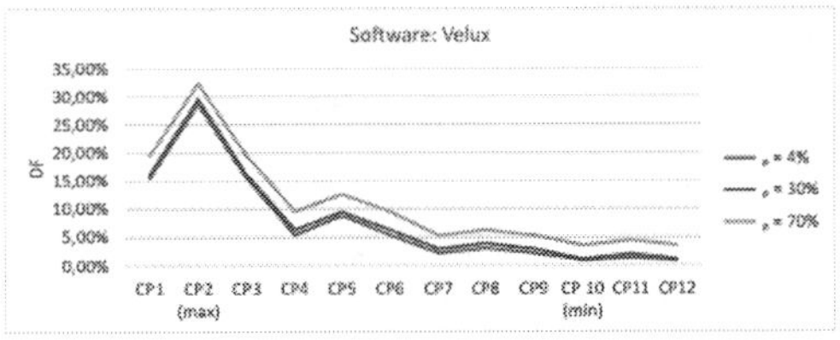

Fig. 13. Room Reflectance Contribution to DF Value (comparison methods).

The comparison of the daylight factor values with different room reflection factors in each method shows at figure 13. As observed above, DF values pattern shows in three room models are similar between those methods. The DF values are significantly increasing in white-painted room surface than others. Another finding is that the DF values increasing are higher at the measurement points away from the opening. The maximum DF in the white-painted room model rose to 26-34% from the original 23-31%, while the minimum DF value rose to 3.5-6.9% from the original ranged from 0.9 to 1.65%.

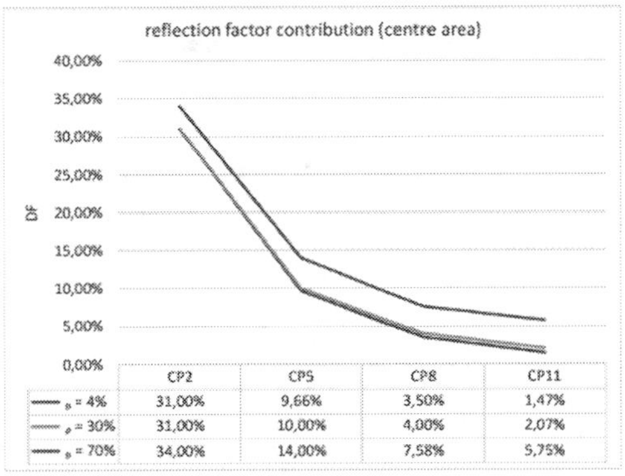

Fig. 14. Room Reflectance Contribution to DF Value at Centre Area.

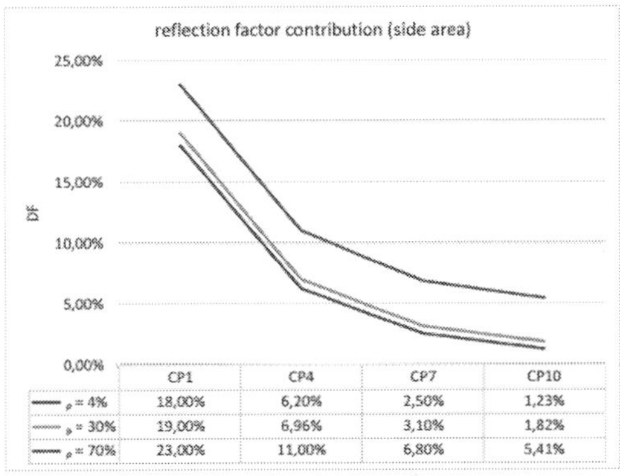

Fig. 15. Room Reflectance Contribution to DF Value at Side Area.

The trial also shows the differences in the contribution of DF values to the measuring point areas at the edge and center of the room. To analyze this, we take data from

Dialux simulation for quick and precise calculation. The illuminance increasing is more significant in the center area than the edge area. This is due to the spatial element that contributes to the increase of the DF value is higher than the edge area. It has a wider reflection element (ceiling, wall, and floor) and a closer position to the reflecting plane so as to obtain a greater reflection value.

The surface changing from gray to white color shows more significant illuminance increasing than the change from black to gray color. Changes from black to gray contribute to increase the DF value up to 48%. While the change of gray to white color can increase the DF value of 10 to 178% in the center area and increased by 21 to 197% at the edge. Thus it can be concluded that the color changing from black to gray is able to increase the illuminance up to 50%, while the color changing from gray to white contributes to increase the DF value much larger that is up to 200%.

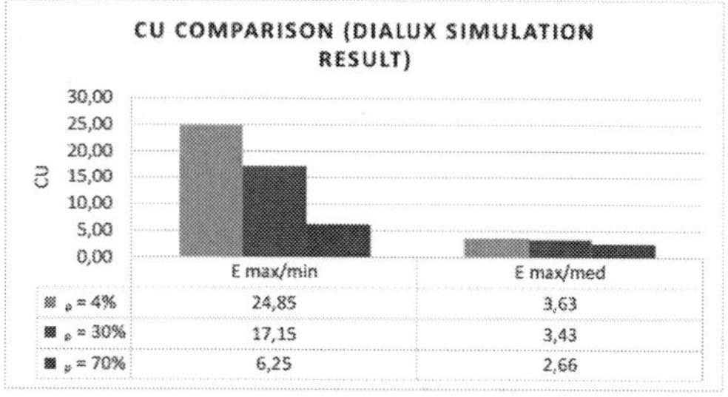

Fig. 16. Room Reflectance Contribution to CU Value (Dialux simulation result).

Figure 16 shows the room reflectance contribution to CU value based on Dialux simulation. As a result, CU value is decreasing (the light is more evenly distributed) at the higher room reflection. The color changing from black to gray contribute to reduce the maximum/minimum illumination value by 31%, while the change of gray to white contributes to reduce the value of CU two times greater.

Conclusions

By comparing the five daylighting calculation methods under overcast sky conditions and analyze the room reflection trial result, we can point several conclusions:

1. All calculation methods display similar light distribution patterns, but have varied results in DF and CU values.
2. Overall, the SC / DF maximum and DF average values by the SNI method are close to the computer simulation results. While the artificial sky measurements are close to the result with the BRE calculation method.
3. In the CU calculation, Velux software has a higher results than the rest. While others are relatively similar.
4. This observation confirms the results of previous studies on the contribution of room surface to the DF and CU value. Higher reflection factor contributes to increase the illumination and lowering the CU.
5. The contribution of DF values increasing is more significant seen in/at:
   - The gray to white painted model than the black to gray one,
   - Measuring points away from the window that near the window,
   - The Centre than the edge of the room.
6. White painted room (ρ =70%) could increase the DF value up to 3 times and lowering the maximum/ minimum illumination value up to 75% than a black painted room.

References

Acosta, Ignacio, et al (2011), Towards an Analysis of Daylighting Simulation Software, Energies 2011, 4, 1010-1024; doi:10.3390/en4071010 published 29 June 2011

Badan Standarisasi Nasional Indonesia (2001), SNI 03-2396-2001 Tata Cara Perancangan Sistem Pencahayaan Alami Pada Bangunan Gedung, Jakarta

D. H. Koenigsberger, et al (1975), Manual Of Tropical Housing And Building, Universities Press: Hyderabad, India

Dialgmbh (2011), Dialux version 4.9-The Software Standard For Calculating Lighting Layouts-User Manual, Ludenscheid

Direktorat Penyelidikan Masalah Bangunan (1981), Penerangan Alami Siang Hari Dari Bangunan, Bandung: Yayasan Lembaga Penyelidikan Masalah Bangunan

Direktorat Penyelidikan Masalah Bangunan (1989), Tata Cara Perancangan Penerangan Alami Siang Hari Untuk Rumah Dan Gedung, Jakarta

Fullermore (1985), Concept And Practice Of Architectural Daylighting, Van Nostrand Reinhold Company: New York

Geebelen, Benjamin (2003), Daylighting Comuptation Methods (From Do Chart To Digital Simulation), K.U Leuven

Iversen, Anne Et Al (2013). Daylight Calculation In Practice: An Investigation of the Ability of Nine Daylight Simulation Programs to Calculate the Daylight Factor in Five Typical Rooms,. Danish Building Research Institute: Aalborg University Copenhagen

Kota, Sandeep; Haberl Jeff.S (2009), Historical Survey Of Daylighting Calculations Methods And Their Use In Energy Performance Simulations, Proceeding Of Ninth International Conference For Enhance Building Operations, Texas November 17-19,2009

Lechner, Norbert (2007), Heating, Cooling, Lighting. Terjemahan Oleh Sandriana Siti. Jakarta: Pt. Rajagrafindo Perkasa.

Lechner, Norbert (2015), Heating, Cooling, Lighting. New Jersey: John, Wiley& Sons

Mandala, Ariani (2016), Komparasi Metode Perhitungan Pencahayaan Alami. Research Report LPPM Universitas Katolik Parahyangan: Bandung

Mandala, Ariani (2017), Komparasi Hasil Kalkulasi Pencahayaan Alami Dengan Metode Manual, Simulasi Maket, Dan Simulasi Komputer. Research Report LPPM Universitas Katolik Parahyangan: Bandung

Phillips, Derek (2004), Daylighting: Natural Light In Architecture. Oxford: Elsevier Ltd

Ecology: Creative Tool in Understanding Interrelated System

Diandra Pandu Saginatari[1,*], and *Paramita* Atmodiwirjo[1]

[1]Department of Architecture, Universitas Indonesia, Indonesia

Abstract. Moving towards a world with higher environmental awareness, the concept of 'ecology' has become particularly relevant for current architectural discourses. Studies about ecology sometimes are directly translated to sustainability because of its focus on efficiency of material and energy process. This paper argues that understanding ecology requires an ability to reveal the intro/interrelationship within the complex yet flexible ecological system. Demonstrating how 'ecology' is incorporated in the first year of architectural education, this paper aims to explore how ecology could be the basis of in developing an analytical tool towards the grounded understanding particular issue of the ecological system and its role as input in creative processing architectural education. The result of this study suggests how ecology has the potential as an analytical tool for reading the interrelated system.

Keywords: ecology, design tool, creative process, interrelated system.

Introduction

There has been a significantly increasing concern on environmental issues in architecture. Architecture starts playing a role in 'correcting' ecological damage caused by the modern society that is shown through the tendency of architects to incorporate environmentally friendly strategies such as using local and energy efficient material, energy harnessing devices, and living component, such as green roof, in their designs (Sykes, 2010). Responses toward these issues are undoubtedly critical in architectural discourse since it also derives the direction of architectural education. It is presented through an inclusion of ecological discourse as the center of architectural education.

Ecology now becomes a common term in architectural education; terms such as ecology-based concept, performance ecology, and ecology and sustainability are increasingly used in architectural education (Goodbun, 2010). It apparently shows that ecology emerges as a complex discourse that closely relates to advanced practical and technical aspects of architecture that tends to be addressed at the later stage of the architectural education. However, this paper argues that the discourse of ecology is essential to be incorporate even in the early stage of architectural education. It does not suggest that we have to push the ecology in advanced practical and technical sectors to be studied in the first year. This paper believes that there is a way to transfer the knowledge of ecology through a creative process that is usually explored in early stage of architectural education.

One of the ways to transfer the knowledge of ecology through creative process is through an understanding of the ecology itself. This paper explores how the understanding of ecology that reveals the essential core of ecology which is the reading of interrelated systems could be a trigger for creative process. This creative process could become the medium in which ecology is explored, learned and understood by first-year architecture students of the Department of Architecture Uiversitas Indonesia. This paper will use examples of student projects to show how the understanding of ecology could be incorporated in the creative learning process of the first year of architectural education.

Ecology in Architectural Education

In past decades, as a multifaceted discipline, the progress of architecture is determined by multiple themes that are emerged within the discipline such as the advancement of technology as new modes of design, environmental issue, everydayness and the phenomena of the "starchitect" (Sykes, 2010). The importance of the environmental issues and the emergence of ecology in the discourse revolves around the tendency of architecture to 'betrays nature' in its every process since the historical time (Ingersoll, 2012). And now, we could easily see the effects such as in the form of climate change. In response to overhaul the ecological damage, ecology becomes the core of architectural discourse.

Ecology is a branch of Biology dealing with the science of the relationship between the organism and its environment. Lately, as we are moving towards broader questions of environmental awareness, ecology emerges across disciplines and seems to signify a paradigm rather than a discipline (Goodbun, 2010). The use of 'ecology' as an additional term in specific disciplines refers to an additional perspective that somehow tries to reveal the way system works by seeing relation and interaction of

* Corresponding author: diandrasaginatari@gmail.com

parts within the system. Industrial Ecology, Urban Ecology, Urban Political Ecology, Social Ecology, Human Ecology and more are examples of the terms using ecology as a suffix, and it indeed indicates a discourse that focuses on revealing relation and interaction within the specific system. For example, industrial ecology, even though it is normative, tries to see how nature, as structure and function, works to be applied to existing industrial system that aims for efficiency; meanwhile Urban Political Ecology (UPE) tries to reveal nature-society relationship as dynamic and networked circulations that reorganize social and physical environments into socio-cultural assemblages (Newell and Cousins, 2014).

The use of 'ecology' as an additional term is also used in architecture and architectural education. Terms such as ecology-based concept, performance ecology, and ecology and sustainability are some of the currently used theme examples in architectural education, showing that architecture as a discipline, need to accentuate the inclusion of this topic in the education.

Catala (2010) stated that:

"Ecology is generally and perhaps too immediately linked to sustainability. The etymology of the word ecology (from the Greek Oikos, 'house' or 'living relations' and logia, 'study of') however, does not necessarily imply the prioritisation of energy-efficient solutions, nor does it favour the notions of 'green' or of 'good' for that matter. Ecology refers to the interdependent (enduring or temporary) relationship between the physical and biological components of an environment that function as one whole ecological unit." (p.89)

Catala's statements are also reflected in architectural education. In architectural education, studies on ecology are translated to sustainability through a focus on efficiency of material and energy process. This focus generates phrases such as 'green architecture' or 'zero energy building'; with emphasis on quantitative methods within these studies. It directly relates to practical and technical aspects of architecture which include practical and technical complexity that will construct advanced skill and knowledge for the architecture student. It seems to indicate that knowledge of ecology will then imply to the implementation of technology to solve and reduce ecological damage. However, this paper believes that the knowledge of ecology in architecture is not merely a problem-solving tool through a practical and technical application that requires quantitative approach. Knowledge of ecology in architecture has the potential to be an analytical tool, as ecology comprises relation and interaction in the complexity of systemic nature such relation and interaction construct complexity which becomes the central idea in ecology.

Ecology and Complexity

"Ecological systems are complex assemblages of interacting organisms embedded in an abiotic environment" (Hartvigsen et all, 1998, p:427), then, ecology is about reading relation and interaction in complexity. It is undoubtedly true that ecological system consists of various parts or elements. Relation and interaction imply connectivity among parts or elements which together constructing the system. As illustrated in Figure 1, there is an issue about the relationship between parts and wholes that is central in ecology (Proctor and Larson, 2005). How parts construct wholes then generate complexity. Hartvigsen et all (1998) stated that complexity in ecology is emerged from the interspecific and intraspecific interaction among parts or elements, and this interaction exists in space and through time. We could conclude that we could see interaction in ecological system as events that exist specifically in particular context. This means that context plays a significant role in determining the event, and it happens through time which usually indicate the dynamic quality.

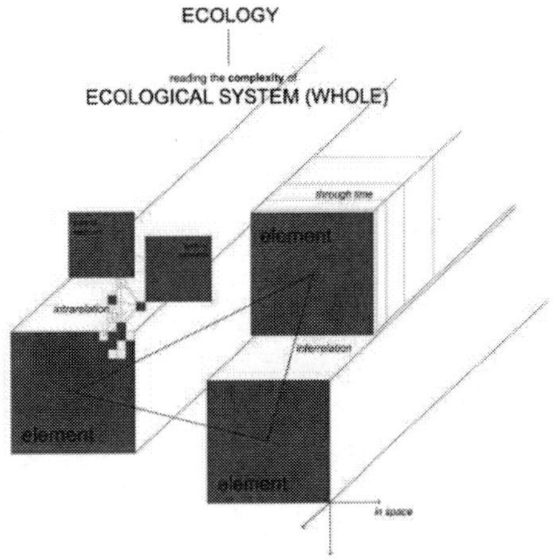

Fig. 1. Diagram of Ecological System.

The complexity in ecological system implies that ecology, which studies about an ecological system, is able to reveal complexity. Loehle (2004), in his paper titled Challenge of Ecological Complexity, suggests that complexity in ecology has at least six types: spatial, temporal, structural, process, behavioural, and geometric. In ecology, spatial complexity refers to vegetation patterns and species distributions. Spatial complexity will cause temporal complexity along with population dynamic and fluctuating climate and weather. Structural complexity deals with relationship and network within an ecosystem and relates to processing complexity that refers to the process which contains many steps or components. Furthermore, since ecology is a study about living organisms in its environment, behavioural complexity arises because of the behavioural response of the living environment to the environment. The living things and the environment also consist of distinct ecological objects that have geometric aspects which will influence an ecological system.

From Loehle's types of complexity in ecology, we could deduce how these complexities construct the

 https://doi.org/10.1051/shsconf/20184106002

ecological system that comprises relation and interaction among elements in space and time. Complexity such as spatial, structural, and geometrical could relate to space within which the system exists. Temporal complexity implies that time is an aspect that will always significantly influence as time causes temporality in the first place. Some complexities such as structural, process, behavioural seem to have a role in composing the relation and the interaction in the ecological system itself. Meanwhile some complexities such as geometrical and behavioural could also refer to the aspects that create the elements or objects.

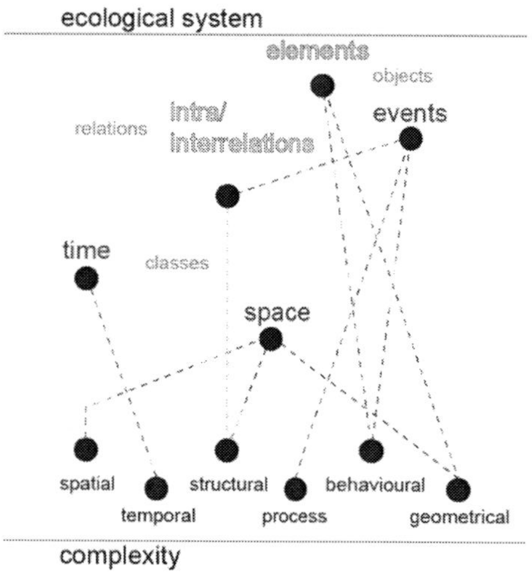

Fig. 2. The Complexity of Ecological System Diagram Based on Loehle (2004) and Hartvigsen et all (1998).

As shown in Figure 2, these types of complexity could be directly related to the essential nature or mechanism of the ecological system which, as stated above and based on Hartvigsen et all (1998), consist of relation and interaction among elements in space and time. To help to reveal the complex ecological system, Lohle (2004), suggests that we could break down the system into objects, classes, and relation. Objects refer to the elements of the system, while classes could refer to the organisational and hierarchical structure in the system, and relations could refer to the nature of the connectivity among elements that construct the system. This could structure our analysis toward an ecological system. By structuring the understanding of the complexity of ecological system we could generate an 'ecological' analytical tool that is substantially relevant in architecture.

Ecology and Creativity

As we unravel the nature and the complexity of an ecological system, we could strongly relate the vocabulary used in the discussion with the vocabulary that we use in architecture. The use of vocabulary such as spatial, temporal, geometrical, relation, interaction, and even the concept of space and time imply that somehow the ecological system is constructed as a structured spatial relationship. This 'ecological' analytical tool suited well with the analytical process in architecture, since in architecture we are expected to use structured analysis to analyze spatial settings or issues. This is what this paper suggests, ecology in architecture is not merely about a sustainable solution, but it could be used as an analytical tool that could be input at any level of an architectural design process.

The use of 'ecological' analytical tool in architectural education is one of the ways to embody ecological knowledge and understanding in architecture that could be implemented even in the first year of architectural education. This paper will illustrate several student projects in Basic Design Studio 1, in the first semester at the Department of Architecture Universitas Indonesia. It comprises basic design and communication knowledge and basic visual art that become the basic competencies for architecture students before they start their Architectural Design Studio in later semesters. This studio uses the triggers that are closely related to creative architectural practice, and one of which is the implementation of this 'ecological' analytical tool as trigger in a creative process.

Light and Shadow is the title of the final project in Basic Design Studio 1. In the final project, the students are challenged to develop their creative ideas in creating two and/or three-dimensional work based on sensitivity and critical thinking on light and shadow exist in various possible context through demonstration of creative exploration of media and techniques. This project offers 5 different units, one of which is titled ***The landscape of Light and Shadow***. This particular unit offers the opportunity to explore the system and relationship between light and shadow particularly in a natural setting where Sun will be the main source of light. Students are encouraged to conduct visual and/or spatial experiments in a natural setting, that consists of various surfaces such as ground surface, water, vegetation, and even exterior of a building, to reveal the relationship of light and shadow as a dynamic spatial quality of the landscape itself. The term "landscape" in Landscape of Light and Shadow is not merely referring to the natural landscape. It refers to how light and shadow is being understood as a *scape* within which ecological relationship among the elements construct the system of light and shadow itself.

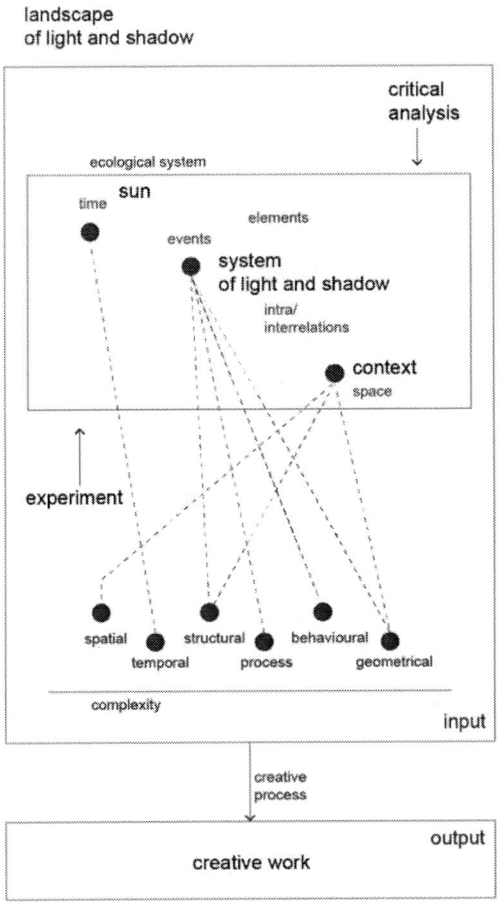

Fig. 3. Mapping the Brief of *Landscape of Light and Shadow* in the Complexity of Ecological System.

The short brief if the project becomes the trigger and basis for the students to develop their creative ideas. As shown in Figure 3, the brief itself has the potential to be an input of creative process that could implement an 'ecological' analytical tool. In the next part, this paper will discuss some of the students' works to unravel how the project demonstrates ecological analysis as input for a creative process. The unravelment of the demonstration is conducted by mapping the process of each project according to the vocabulary of complexity in the ecological system.

Project #1: 2500 Shades of Shadow. This project explores the composition of umbra and penumbra of shadow. It began with research on how umbra and penumbra are formed in the shadow that leads to the understanding of umbra as the darker part of shadow and penumbra as the lighter part of the shadow. Then, an experiment was conducted to know the aspects that influence the degree of umbra and penumbra on the shadow. The experiment found that the height of the object from the surface of which the shadow existed affected the value of the colour of umbra and penumbra. The shorter the height, the object created the darker value of umbra and penumbra, and the value get lighter as the object get higher. Then, based on this experiment, further observation was conducted to collect umbra and penumbra from objects with different heights and different surfaces of which the shadows were formed.

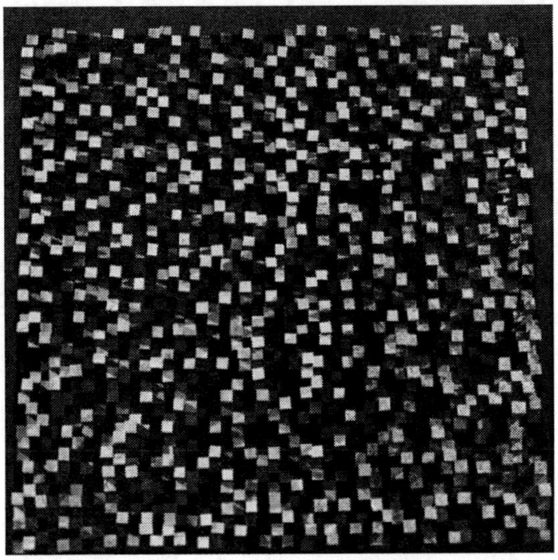

Fig. 4. *2500 Shades of Shadow* by Adwani and Triarianti, 2016.

The output of this project is a catalogue of umbra and penumbra presented in a collage of 2500 1cm x 1cm squares of shadows which are collected from photographs of umbra and penumbra in the observation. Each square of shadow has a different height from the surface. Referring to the findings, the higher the square, the lighter the value of the shadow shade. This work presents an artistic composition and the detailed data of umbra and penumbra that are based on ecological analysis that derives understanding about light and shadow, especially umbra and penumbra.

Project #2: Kridaprana Andakara. The title is derived from Sanskrit Language meaning *change of sunlight reflection on vertical surface.* This project explores the shifting position of sunlight reflection. It began with the basic understanding of reflection that consisted of elements such as light source and flat reflective surface. As a system reflection happened when incident ray with a particular angle of incidence hit a flat reflective surface and created reflected ray with the same certain angle of reflection as the angle of incidence. So, one of the aspects that influenced the reflection was the angle of the ray. An experiment was then conducted to explore how the reflection of sunlight changed through time. By using several mirrors and vertical surface as a canvas for the reflection, the experiment captured that every 10 minutes, the reflection moved up or down around 2 cm depending on the position of the mirror. In the determined duration of time, the experiment produced a mapping of changing reflection of sunlight.

https://doi.org/10.1051/shsconf/20184106002

Fig. 5. *Kridaprana Andakara* by Agusiwi and Afifah, 2016.

The output of this project is a mapping of shifting reflection of sunlight presented as a three-dimensional work that consists of vertical surfaces with colour marks showing the changing position of sunlight reflection. Complementary colours of red and blue are used to signify the different direction of the shifting position of sunlight reflection. This work presents a creative map showing the dynamic flow of sunlight reflection based on an experiment that includes ecological analysis on the phenomena of reflection.

Project #3: Distorsi Dalam Waktu (Distortion in Time). This project explores the changing shape of shadow from a man-made object through time. It began with a basic understanding of how the shadow of a man-made object in an outdoor setting was formed and changed through time. Then observation and experiment were conducted to explore how the shadow of man-made object in outdoor setting changed, both its position and its shape. The observation and experiment found that as the sun moved, the position and shape of the shadow changed. The shape of the shadow of the object get distorted as it changed because of the movement of the sun. The intensity of the sunlight also played a role in forming the shape of the shadow as the higher the intensity, the sharper the shape of the shadow that was created.

The output of this project is an abstraction of shapes of shadows presented in three-dimensional work that consists of horizontal layers of changing distorted shapes of shadow. Different distorted shapes of shadows are stacked layer by layer based on the timeline of the observation. The work is displayed on the ground, as of how the shadow of the object is, to help make sense the idea behind it. This work presents the findings of observation and experiment, which requires an ecological analysis on how sunlight could create ever changing shapes of shadow, which are creatively processed through abstraction being delivered as an artwork.

From the explanations above, each project has its specific understanding of light and shadow. Each of them is applying 'ecological' analysis in the process as we could always relate the development process with the essence of the ecological system and the complexity.

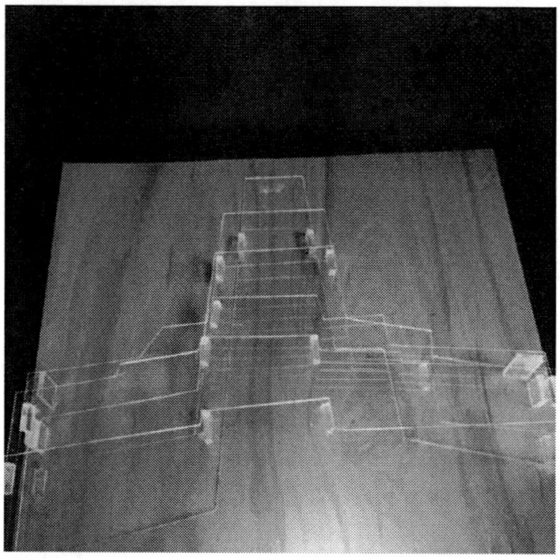

Fig. 6. *Distorsi Dalam Waktu* by Ratriananda and Kusumowidjojo, 2016.

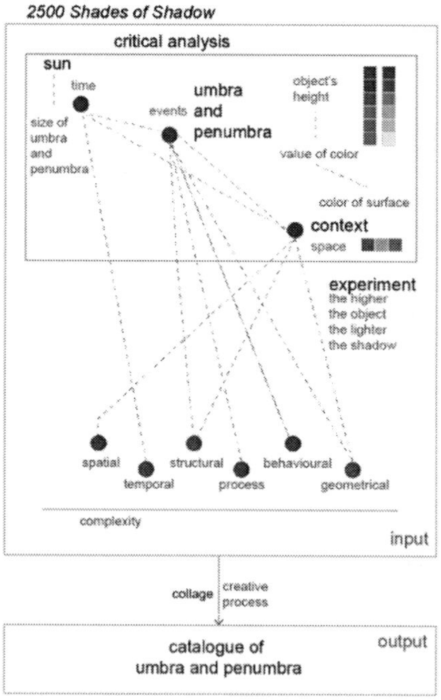

Fig. 7. Mapping *2500 Shades of Shadow* in the Complexity of Ecological System.

Figure 7 shows how *2500 Shades of Shadow* has umbra and penumbra as the main 'ecological' event that consists of intra/interrelated elements, exist in particular settings, and has the sun as the source of light that signifies time. When we discuss the system of umbra and penumbra, there are structural, process, behavioural, and geometrical complexity within. Then, settings of which umbra and penumbra exits will have spatial, structural, and geometrical complexity.

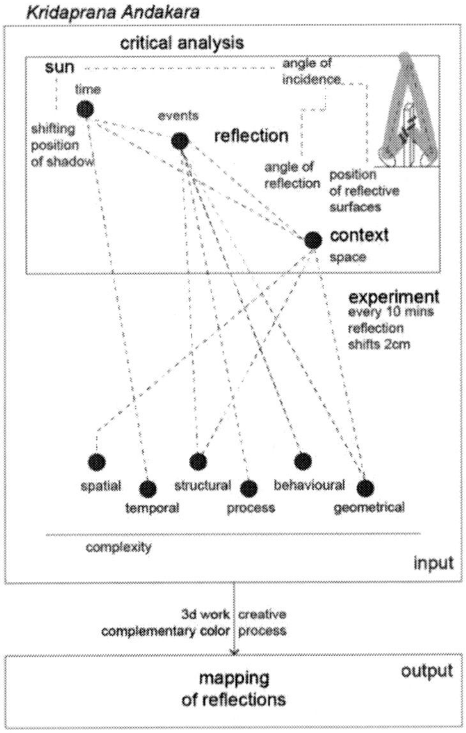

Fig. 8. Mapping *Kridaprana Andakara* in the Complexity of Ecological System.

The process is also applied in *Kridaprana Andakara*, shown in Figure 8. This project has an understanding of reflection as the 'ecological' event that has structural, process, behavioural, geometrical complexity within. Elements that are existed, such as the reflective surfaces will have their behavioural and geometrical complexity. Spatial, structural, and geometrical complexity arise in the process of mapping the experiment of shifting position of sunlight's reflections.

As in *Distorsi Dalam Waktu*, as shown in Figure 9, the main 'ecological' event is the shadow of a man-made object in outdoor settings. Discussion on how this happened has spatial, structural, behavioural, and geometrical complexity. Elements of the event have at least behavioural and geometrical complexity that could relate to spatial and structural complexity too.

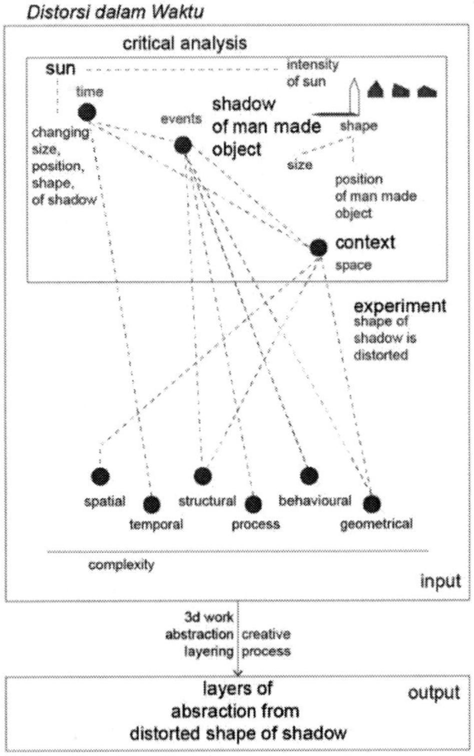

Fig. 9. Mapping *Distorsi Dalam Waktu* in the Complexity of Ecological System.

The three projects above are under the influence of time because they have the sun as the primary source of light. This comprises temporal complexity in the system that complements the relationships among elements.

Ecology and Dynamicity

In Landscape of Light and Shadow Project, the students are directed to learn about ecology through a creative process. The idea of ecology that is studied and explored is focusing on how to reveal interrelated system by using 'ecological' analytical tool. It shows how the understanding of ecology could be embedded in the early stage of architectural education and could derive creative works. If we reflect on the discussion, complexity in ecological understanding provides an analytical framework in understanding interrelated system. The analytical framework is formed by the nature of every element in the complexity that always has relation and interaction among one another for the purpose of maintaining a system. In other word in using this ecological understanding, analytical process is not limited to the relation and the interaction between elements, but extended to seeing the result of the relation and the interaction. Various elements and complexities will create diverse combinations of relations and interactions that will lead to different results. Furthermore, any changes of internal and external properties will affect elements and the relations within.

This fact shows that ecological system is dynamic; it changes and fluctuates in time depending on the condition and the relations that occur.

Goodbun (2012) suggests that, based on Gregory Bateson, the ecology that is required to understand a complex system is the ecology that prioritises flexibility. This statement shows that flexibility, that entails dynamicity, is important in understanding a complex system. This dynamicity in ecological system, arises because of the complexity itself. Several types of complexity that is proposed by Loehle (2004), such as temporal, behavioural, spatial and geometrical complexity, already imply dynamicity within. So, it can be concluded that, beside complexity, dynamicity is also the nature of an ecological system. Dynamicity in ecological system, as an analytical tool, plays a role in allowing different perspectives to emerge, and in relation to architectural education, dynamicity in ecological system is beneficial and aligned with the spirit of creative process.

In the process of the final projects, dynamicity emerges through various experiments that are conducted by the students. The experiments examine specific issues and contain manipulation of elements that are involved within. The experiments also involve independent tools and devices to test the issues. The experiments lead to diverse yet specific findings to understand the ecological events of light and shadow. Concerning the creative process in architectural education, dynamicity and complexity of ecological system are undeniably essential in supporting it as 'ecological' analytical tools in creative process.

Conclusion

Ecology as a discipline is strongly relevant to architectural education. Ecology is utilised to learn about sustainability through a rather quantitative approach to achieve efficient and environmental friendly problem solving. This paper particularly addresses this issue by presenting an alternative possibility to integrate the idea of ecology in architectural education. The idea of ecology as a study on interrelated systems could be transformed into an analytical tool that could reveal complex systems and issues. Its nature of having complexity and dynamicity makes ecology a complete analytical tool that could lead to a deep understanding through an explorative process. Beside as an analytical tool, the understanding of ecology also reflects spatial understanding. As an event that exists in space and through time, relation and interaction in ecology also leads to the understanding of spatial relation. To think ecologically means to think spatially. This is also one of the potentials of ecology that could be used in architectural education.

In the Department of Architecture Universitas Indonesia, analytical skill and creative thinking are considered essential in the educational process. Since the early stage of the education, we always incorporate analytical skill in the process, aiming for critical thinking. Together with creativity, analytical skill will be an adequate instrument to determine creative design work with a strong analytical background addressing specific issue. Because of this consideration, we, as teacher, get opportunities to explore various teaching methods that will be beneficial for the progress of our architectural education. The Light and Shadow project is one of the example of our exploration in making an integrated exercise based on our proposition and understanding towards ecology, that will give the students opportunity to understand issues in a holistic way and to create creative works based on a through analytical process.

Acknowledgement

The Works included in this paper are the work of Architecture Students, from the Regular and International program, Universitas Indonesia, class of 2016. The works are part of Landscape of Light and Shadow project as the final project for Basic Design Studio 1 in 2016.

References

Catalá, A. (2010). Numerical Ecosystems. *Architectural Design*, 80(6), pp.86-93.

Goodbun, J. (2012). Flexibility and Ecological Planning: Gregory Bateson on Urbanism. *Architectural Design*, 82(4), pp.52-55.

Goodman, J. (2010). Gregory Bateson's Ecological Aesthetics - an addendum to Urban Political Ecology. *Field: a free journal for architecture*, 4(1), pp.35-48.

Hartvigsen, G., Kinzig, A. and Peterson, G. (1998). Complex Adaptive Systems: Use and Analysis of Complex Adaptive Systems in Ecosystem Science: Overview of Special Section. *Ecosystems*, 1(5), pp.427-430.

Ingersoll, R. (2012) The Ecology Question and Architecture. In: Cattaneo, E., Ardesio, G., and Andolina, M. (2013) *Loaded void: city theory since 1956 : falling modernism and arising landscape urbanism.* Santarcangelo di Romagna: Maggioli. pp. 786-802

Loehle, C. (2004). Challenges of ecological complexity. *Ecological Complexity*, 1(1), pp.3-6.

Newell, J. and Cousins, J. (2014). The boundaries of urban metabolism. *Progress in Human Geography*, 39(6), pp.702-728.

Proctor, J. and Larson, B. (2005). Ecology, Complexity, and Metaphor. *BioScience*, [online] 55(12), pp.1065-1068. Available at: http://www.jstor.org/stable/10.1641/0006-3568(2005)055[1065:ecam]2.0.co;2 [Accessed 18 Sep. 2017].

Sykes, A. (2010). *Constructing a new agenda*. New York: Princenton Architectural Press

Mapping of Affordance and Activity as the Biophilic Design Principle of Blue Lagoon Tourism Area Yogyakarta

Hastuti Saptorini[1,*] and *Dea* Viviani[2]

[1]Lecturer at Department of Architecture, Universitas Islam Indonesia, Indonesia.
[2]Student at Department of Architecture, Universitas Islam Indonesia, Indonesia.

Abstract. Bathing and playing in the river is not a new phenomenon for some people. But this experience creates a recreational spirit on the Tepusriver which is now better known as the Blue Lagoon. This area is a term for the settlement that is split a piece of the river in the dusun Ndalem Ngemplak Sleman. Its tributaries are clear and blue, surrounded by some springs with natural bamboo groves and old trees that invite local and around communities for recreation. The local community has anticipated it by responding to the communities' demands as well as the recreation area. Appear sporadically *"warungs"* and recreational activities follow-up though not yet grown in conceptual. This paper aims to study the map of affordance and activity of previous research findings and could be used to the biophilic design to afford the health, productivity, and wellbeing in the tourism area. The method is by synthesizing the previous research findings in 2016, some relevant urban design theories, and biophilic design principle. The conclusion is paid attention to two main principles. The first principle is maximizing the utilization of existing natural properties and the existing cultural skills into its development to nourish visitors both physically and psychologically. The second principle is the Government, and Non-Government Organizations (i.e., expertise, academics, universities, investors) support, both policy and financially, in many sectors: tourism, environment, and infrastructure.

Keywords: affordance and activity, biophilic tourism design principle

Introduction

Blue Lagoon is a popular designation for the Tepus River cutting point that develops as a "river recreation" place in a residential neighborhood in the *dusun* Ndalem, Ngemplak, Sleman, Daerah Istimewa Yogyakarta. The water is clear and blue, surrounded by a number of springs with natural bamboo groves and old trees that invite local communities and around to perform a number of recreative activities, both individually and communally.

Recreational activities grow in the river and banks, and in its surroundings. The author, in a previous study, found the behavior of visitors in the water nature in three places, namely river baths, sendhang wadon, and sendhang kakung. Swimming, bathing, playing theater, race on tires, etc. are activities on the river and invites other visitors. The phenomenon of the audience has not been directed and controlled because the infrastructure as a tourist area has not been prepared by default. The physical condition of nature is still pure and has not been touched by development. The undeveloped infrastructures is potential unsecure and unsafe for visitors.

Similarly, the activity conducted on the riverbanks: fishing, playing, watching theater, etc. is a phenomenal activity, both by teenagers and children. Security and safety are a threat because infrastructure has not yet constructively sustained for the public good.

This condition has not responded to Article 26 paragraph (d) in the regulation of tourism, Law no. 10 of 2009 which emphasizes the certainty of comfort, hospitality, security, and tourist safety (page 17). The basis of the regulation is also contained in the Minister of Tourism Regulation No. 14 of 2016 on Guidelines for Sustainable Tourism Destination. Clause (a) on page 14 confirms the existence of a sustainable destination strategy indicator is to address issues of health and safety issues, in addition to the general, economic, and social environment.

According to The Workers Compensation Board (WCB) of British Columbia (WCB: 2002), the "Health and Safety Program" in Yudhisthira emphasized that all activities that develop and need the planning process take into consideration 6 things as risk management. Of the six things in question, one is to emphasize "tourism planning by enacting procedures that will ensure the safety of visitors" (2012: p. 21). This is explicitly reinforced by the results of his study which found 10 parameters of tourism quality indicators in terms of safety. One of the parameters associated with the Blue

* Corresponding author: hastuti.saptorini@uii.ac.id

Lagoon phenomenon is the sixth parameter, namely the importance of infrastructure for visitor safety (page 22).

Despite the anticipatory consideration of security and safety for visitors, the Blue Lagoon needs a development concept as it is a potential tourism area. This is explicitly stated in the Minister of Tourism Regulation No. 14 of 2016 on Guidelines for Sustainable Tourism Destination. Clause (c) on page 15 confirms the existence of an indicator of a sustainable destination strategy as evidenced by evidence supporting the plural year development strategy / plan.

Theoretically, the tourism development strategy/ plan in question is based on facts and valid information of the region. This is as a matter of Godfrey K. et al (2000), that tourism strata/plans are structured through a three step process:

1. Identify opportunities and barriers based on potential and demand,
2. Establishing issue-based development objectives in the short, medium, and long term,
3. Define a stage of action designed to achieve goals in a specific and measurable timeline. Thus, the identification of facts and information, Blue Lagoon is an early stage that needs to be done to underlie the design process in its development strategy.

To date, the Blue Lagoon has not been widely identified, researched, studied, or published scientifically. Some non-scientific studies can be found, but not written and published by the Authorized Publishers. In 2016, found 2 researchers about this tourist area. First is Susi Febriani who is researching for her thesis, entitled "Participation Level of Society Tourism Activities in Blue Lagoon Tourism Village Dalem Widodomartani Ngemplak Sleman. The research aimed at 2 points, that is knowing the factors, and the level of community participation in this area, using probability random sampling technique. Researchers have interviewed 57 households of 130 household population. Data were analyzed by qualitative descriptive method. The result found 2 things. First, that the factors of community participation are influenced by age, gender, education level, occupation, and number of family members. Secondly, the level of community participation is in the medium category of 35.09%.

From the research can be described that public participation is not maximized. This condition may be due to the unavailability of the whole Blue Lagoon tourism system, so that community involvement is still "sporadic and fragmentative".

The second researcher was Dea Viviani who conducted a mini research for his Scientific Writing with the title "The Behavior of Dalem Villager in the Tirta Budi Blue Lagoon Yogyakarta Tourism Object". This study aims to find the typology of dusun nDalem people's behavior towards the development of Tirta Budi Blue Lagoon tourism object. Behavior data is collected by observation and processed by way of mapping. The result is that there are routine and incidental behaviors around the river and its banks. This second study does not study affordance in terms of designing biophilic, so it has not yet identified comprehensively for environmentally-based tourist areas.

Affordance is an object/property in an environment that is likely to be utilized by both human and animal beings (Gibson, Eleanor J, et al, 1999). In the context of the Blue Lagoon, as a natural area for tourist setting, it actually has a productive biofilic environment. This 130-household inhabitant still has the affordance that needs to be explored to create a biofilic activity. In principle, biophilic design is to create a healthy, productive, and wellbeing environment for its inhabitants (Abby Lerner and Mike Stopka, 2016). Biofilia refers to the existence of an instingtive bond between human beings as a species responsive to natural forms, natural analogies, and the nature of a space (Priatman, 2012: 35). This gap is a consideration of this paper. So, the objective of this paper is mapping the affordance and activities as the basic concept of biophilic design principle of Blue Lagoon tourism area Yogyakarta.

Theoritical Review

Affordance

Affordance are perceived through an invariant combination of properties, which are meaningful to a person and animals (Gibson, Eleanor J, et al: 1999), and have opportunity to creat activities in some objects of the environment (Soegaard, M: Wikipedia, https://www.interaction-design.org/literature). On the other hand, Chemero, A (2003) argue that affordance do not dissappear when there is no local animal to perceive and take advantage of them (p.193).

In terms of territory context, affordance is defined by Laurens (2004: 104) as the benefit of an area for individuals or communities to demonstrate power, defense, exclusive use, ownership, personalization, needs satisfaction, and identity through signification. This meaning is almost in line with Kytta's theory in Bobby (2011: 609). There are 3 categories in affordances, namely: perceived, utilized means 'used', and shaped means 'formed.' "As for Lindberg, P (2014) affirms, affordance is a potential activity that the environment offers to users (p. 14).

Blue Lagoon Previous Research Findings

Anonymous researcher (2014) found that Blue Lagoon has fauna properties in moderate category. There are 399 birds totally. They consist of 22 kind of birds, 20 families, 14 suku. The most numbers of Suku are Alcedinidae and Cisticilidae. From this number, there are 4 kind of birds that preserved Government related to *Peraturan Pemerintah No.7/1999 Pengawetan Jenis Tumbuhan dan Satwa. Jakarta : Departemen Kehutanan.* There are *Cekakak Sungai, Raja Udang Meninting, Burung Madu Sriganti, and Burung Madu Kelapa.* Types of high abundance birds are Bondol Jawa, and Walet Linci, while 20 other species are low in abundance.

Viviani (2016) found that there are many behavioral tourists in Blue Lagoon. Based on research conducted on Tirta Budi Blue Lagoon tourist attraction, found two categories of tourist behavior that is routine and incidental. Both categories of behavior are carried out within the river and its banks, as well as around the river. The routine behavior of visitors conducted in the river and banks is bathing, fishing, swimming, and mutual help. They are visitors who come from inside and outside the village. The local community sells a variety fast-food, heavy or light meals at a *warung* stall spread across the open courtyard inside the area.

Incidental behavior that develops in rivers and banks, as well as around them, generally has a high cultural value and characterizes this area. This incidental momentum is visited not only by local communities also outside visitors, even the Sleman regent and the Governor. The behavioral typology developed in this region tends to center. The visitor's point of interest has been centered on the Tepus riverbed used as the main bath as can be seen from Figure 1. The behavior of other visitors is more widespread around a river in the Blue Lagoon area.

Fig. 1. Tourists' behavior in general bath as a central tourism area.

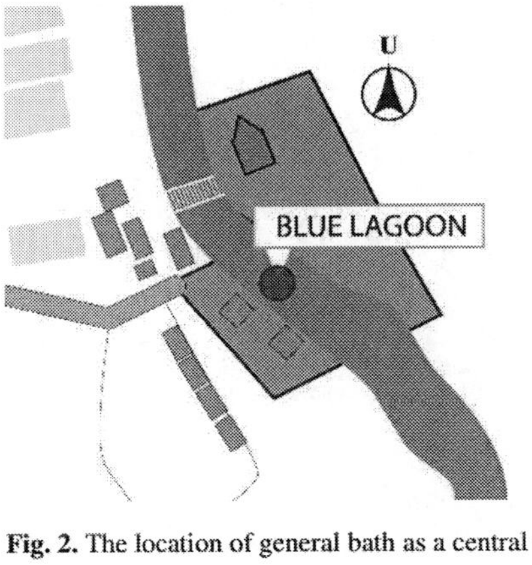

Fig. 2. The location of general bath as a central tourism area.

By collaboration with Author, the latter study was written in the context of a biology science writing contest in 2017. The title is Tirta Budi Blue Lagoon Tourism Object Development Strategy As A Potential For River Tourism and Sustainable Watershed Area. The paper concluded that Tirta Budi Blue Lagoon has a potential community that can create activities economically. This requires the design of the area in accordance with the existing potential and Government regulations on the development of river tourists area. The proposed development strategy/the finding is to emphasize the local potential of the area, both physically/environmentally and non physically/culturally (Saptorini, 2017). Here are some theories relevant to the proposals/findings.

Urban Desin Process and Tourism Policy Development Godfrey and Clarke (2000) emphazised that tourism development strategy has 3 steps. That are:

1. Identifying opportuniies and constraints (based on an evacuation of supply and demand),
2. Setting development goals and objective,
3. Define a series of action steps (Designed to achieve the goals and objectives within some specified time frame).

This theory is technically reinforced in terms of urban design by Hamid Shirvani. He examined that there are 8 eight components for considering the output of urban design (1985:p.7). These are: land use plan, mass layout, circulation, open space, pedestrian, supporting facilities, infomation, and preservation. In terms of tourism area context, Smith (1994) focussing in 5 elements and could be manifested in tourism products. Those are: physical plant, services, hospitality, freedom of choice, and involvement (p.587). This concept was figured as the circle, which the physical plant as the core.

This concept is affirmed legally through regulations issued by Tourism Minister Republic Indonesia. The regulation No.014/2016 establishes a number of criteria in the guidelines of sustainable tourism design. In principle, the scope of the guidelines for sustainable tourism destinations includes 4 verses. There are management of sustainable tourism destinations, economic utilization for local communities, cultural preservation for the community and visitors, and environmental conservation (p.3). This regulation supports the potential of Blue Lagoon in realizing as a tourism destination that prioritizes the culture and environment. The biophilic design concept is an appropriate and contextual choice with the area's condition.

Biophilic Design Principle

Biophilic design is the designing for people as a biological organism, respecting the mind-body systems as indicators of health and well-being in the context of what is locally appropriate and responsive. Good biophilic design draws from influential perspectives – health conditions, socio-cultural norms and expectations, past experiences. Those are to create spaces that are inspirational, restorative, healthy, as well as integrative with the functionality of the place and the ecosystem to which it is applied. The spirit of biophilic design must nurture a love of place (p.13).

This theory is formulated by Terrapin Bright Green, LLC. 2014 in 14 Patterns of Biophilic Design. The principle of this strategy is still relevan to improve health, productivity, and well-beeing. The strategy is organized by 3 categories: nature in the space, natural analogues, and nature in the space as follows.

Nature in the Space

Nature in the space includes plant life, water and animals, as well as breezes, sounds, scents and other natural elements. Common examples include potted plants, flowerbeds, bird feeders, butterfly gardens, water features, fountains, aquariums, courtyard gardens and green walls or vegetated roofs. The strongest nature in the space experiences are achieved through the creation of meaningful, direct connections with these natural elements, particularly through diversity, movement and multi-sensory interactions (p.9).

Natural Analogues

Natural Analogues addresses objects, materials, colors, shapes, sequences and patterns found in nature, manifest as artwork, ornamentation, furniture, décor, and textiles in the built environment. mimicry of shells and leaves, furniture with organic shapes, and natural materials that have been processed or extensively altered. The strongest natural analogue experiences are achieved by providing information richness in an organized and sometimes evolving manner (p.10).

Nature of the Space

Nature of the Space addresses spatial configurations in nature. This includes our innate and learned desire to be able to see beyond our immediate surroundings, our fascination with the slightly dangerous or unknown; obscured views and revelatory moments; and sometimes even phobia inducing properties when they include a trusted element of safety. The strongest nature of the space experiences are achieved through the creation of deliberate and engaging spatial configurations commingled with patterns of nature in the space and natural analogues.

Method

The study method in this paper is based on a map of previous research findings, which is then linked with a number of relevant parameters from the affordance theory, the designated tourism design teory and biophilic design. The map of previous research findings is a category of visitor behavior found routinely and incidentally in rivers and banks, as well as nearby ones. The map of this finding then identified its potential properties used in biophilic design towards healthy, productive, and wellbeing areas. Diagramatically, the method of discussion of this paper as follows Fig.3.

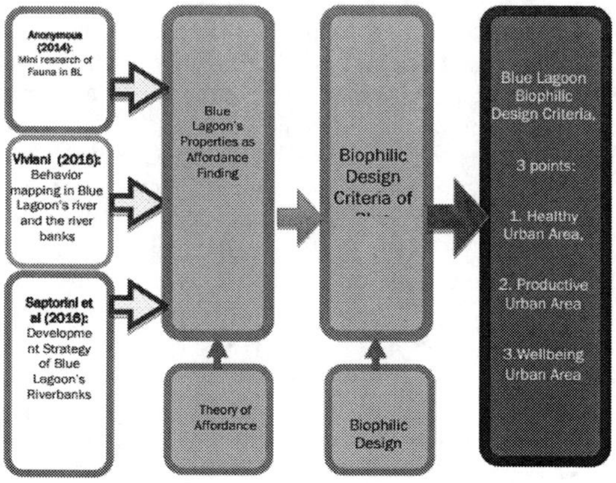

Fig. 3. Method

Result And Analysis: Map Affordance and Activity to Biophilic Design Principle of Blue Lagoon Tourism Area

This section is the result of the mapping of affordance and activity in the Blue Lagoon, and its analysis to formulate the biophilic design principle, as the purpose of this paper. By syntesizing the map of previous research findings, the theory of affordance, and the biophilic design principle, the result would be examined into 5 elements to manifesting the generic of tourism design product. These are physical plant, service, hospitality, freedom of choice, and involvement. Each point may related to the 8 urban design elements which offers the design strategic criteria to increasing the healthy, productivity, and the well-being of Blue Lagoon as the biophilic design tourism area.

The Map of Previous Research Findings

Previous Research Findings has found a map of visitor behavior that is routinely and incidentally done in rivers, banks, and its surroundings as shown in the Fig.4 - Fig.11.

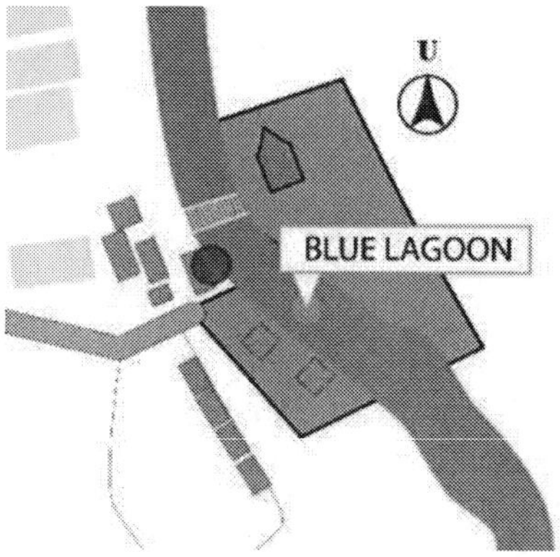

Fig. 4. Location of Routine Visitor Behavior: In The River and Banks.

Fig. 5. Fishing, one of the routine Visitor behavior In The River and Banks.

Fig. 6. Fishing, one of the routine Visitor Behavior In The River and Banks.

Fig. 7. Fishing, one of the routine Visitor Behavior In The River and Banks.

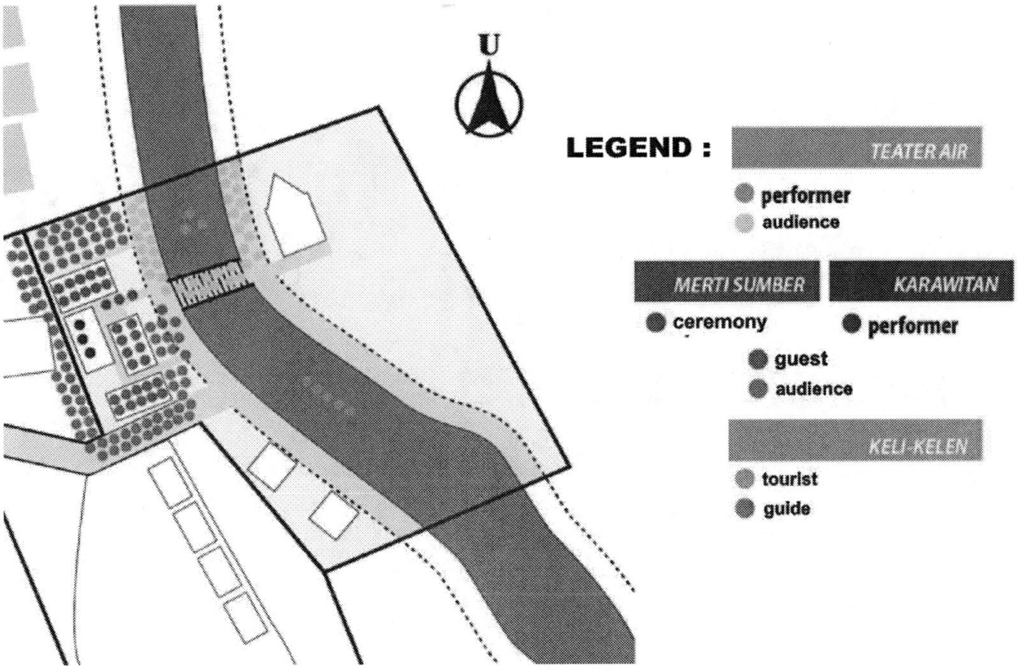

Fig. 8. Insidental Visitor Behavior In The River and Banks.

Fig. 9. Water Theatre, One of The Insidental Visitor Behavior In The River and Banks.

Fig. 11. Camping, One of The Insidental Visitor Behavior In The River and Its Surroundings.

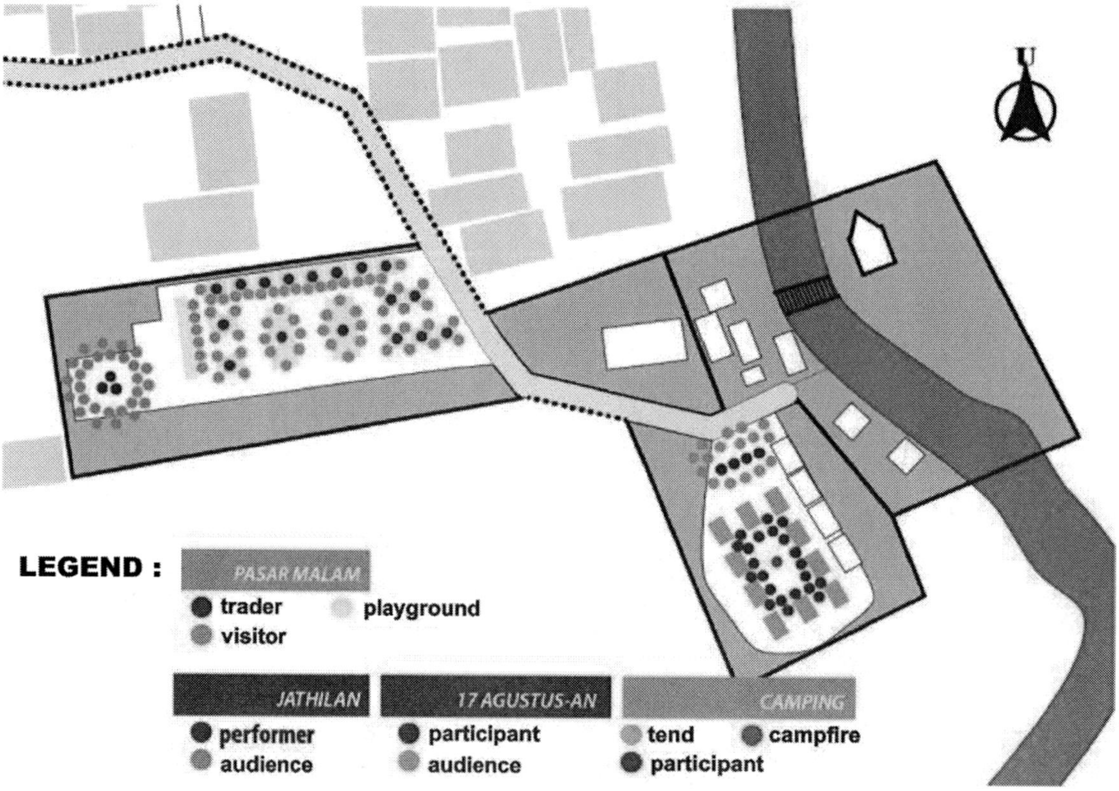

Fig. 10. Routine Visitor Behavior In The River and Its Surroundings.

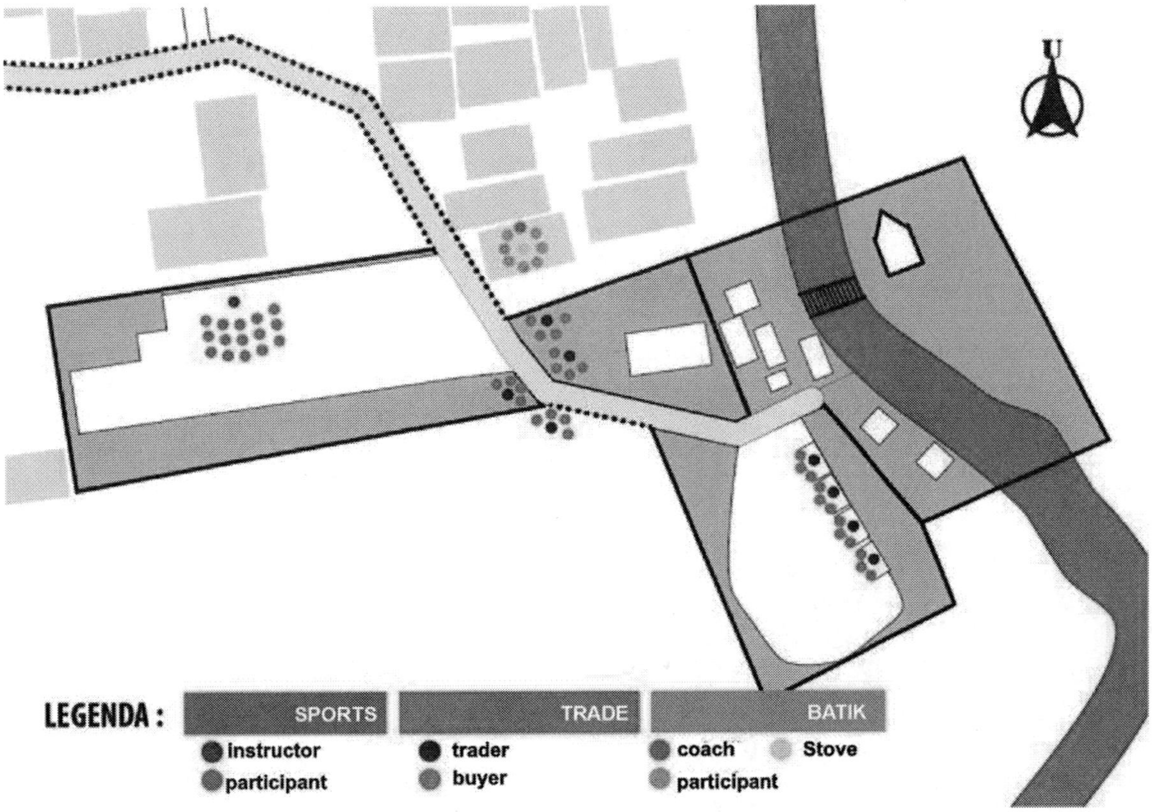

Fig. 12. Incidental Visitor Behavior In The River and Its Surroundings.

Mapping of Natural Properties and The Activities

The affordance mapping and its activities are translated as properties in the setting area that generate activity. The results are linked to the principle of biophilic design. This study is packaged into 5 elements of Smith's theory in Table 1 - Table 5 as follows.

Physical plant

The main affordance related the physical plant are the water, plant, and birds. Water is the main properties where classified in two locations: in the river and banks, and the surroundings' river. The river water is clear and blue, so tourists are interested in using pool. Swimming, swimming contest, fishing, bathing are the phenomena activities in this property.

The surrounding water are spring waters. The main spring waters are "man spring water" and "women spring water". The sub/another/child spring water are scattered throughout the area and are used to create gurgling sound sources and water plants.

Bamboo grove as we can seen in Fig.13 show that this is the dominant plant. This vegetation grew hanging over the Tepus riverbed. It is use tourists for overshadow who are swimming, swimming contest, theatre, fishing, etc. Table 1 is the mapping of main natural properties and the activities. These properties and activities are components that can be taken into consideration of the biophilic design area. As the Shirvani theory, zonification is the beginning of the design.

Fig. 13. Green bamboo, the dominant plant as generator tourism activities.
Source: https://www.jejakpiknik.com/blue-lagoon.

 https://doi.org/10.1051/shsconf/20184106003

Tabel 1. Mapping of main natural properties: water, plant, birds.

Natural Properties			Activities	Biophilic Design Principle
Water	In the river	General baths	Swimming, - contest, bathing, fishing, theatre	Minimizing landuse change and maximazing to keep up the natural assets by connecting with natural system
	Surroundings	Water springs: man, female	Bathing, washing,	
Plants	Dominant	Bamboo grove, bread fruit.	Perched birds	
	General	Plant seeds	Perched birds	
Bird, goat	Protected PP 7/1999	Cekakak Sungai, Raja Udang Meninting, Madu Sriganti, Burung Madu Kelapa	Eating grains, whistling, investation	

Fig. 14. Activities in the river: Swimming, bathing, fishing, theatre.

In the bamboo surroundings, there are stone hole where tourists using them for sitting while listening the frozen bamboo friction. These properties are natural resources which potential vegetation to be applied in landscape design principle for maintaining the area ecosystem chain.

Fig. 15. Cekakak sungai bird, the protected birds.
Source: https://id.wikipedia.org/wiki/cekakak_sungai.

Zonification Design

Zonification of properties and its activities is the design proposal for this area as shown in Figure 16. The natural properties are the main assets that should be keeped up and maintained as a part of biophilic design area. By minimizing land use change is the main principle of the design area.

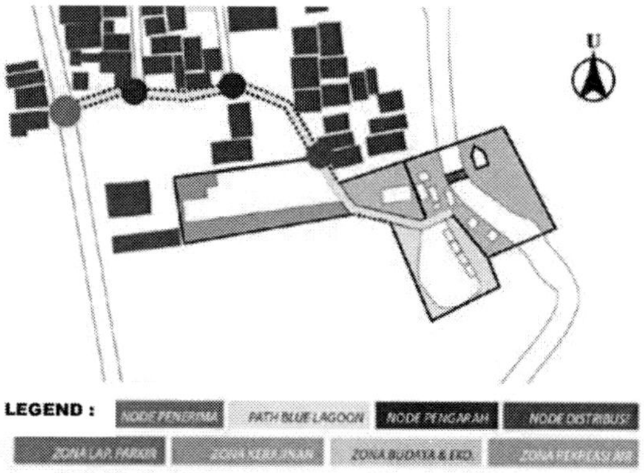

Fig. 16. Zonification Design Based On Properties and Activities.

Services

The main affordance related the services are the performance of specific task required to meet the need of tourists (Tabel 2). The tourists' need are homestay, rest area, *warung makan*, fit room, first aid place, prayer room, etc

Tabel 2. Mapping of the main services properties: water, plant, birds For Tourism Services.

Property		Activities	Biophilic Design Principle
Natural	Home made		
Stone		sitting	Designing building layout according the similar caracteristic activities. The principle design is this building services could be spreadly distributed in the area.
	Bence	sitting	
	Gubug	massage, getting a rest,	
	Rental camera	taking picture	
	Warung	drinking, eating, chatting	
	Communities home	sleeping, training painting, cooking	

Biophilic Design Principle

Strategy to meet the tourists' needs, in principle, is setting the building layout evenly distributed. To express a natural impression, the building appearance is directed using the principle of natural element analogies. The natural asset of water, rivers, bamboo, birds, is a natural element of the Blue Lagoon that can be adopted of its form, behavior, and its color into the expression of the building.

Hospitality

The main affordance related the hospitality has been given by clear the access, walkable pedestrian, lighting, multy natural sensory: visually, auditorically, odour sensory, etc (Tabel 3).

Biophilic Design Principle

The strategy to achieve hospitality for tourists, in principle, is to set the landscape that provides many benefits for visitors, both physically and psychologically.

Physical benefits when Blue Lagoon landscape facilitates comfort movement. This could be manifested in the form of circulation pattern arrangement in site area that flows with standard circulation. The design guidelines are sufficient road widths, constructive road surfaces, enough lighting, equipped with street furniture. The selection of shady vegetation, spread along the pedistrian pathway will overshadow tourist so thermal comfort is achieved.

Psychic benefits, when Blue Lagoon landscape provides visual comfort, auditory, and a pleasant smell for visitors. The selection of colorful shrubs grown on the knot of the pedistrian knot will attract the attention of tourists so that visual comfort is achieved. The selection of plant movement plants that produce grain seeds, will bring birds and the sound of birdsong will entertain tourists.

Filling shrubs perfume along the circulation path will produce a fragrant aroma, so visitors can be treated naturally.

Tabel 3. Mapping of Blue Lagoon"s Properties and Activities For Tourism Hospitality.

Property		Activities	Biophilic Design Principle
Natural	Man made		
	Pintu gerbang	signing the area/landmarking	Combine vegetation and manifacture elements for main gate area
	Pedestrian	moving in the area	Construct a standard wide pedestrian for tourist walkability
Vegetation		producting odour, color, and seed	Choose full color flower, seeded, and aromatic vegetations for visual, whistling, and auditory comfortabilies
	lighting	Visual, atmospher character supporting	Maximize natural light into the area to keep the visitor healthy and help visual comfort in enjoying its natural property

Freedom of choice

The main affordance related the freedom of choice is in the form of tourism facility that give many optional properties for tourists (Tabel 4). For this purposes, Blue Lagoon gives optional attraction for familiy, individually, and communally. Doing in water, training some cultural skills, sensing the natural atmosphere, are the choice which be given to the tourists. Doing in water has many choices both doing bathing/swimming, and fishing in the main pool. Doing some washings (dishes and or clothes), bathing, and so on in the water springs both man and women. Training some cutural skills are choices for tourists in the form of many tranings e.g.: paintings, traditional dancing, local food cooking, and so on. Sensing the natural atmosphere has many choices both hearing the birds whistling, watterfall, visual green area, etc.

Tabel 4. Mapping of Blue Lagoon"s Properties and Activities For Tourism Freedom Of Choice.

Property		Activities	Biophilic Design Principle
Natural Attraction	Sosio-Cultural Attraction		
Water attraction		Theatre, swimming, fishing, kelen kelen, contests	Maximize local material for river bank construction
	Camping area	Renting the camping facilities	Maximize community involvement in routine and incedental events
	Karawitan area	Cultural performing	
	Village attraction	sosio living in settlement	
	Merti sumber	Cultural festival performing	

Sosio-Cultural Attraction

Here are some design guidelines to develop a socio-cultural attraction for the Blue Lagoon's community to be more productive and improve their economies.

a. Renting the camping area with a standard quality. The design strategy is to arrange with a sufficient area, the shady floor surface, protected by many shade trees, sufficient quantity bathing place and decent quality, equipped with a safe and clean public kitchen for nourish the user.
b. Providing a performance area for karawitan so that the mother community can perform periodically to entertain visitors.
c. Strengthen community skills in cooking local culinary.
d. Inviting visitors to tour the neighborhoods with local culture in the neighborhood of local architecture.
e. Having a systemic promotion management to invite regularly periodical visitors in annual merti festivals event.

Biophilic Design Principle

Water Attraction

The strategy to strengthen river cliffs used for water attractions, in principle, is to maximize the safety and security of tourists. The design guideline is maximizing the use of rocks as a construction material for river bank booster for tourist safety and security. Stone adhesive is used a strong rooted plant to knit the retaining stone.

Tourists tribune for watching the attractions allocated to a space of sufficient size and comfort visually and thermally. For that, the terracing design is proposed as the seat model. Its location is oriented under shade trees so thermal comfort environment is achieved.

Involvement

The main affordance related the involvement is the stakeholders who mandatory and potential involve (Tabel 5). They are the local community, Government, Non Government Organisation (NGO), and Investors. Their involements a re in the form of both physical and or non physical development.

Tabel 5. Mapping of Blue Lagoon's Properties and Activities For Tourism Involvement.

Property			Activity	Biophilic Design Principle
Gov	NGO	Comm		
Regional relationship			Sleman regent is present in a number of event	Enhance the stakeholders' development networking
Provincial Relaionship			The governor was present at the merti source ceremony	
	Investor, Academicions		Supporting finance and ideas	
		Gerakan community: fathers, mothers, teenagers, children	Organizing some culturalevents and area development	

Biophilic Design Principle

Relationships with networks of development actors, both governmental, non-governmental, and community are the assets that Blue Lagoon already has. Therefore, communication and cooperation should be scheduled regularly and systematically. Involvement of each actor, according to the role, intensified towards the Blue Lagoon as a national tourist area, as well as international.

Conclusion: Toward A Healthy, Productive, and Well-being Blue Lagoon Tourism Area

The development of Blue Lagoon into a healthy, productive, and well-being tourist area when pay attention to the following principles.

1. Maximizing the utilization of existing natural properties is a strategic principle for the Blue Lagoon to nourish visitors both physically and psychologically.
2. The addition of plants that fit the criteria of physical and psychical comfort can be developed through the arrangement of the Blue Lagoon landscape and the building performance.
3. Capacity building of local communities is needed, especially in developing tourism development paradigm. potents should be enhancement. Cultural skills that they already have still need assistance in terms of innovation and technology.
4. Government support, both policy and financial is indispensable for the Blue Lagoon development sustainability. Not only in the sectors of tourism, environment, and infrastructure, but also in the economic and banking sectors.
5. The Non Government Organisation, such as expertist, academicions, universities, investors and also potentials stakeholders may be involved to assisst and or to advocacy the future development.

References

Febriani, Susi. 2016/09 November 2016. "Level of Public Participation in Tourism Activities In Blue Lagoon Tourism Village Dalem Widodomartani Ngemplak Sleman. "Thesis Digital. Akses 23 September 2017. URL:http//eprints.uny.ac.id/id/eprint/43435.

Chemero, Anthony. 2003. An Outline of a Theory of affordances. *ECOLOGICAL PSYCHOLOGY*, 15 (2), 181-195.

Gibson, Eleanor J., et al. 1999 *"The MIT Encyclopedia of the Cognitive Sciences"*. Lokasi terbit : Nama penerbit.

Godfrey, Kerry et al. 2000. *The Tourism Development Handbook. A Practical Approach To Planning and Marketing*. ISBN 0304704512, 9780304704514 Cassel. Madison.

https://dokumen.tips/documents/miniriset-keanekaragaman-dak-kelimpahan-jenis-burung-di-obyek-wisata-blue-lagoon.html

https://www.jejakpiknik.com/blue-lagoon

Laurens, Joyce Marcel. 2004. *Arsitektur Perilaku*, Grasindo. Jakarta.

Law of the Republic of Indonesia Number 10 Year 2009 About TOURISM

Lerner, Abby et al. 2016. The Financial Benefits of Biophilic Design in the Workplace: A Review and Summary of Current Research, MIST Environment.

Lindberg, Paivi. 2014. In search of affordances and Visual Quality. Interpreting environments of children aged under three inseven finish day-care centres.

National Institute For Health And Welfare: Juvenes. Terveyden JA. Hyvinvoinnin Laitos. Finland.

Priatman, Jimmy. 2012. Biophilia Design Concept "as a Green Dimension on Empathic Architecture. National Seminar-Towards a Great Architecture. Jurusan Arsitektur Universitas Kristen Petra. Surabaya.

Regulation of the Minister of Tourism of the Republic of Indonesia Number 14 Year 2016 About GUIDELINES OF DESTINATION OF SUSTAINABLE TOURISMShirvani, Hamid. 1985. *The Urban Design Process.* New York: Van Norstrand Reinhold.

Saptorini, Hastuti et al. 2017. *"Tourism Development Strategy Of Tirta Budi Blue Lagoon Yogyakarta, As A Rich Tourism Potency And Sustainable Derivative Area".* Compiled to Follow the Scientific Writing Competition Biological Week 2017. Unpublished.

Smith, Stephen LJ. 1994. *The Tourism Product.* USA: Elsevier Science Ltd.

Soegaard, M: Wikipedia, https://www.interaction-design.org/literature.

Terrapin Bright Green, LLC. 2014. *14 Patterns of Biophilic Design.* New York: Terrapin Bright Green llc.

Yudistira, Anom IGA., et al. 2012. *Rancangan Sistem Penilaian Keselamatan Pengunjung Tempat Wisata.* Jurnal Teknologi. Tahun 29 Nomor 320, hal 21-22.

Eartquake Vulnerability Level of Reconstructed Houses, Lesson Learned after Ten Years Java Eartquake 2006

Noor Cholis Idham[1,*] and *Munther* Mohd[2]

[1]Universitas Islam Indonesia, Indonesia
[2]Nablus University, Palestine

Abstract. Ten years after devastating earthquake May 27 th, 2006, housing reconstruction in the affected area has been done and development is still undergoing. Since then, houses are constructed in various forms and quality. Unrestrained housing development contributes uncertain quality of the buildings in term of building safety. This study examines the level of earthquake vulnerability on the reconstruction houses in the area according to the structural firmness. The method used is by assessing the structural completeness, especially the upper part of building elements, synchronized by the five damages classification for grouping the building affected by the quake. This method is aimed to predict damage level for the future similar earthquake. The outcome of this study shows the level of earthquake vulnerability in the every sub-district. It does also compare the level of building safety differences

Keywords: reconstruction development, earthquake vulnerability assessment, structural safety level, damaged prediction

Introduction

A strong earthquake of magnitude 6.3 struck Java Island, Indonesia, on May 27, 2006 at 5:54 am local time. The epicenter was located at 7.962°S, 110.458°E, about 25 km south-southeast of Yogyakarta with a fairly shallow focal depth, about 10 km [1].

The quake had affected two provinces Yogyakarta and Central Java of Indonesia. Due to this earthquake, 5,751 people loss of life and 29,473 injured, 145,929 houses were completely collapsed and 190,843 others were damaged [2]. This left 1.6 million people homeless [3]. The worst area destroyed was Bantul district of Yogyakarta province.

Table 1. The list of earthquake victims and houses damages (source: BAPPENAS, 2006a/the latest data, taken from Media Centre, June 17th, 2006[2].

Location	Death	Bad Injury	Light Injury	House Collapsed	House Severe Damaged	House Moderate Damaged
DI Yogyakarta	4,688	10,975	7,862	**115,170**	**124,748**	**170,543**
Kab.Bantul	4,143	8,673	3,353	**71,763**	**71,372**	**73,669**
Kab.Sleman	234	689	,539	19,113	27,687	49,065
Kodya Yogyakarta	204	245	73	7,186	14,561	21,230
Kab.Gunung Kidul	84	1,086		12,581	5,950	18,078
Kulon Progo	23	282	1,897	4,527	5,178	8,501
Central Java	1,063	18,498	-	30,759	66,095	103,136
Klaten	1,045	18,123		29,988	62,979	98,552
Magelang	10			386	386	546
Boyolali	4	300		307	696	708
Purworejo	1	4		10	214	780
Sukorharjo	3	67		51	1,808	2,476
Wonogiri		4		17	12	74
Total	**5,751**	**29,473**	**7,862**	**145,929**	**190,843**	**273,679**

* Corresponding author: noor.idham@uii.ac.id

https://doi.org/10.1051/shsconf/20184106004

Fig. 1. Destructed houses of masonry-brick bearing wall.

Soon after implementation of emergency period, Indonesian government supported by various relief organizations redeveloped the destroyed area with reconstruction program. The program was successful to rebuild 279.000 houses and to restore 253.000 others within two year [4]. There were Eighty-five agencies who participated within the UN-OCHA humanitarian-cluster system redevelopment and 546 organizations recorded as delivering assistance. Budget confirmed allocated and distributed 5.4 trillion Rupiah (580 million USD) [3] from total plan about 15.3 trillion Rupiah (1.64 billion USD) for housing sector only [5].

Fig. 2. Reconstruction house and its upper roof reinforced concrete frame structural system.

Reconstructing the collapsed houses were involving many donors in various ways. Houses with any form, size, and structural type were constructed. The people were also given a chance to rebuild their houses themselves as a program so called the community based settlementa. Two reputable universities from Yogyakarta and Semarang soon after 2 years, when reconstruction program had finished, reported that all the houses are technically earthquake-proof with good quality [4]. However, there is no guarantee in building performance since reconstruction house were done in various schemes starting from providing complete house, core house, or only giving materials and funds in various quality level [7]. As a result, there are many types of new houses starting from simple to complex structural system, although reinforced concrete framed with brick wall houses are still the majority. Among the reinforced concrete framed houses themselves, the variety of structural system and construction is also wide. The people are also build or extend the houses autonomously afterward.

Even though located in the ring of fireb with frequent destructed earthquakes, Indonesia has less experience in post-quake reconstruction. Just like in any other post-quake reconstruction, it is almost impossible to build immediately a better housing in the affected area [6]. The most important action in such circumstance is to aid as soon as possible the people by means of appropriate ready-use shelters and houses. Trough the time, housing development must carry on in the areas with less attention than before. Lack of supervising and consultation service provided for the people turns the housing development in uncertainty. Though some public guidance in earthquake proof house were available, preliminary surveys indicate that the community members are still unable to afford quality of building without

[1] The re-housing scheme which gives the people some money (limited) and an autonomy to manage the reconstruction project by themselves with facilitators supported by government. It has an advantage for the people to have a chance to rebuild their houses according their preferences but has lack of quality since the limited availability of facilitator and the massive number of the project. Misappropriated in managerial issues also found in many steps of reconstruction starting from data collection to reconstruction process [8]

[2] The line of continental crushes as a result of the meeting boundaries; Indonesian territory surrounding by many ring of fires between Eurasia, Australia, Pacific, and Philippine plates.

professional skills involvement in order to build seismically resistant housing [9]. Building vulnerability is very essential issue since collapsed building has about 80 % of human fatalities affected by earthquake [10] and strong concern about vulnerability in the affected area is still in high level.

The Significant Role of Upper Structure Examination

From the previous 2006 earthquake experience, the failure of upper parts of the houses was the most significant aspect in contributing severe casualties. In fact, it was a very view cases involving the sub structure system issues such as house foundation and soil failures. Poor quality of materials and lack integrity among upper building components are the most causal factor to the collapse of the houses [13]. When the quake affects the defect upper parts, the houses were easily damaged and even easily collapsed without any notification. One of the most reasons was because of weak masonry brick wall or improper reinforced concrete structural system used mostly by the people.

The new houses in general are constructed by bricks wall and reinforced concrete frame with relatively better quality of materials. Unfortunately, houses with timber structure only found in view number. This is opposite way if compare to Japan where almost 90% of building stock in earthquakes prone made from timber [14]. Concrete framed core house with pitched roof type was majority constructed for replacing the collapsed houses. People were expected to extend by themselves afterward according to their need and ability. This was decided due to the limited budged and the large scale impact of disaster.

Concrete materials (Portland cement, sand, and gravel for 1:2:3) with reinforced 12mm diameter steel bar are used for the reconstruction. 60-70cm deep and 50cm wide stone foundation is also being used, while mortar usually contained 1 PC: 3 lime powder: 5 sand or 1 PC: 6 sand (without lime powder). The materials used for mortar are much better than previous collapsed houses which mostly very brittle [13], while roof uses mostly ceramic roof tiles. The whole construction seem well done since in the every sub-district had their own facilitators to guide the construction process.
Major aspects of building stability from earthquake are building configuration, structural collapse mechanism, beam-column joints, and material-construction quality [15]. Since the quality of materials is better than used before and the form is relatively simpler, house's structural collapse mechanism combined with structural integrity by the joints have important role to determine the level of structural stand-ability under quake.

The concept of 'roof first' actually is not merely emphasized on the function of roof as space giver but also the capability to protect inhabitant bellow since it was seen as considered element for causing a big number of casualties. In regard to the collapse mechanism in a simple form building, the upper parts are the most significant aspect to be evaluated. Most of the post-quake building mitigation was focused on column-beam joint appropriateness [15], yet roof frame and its connections to the main frame is less considered. For these reasons, re-evaluating upper structure focusing on main roof elements is essential.

Field Data

The 2006 Javanese earthquake stroke the middle part of Java Island. Bantul district of Yogyakarta province is one of the most affected areas beside Klaten region in Central Java. From the two areas, 72% of the total housing stock was destroyed resulting 95% of total fatalities and serious injuries [8]. More than 50% of houses destructed in Bantul was inhabitable (see table).

Table 2. Statistical Data of Houses Affected by 2006 Java Earthquake in Bantul region (source: updated from BAPPENAS, 2006a/the latest data, taken in June 17th, 2006[2]).

Sub-districts	Total Houses*	Collapsed	Heavily Damaged	Lightly Damaged	Total uninhabited	% uninhabited **
BAMBANG LIPURO	8853	6587	2732	816	9319	75.19
BANTUL	12403	4708	7378	3295	12086	69.60
IMOGIRI	11480	5664	5353	4471	11017	68.55
PLERET	8821	8139	2322	1438	10461	84.71
JETIS	10352	11195	2532	655	13727	94.72
KRETEK	6838	1081	4665	2486	5746	60.02
PANDAK	9576	2886	5185	4065	8071	60.20
PIYUNGA N	9916	5514	4801	3135	10315	74.30
PUNDONG	7297	6793	1903	500	8696	85.12
SEWON	20976	8281	8496	6004	16777	57.13
BANGUN-TAPAN	20009	5557	8232	7452	13789	49.22
DLINGO	6501	1377	3380	4720	4757	52.27
KASIHAN	19735	1790	4657	12103	6447	23.33
PAJA-NGAN	7344	1228	2216	2610	3444	33.50
SANDEN	7135	97	2052	4650	2149	21.51
SEDAYU	10266	243	1800	4591	2043	14.21
SRANDAK AN	6544	342	3054	3506	3396	37.07
total	**184046**	**71482**	**70758**	**66497**	**142240**	**55.20**

*Based on year 2003 census statistical data (podes 2003)
*** Ratio of total uninhabited to baseline data by 1.4 growing level assumptions in 3 years*

Ten sub-districts having higher percentage of deserted houses in Bantul region of Yogyakarta province had been selected for field samples. About 300 houses had been picked randomly by observing 30 houses in each sub-district. The reconstruction houses with additional room or building are examined for their completeness of main upper and roof structural system. In order to validate the sample, the reconstruction houses which are taken as samples should be observed visually.

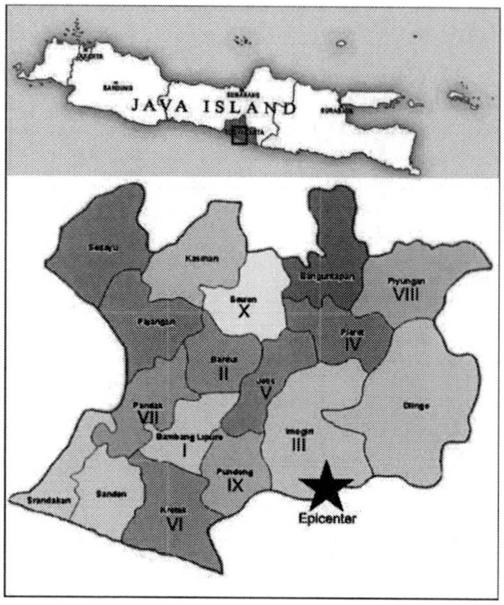

Fig. 3. Data gathered from the most affected areas ten Sub-districts of Bantul region in Yogyakarta Province of Indonesia.

This study offers the simple way to evaluate the earthquake vulnerability of the houses by visual method developed for the case of Java Indonesia. The similar procedure was proposed by FEMA 154 [11, 12] and followed by various scholars in some countries. The method assess a simple structure performance and will not involving detailed structural investigation nor complicated consideration. Only visual applicable aspect of the houses will be examined.

The most problem to observe the reconstruction houses especially for these 'semi-vernaculars housing', is the limitation of technical information about houses e.g. the exact use of material and technical construction system used. For this reason, the best way for visual assessment is by inspecting the comprehensiveness of upper parts of the houses and roof structures. The presence or absence of appropriate structural elements is vital. Incomplete, incorrect, or misplaced part of structural elements represents the improperness the system as a whole. Identification will be listed in order to classify in which grade the houses will be. Although this is simple technique, since it does contain the most important aspect of building stand-ability from the quake, the study is worth enough to decide either the building is structurally acceptable or not. In order to confirm the classification, the grading method is proven by utilization of SAP 2000 program.

Fig. 4. Roof Structural failed both for bearing brick wall and reinforced concrete frame.

Grading System

Method used for grading system is by giving safety values in the levels of structural integrity. The minimum requirement for surviving from quake should be taken as starting point as a 'minimum standard' level, by means that the less should be mark as negatives and the more as positives. According to the five levels of damages classification used by Indonesian authority, damages are classified starting from very slight, slight, moderate, severe, and very severe. In this regard, moderate damage level is the minimum circumstance of the building after earthquake that still can be emergency occupied and expected to be still stand up without affecting the inhabitant's safety at the moment. In order to link with structural safety, the moderate damage level is then correlated with a minimum requirement stated as sufficient level of structural system in facing quake (see table2).

Five grades for structural classification starting from strong, less strong, sufficient, weak, and very weak then parallelized to the five damage levels of very slight, slight, moderate, severe and very severe (figure 8). To be having positive or negative values, the middle level of minimum accepted level is given by 0 values. The other higher accepted for strong and less strong structure are grouped as positive one and two (+1 and +2), while lower groups as rejected level or weak and very weak structures are stated as negative one and two (-1 and -2).

Table 3. Damage classification level used for vernacular house (developed from Dirjen Ciptakarya PU 2006 [16]).

Damage Level	Meaning	Building Condition	Building Items Affected
1	No or Very slight damage	No or Small crack in non-structural parts of building	Non structural wall Roof tiles Openings
2	Slight damage	Some cracks in non structural members	Non structural wall Roof tiles, floor tiles Openings
3	Moderate damage	Crack on structural members such as bearing wall, column, beam and connection	Structural wall Column, beam Ceiling, Roof frame
4	Severe damage	Severe cracks or lost connection between main structural members	Structural wall Column, beam, ceiling Roof frame, Foundation
5	Very Severe, Collapse	Some parts or whole building collapsed	Structural wall Column, beam Roof frame, Foundation

In order to apply the point system in the structural integrity, the level from strong to very weak then translated to the structural grading system by grouping the most complete to the least one. In case of reinforced concrete frame system, the rigidity level which is directly related to stand-ability is judged by the structural comprehensiveness which is confirmed by the presence and absence of structural members and its proper connection. Since the most type of reconstruction houses is pitch roof, the most concern then focused to gable system and its connection with main frame underneath. The presence of brick wall inside the frame is also taken into account since it affects significantly the work of frame system.

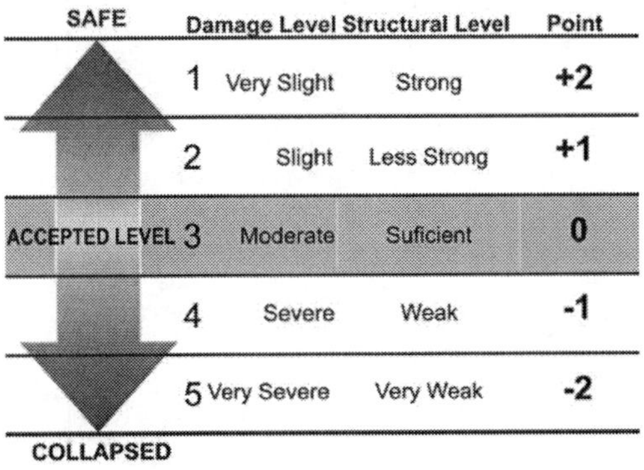

Fig. 5. Point system applied for structural levels paralyzed with damage level chart.

Similar grading technique is also used for wooden and reinforced concrete frame-wood combination structural system regarding the flexible joint system principles. Since the wooden frame for gable roof is easily collapsed in front-rear direction, strengthening between the frames is most concerned beside the frame configuration properness as a main factor that easily affected by earthquake (figure 6).

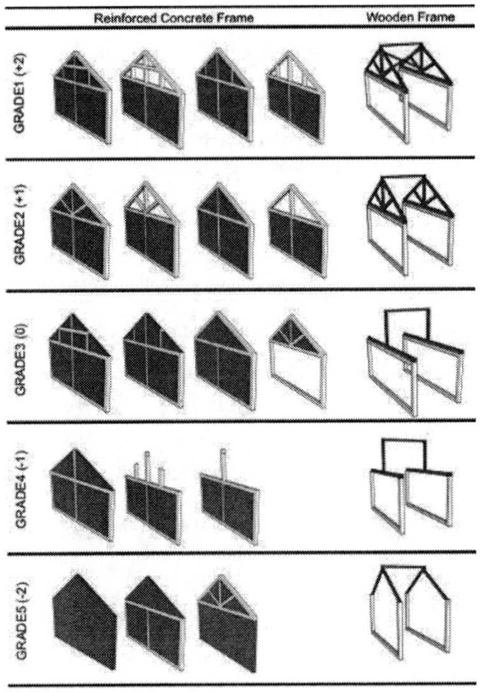

	Structural Definition
GRADE 1 (+2)	**Strong Connection** Completed roof frame with more than three connection to the main frame either with or without walled brick in between. While for wood has 3 D connection to other frame. When earthquake strike it will stand completely
GRADE 2 (+1)	**Less Strong Connection** Completed roof frame with less connection to the main frame. For concrete categorized as upper group as long as has rigid connection (3D connection) to other frame, for wood less connect. When earthquake strike it will little affected and still stand completely
GRADE 3 (0)	**Sufficient Connection** Uncompleted roof frame structure or completed one but less sufficient support bellow. For wood has no diagonal element for both sides. When earthquake strike it will severe affected but still stang completely.
GRADE 4 (-1)	**Weak Connection** Uncompleted roof frame structure but still has sufficient connection with the main frame bellow. For wood has no diagonal element for both sides with small horizontal under. When earthquake strike it will affected easily and some part will easily collapsed.
GRADE 5 (-2)	**Very Weak Connection** Incompleted roof frame structure with improper structural system without any connection between roof and main structural system. For wood without horizontal bracing member. When earthquake strike it will affected easily and the whole roof structure will easily collapsed.

Fig. 6. Grading systems criteria for reinforced concrete and wooden gable roof structure types.

Results

The vulnerability level had been analyzed in two ways, firstly by discovering the structural stand-ability level and secondly by estimating the level of unaccepted building in the area. The higher structural stand ability level means the less vulnerability while the level of unaccepted building related directly with the level of vulnerability. Both are needed in order to discover the level of vulnerability in different ways as the level of 'accepted' and 'non-accepted' for original and additional reconstruction houses. Though they are correlated each other, the structural stand-ability explains the general-average level of structural integrity which is taken from point system examination directly (see appendix 1) while the level of unaccepted building uses percentage system in order to show only the distribution of vulnerability in each area (see appendix 2).

Structural Stand-ability

Structural stand-ability level from earthquake shake which is defined by points system from +2 to -2 is applied for all data in the every sub-districts. The outcome shows that all levels are found in the areas. In general, all original reconstruction houses level are always above the additional houses. The levels of original and additional houses are not always parallel, and even some sub-districts such as Imogiri and Pandak have big difference between the two. Point system applied to samples had shown that -1 to 2 point levels had been found in almost the every sub-district. Unfortunately there are level -2 for original reconstruction house in the four sub-districts; Kretek (2 cases or 6.66% data), and Pundong (1 case or 3.33% data). For additional building, the -2 level is also found in Bantul (1 case or 3.7% data), Imogiri (1case or 3.33% data), Kretek (7 cases or 23.33% data), and Pundong (3 cases or 10% data).

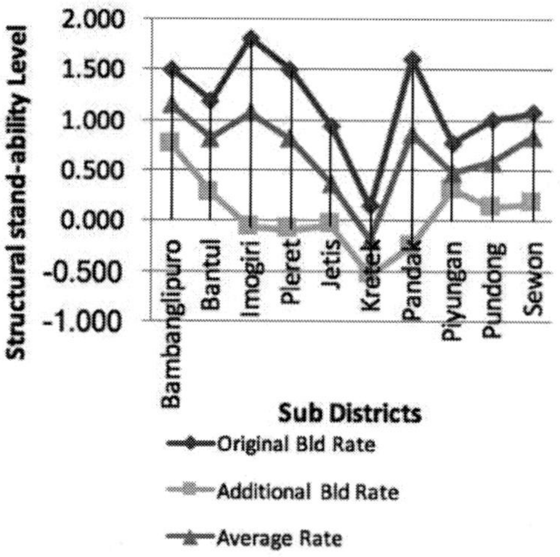

Fig. 7. Result chart of stand-ability level which is directly in opposite with vulnerability level of the houses.

Non-Acceptable Evaluation

Building vulnerability as lower level of acceptable structure or stated as non-acceptable is defined by percentage of failed house from the total sample in the area. Again, as shown in structural stand ability examination, the additional buildings have higher vulnerability. However, this examination considers only the amount lower level quality without concerning the higher as accepted group. This is useful method in order to focus the vulnerability level. The chart shows that the areas having higher level in structural stand-ability are not always have less percentage in vulnerability. Furthermore, only one district (Imogiri) that have 100% of original new reconstruction houses are accepted (0 to +2 levels), however, the rest of sample areas have unacceptable houses. Unfortunately, this is not followed by new additional buildings which are found in every sub-district with the unaccepted percentage starting from 20% to almost 78%. This is showing us that, by this way, the complete situation on vulnerability of reconstruction houses can be identified more clearly.

 https://doi.org/10.1051/shsconf/20184106004

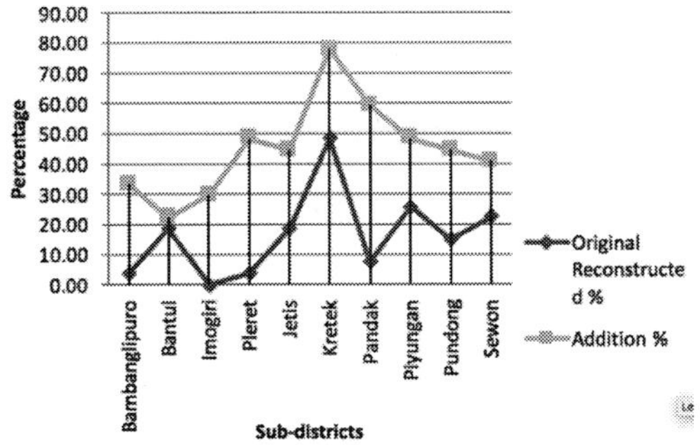

Fig. 8. Percentage of Non-Acceptable Reconstruction houses which directly shows the level of vulnerability.

Discussion

The examination of new reconstruction house and the additional building have no correlation in between. Imogiri, Pleret, and Pandak are sub-districts that have higher-positive level in building safety for original houses, but in the same time, the three are also have higher level in vulnerability. For additional houses are even found below zero as under accepted level (see figure 11). In percentage table, Imogiri, which has all new reconstruction house accepted, have some significant percentages of non-accepted additional buildings (see figure 12).

From this study, some aspects found:

1. Not all the reconstruction houses have similar quality in structural stand-ability to stand with future earthquake
2. Original reconstruction houses have no parallel quality with the additional in every sub-district
3. The percentage of non acceptable level is higher in additional buildings than in original buildings in any sub-districts.
4. In average, almost original construction houses are above the accepted level but still there is below level in one sub-district (Kretek)
5. 100% structurally accepted only found in Imogiri for original buildings, the others still have un-accepted building in certain levels.
6. Half of total sub-districts examined have under accepted level for additional building (Imogiri, Pleret, Jetis, Kretek, Pandak)
7. The highest quality for original reconstruction houses is sub-district Imogiri while the least one is Sub-district Kretek
8. The highest quality for additional buildings is Sub-district Bambanglipuro while the least one is Sub-district Kretek
9. The highest unaccepted percentage found in Kretek both for original and additional buildings

Biggest difference between original and additional is sub-district Imogiri while the smallest one is Sub-district Piyungan.

Conclusions

The ten most affected sub-districts had been analyzed and the outcome is that there is still high in vulnerability level even though new reconstruction houses had been constructed. Some other conclusions from the study are:

1. Not all reconstruction house in the area safe from the future earthquake threat
2. Non accepted building is still found in high percentage for new post-earthquake houses
3. Additional building is the most source of structural failure and there is high possibility to make any casualty in the future similar disaster
4. The trend of structural quality of additional building is going to decrease

By this result, this study has discovered the level of earthquake vulnerability applied to regions after Java earthquake 2006. Damage level from similar future earthquake can also be predicted accordingly in every district examined. The procedures presented here is simple which will facilitate the work of the inspection teams easier and will reduce the time required to complete the job. The next study for this issue is still need to be carried on in regard of other aspect such as structural quality level and relation with the building type, structural type, and materials used. The other aspect may set off with occupant issues such as the relation with craftsmanship and ability level since this vernacular reconstruction deal not only with technical matters.

References

BAPPENAS. Action Plan of the Rehabilitation and Reconstruction for the Post earthquake Affected Areas in the Province of Yogyakarta and the Province Of Central Java. English version by JICA 2006

BAPPENAS. Preliminary Damage and Loss Assessment; Yogyakarta and Central Java Natural Disaster, A Joint Report from BAPPENAS. The Provincial and Local Governments of D.I.Yogyakarta, the Provincial and Local Governments of Central Java, and international partners, 2006

BAPPENAS. Yogyakarta and Central Java Earthquake: Casualties Data, URL: http://rehabyogyajateng.bappenas.go.id/index.php?option=com_content&task=view&id=21&Itemid=60 retrieved on August 17th, 2009

Boen T. Yogya Earthquake 27 May 2006, Structural Damage Report. URL: www.eeri.org/lfe/pdf/indonesia_yogya_structural_damage.pdf retrieved on October 15th, 2009

Dirjen Ciptakarya PU. Earthquake Proof Technical Guidance for House and Buildings Departmen Pekerjaan Umum Jakarta 2006 (in Indonesian)

FEMA (2002) FEMA 154: Rapid Visual Screening of Building for Potential Seismic Hazards: A Hanbook, Aplied Technology Council, second edition, California March 2002

FEMA (1988) FEMA 155: Rapid Visual Screening of Building for Potential Seismic Hazards: Supporting Documentation, Aplied Technology Council, California September 1988

Guchan SN. Observations on earthquake resistance of traditional timber-framed houses in Turkey. Building and Environment 42 2007:840–851

Horie K ,Okimura T, Torii N. Application of Seismic Risk Assessment Method Reflecting Building Damage Levels, URL: http://www.lib.kobe-u.ac.jp/repository/80020082.pdf retrieved on October 15th, 2009

JRF. Two Years after Java's Earthquake and Tsunami: Community Based Reconstruction, Rising Transparency, Ongoing Report Java Reconstruction Fund (JRF) 2008

LODIY. Lembaga Ombudsman Daerah Provinsi DIY. In http://tawonmedia.com/lod-diy.or.id/makalah-dan-artikel/lod-diy-dalam-rehabilitasi-dan-rekonstruksi-pasca-gempa-yogyakarta.html retrieved on August 17th, 2009

Tas N, Cosgun N, Tas M. A qualitative evaluation of the after earthquake permanent housings in Turkey in terms of user satisfaction—Kocaeli, Gundogdu Permanent Housing model, Building and Environment 42 2007: 3418–3431

UN-OCHA. The Humanitarian Cluster Response to the JAVA Earthquake, 2006-2007, One Year Review July 2007

USGS.URL:http://eartquake.usgs.gov/eqcenter/eqinthenews/2006/usneb6/ retrieved on August 17th, 2009

Widodo. Building Damages in Yogyakarta Earthquake May 27th 2006: Failure on Code, Dissemination, or Construction Process, (in Indonesian) paper presented on HAKI 2007

Wu JY, Lindell MK . Housing Reconstruction after Two Major Earthquakes: The 1994 Northridge Earthquake in the United States and the 1999 Chi-Chi Earthquake in Taiwan Disasters, 2004, 28(1): 63–81

Sustainable Customized Consolidation Design of Kuin Riverside Kampong Regeneration in Banjarmasin, Indonesia

Budi Prayitno[1,*]

[1]Department Architecture and Planning, Faculty of Engineering Gadjah Mada University, Indonesia.

Abstract. On the one hand, the hyper-urbanization phenomenon gives numerous advancements in urban living quality, yet on the other hand, it also raises an equal amount of problems. Environmental issues come in the form of density issues, the existence of slums, floods, social inequality, and urban architectural identity. As a city with "City of a Thousand Rivers" as its brand, Banjarmasin now faces a shift in an urban architectural image from a river based city to a land-based city due to the rapid land-based infrastructure development. This resulted in the degradation of environmental and architectural quality of river-based village (kampong) settlement, the main component of Banjarmasin river front city, due to a strong paradigm shift. Kuin, a river side residential area/riverside tourist destination that is currently experiencing a degradation of environmental quality and place identity, is undergoing an urban renewal. Unfortunately, the policy approaches to urban planning that has been implemented are based on general formal guidelines; guidelines which do not take the informal nature of kampong river-side settlements, due to the lack of guidelines based on riverside place identity. This paper aims to explore the characteristics of riverside settlement using architectural image observation method, space syntax method for analyzing settlement configuration genotype, connectivity, interface and interlink territory integration, as well as questionnaire and interview methods to assess the perceptions of residents and municipal authorities. From the results of this research, five compositions forms have been identified: attachment to riverside settlement identity in the form of pilling, spanning, floating, embracing, and ascending with elements such as waterfront alleys, *jukung* (traditional boat), *lanting* (floating houses) as well as floating traditional markets that serves as the frame to the configuration. The identification process is done by using the approach of observing the spaceuse appropriation and the space-user perception on how to consider its sustainability aspect as a means to determine a level of adjustment. The result shows that self organized and self customized kampong residents and tourist are aware towards the river environment, the assets of local floating markets as well as the local social space. On the other hand, municipal authorities gave more attention to formal normative and regulative aspects. This analysis is used as the basis for recommendations for kampong riverside settlement design consolidation, which is done through guided participatory design workshops.
The result of this study is constructed as a concept for urban riverfront composition architecture, amphibious space territory, and urban riverfront settlement identity and expected to be able to further advance the knowledge surrounding the subject of urbanism and territoriality.

Keywords: customized consolidation, kampung regeneration, place identity, sustainability

Introduction

The place identity of an area is strongly determined by the attachment value of its social, economic, and environmental aspects. In its development the attachment value will always change in accordance with the architectural design trends of its era. The identity value of the area is currently at risk due to the strong development of contemporary styles and the complete disregard in the governance of the area. The identity of Banjarmasin as a city of a thousand rivers and the Venice of the East is currently experiencing a visual degradation. This is due to the area's aggressive land-based urban development. Rivers that initially amounted in the hundreds is now only a few dozens. Kuin as a region famous for its traditional floating markets is indicated to experience a decrease in value of its identity. This is indicated by the declining number of floating market traders, the number of incoming tourists, as well as the increase of housing density and the degradation of the quality of the river, which in turn affected riverside settlements. The land-based approach utilized by municipal regulation has side-lined local river-based values.

Current local infrastructure development focuses more on dealing with existing problems due to negative impacts

* Corresponding author: budiprayitno_ugm@yahoo.com

brought forth by previous approaches. The high demand for a fast development as well as the need for a future vision can only be realized by a positive outcome approach in a regeneration concept. There is also a need for various activities to be redefined as mixed-use and hybrids. This water regeneration model was developed in the 90s, characterized by a variety of development models with festivals and celebrating market placed as key magnets in commercial, housing, and recreational areas (Breen, 1994).

Place identity and place are widely viewed from various capital, commodity, marketing, culture and tourism aspects (Malone, 1996; Philo, 1993; Ward, S, 1994). The demand for creativity in devising place making is closely related to place marketing and investment to create jobs and increase space value (Philo, 1993). As part of place's identity, the architectural assets of waterfront areas need to be spatially reconstructed and calls for a reconstruction of their shape and ordering systems (Ward, 1994; Goodwin, 1997; Crilly, 1993). It is intended to answer the need for a customized consolidation of the development of various space uses, user, and territory integration needs as well as a sustainable consolidation on social interaction of productivity and environmental sustainability aspects.

Customized Consolidation

The interface territory of Kuin's riverside kampong settlement area, which will be reconstructed to answer the need for an improved visual quality that supports its place identity, is determined by the space connectivity, integration and appropriation values. The methods used in the reconstruction are 'redefine, reconnect, and revitalize'. The 'redefine' method is intended to define a new meaning for the place identity of Kuin's amphibious architecture. The 'reconnect' method aims to increase the value of connectivity and space integrity to support Kuin's space network and access. The 'revitalize' method is intended to formulate new dwelling, working, and access hybrid functions. Several ways that can be done to support architectural engineering is through territory attachment by manner of floating, pilling, spanning, embracing, and ascending. The reimagining of amphibious architecture is done through the mapping of perception of space identity from the stake-holders and agencies.

Space, territory and identity are composition architecture system that will be analyzed in assessing the level of sustainable consolidation that occurred in the Kuin region as a basis for constructing the amphibious regenerative village approach. Kampong riverside settlement or commonly called amphibious settlement because of its land and water elements, generally are self-organized, spontaneous, and organic in their growth. Processes that occurred by accident are not by design, which often causes social, spatial, environmental, and architecture problems. This is marked by the emergence of several phenomena namely high density, slums, floods, social inequality and architectural identity.

In addition to the negative impacts that emerged, commonly there are also hybrid and mix-used phenomena of various activities. The types that generally appear are based on the need for livelihoods resilience. Livelihood resilience, which is caused by economic development and creative trends, spawns hybrid types of mixed dwelling functions, mixed hybrid users between residents and tourists, hybrid bridges that synergize physical infrastructure assets, land status, and infrastructural assets environment.

The architectural methods to reinforce and respond to the quality degradation of place identity can be achieved through designing architectural tectonics. The architectural tectonics which can be processed in an attempt to re-imagine the degradation of identity quality are able to use constituent construction materials, as well as space composition and form composition concepts. 'Lanting' houses (floating houses), stilts homes, catwalks, jukung (traditional housing), and boat mooring docks are strong elements in the establishment of place identity in the Kuin area.

The reinforcing of place identity can be done on a local scale (area-based or citywide-based). Such spatial reinforcing methods are commonly done through space syntax simulations to assess connectivity and integration on a regional scale and the types of spatial relationships in building genotypes. The space syntax method can be used as a basis for establishing intra-network (area-based network) and inter-network (citywide-based) place identity reinforcement. This network configuration is determined by the space appropriation of various activities, space use appropriation and space intelligibility.

Research Methods

This research uses case study method with rationalistic paradigm. Research area with delineation of residential area and traditional floating market in Muara Kuin Village is on the banks of the river Kuin. The research method used is visual observation, space use mapping, questionnaire, and guided participatory design workshop. Visual observation observes objects forming the identity of an area. Space use mapping is done by using space syntax method. The assessment of the perceptions of space users and space management actors are put through the questionnaire method that the results are validated through guided participatory design workshop. The subjects for the questionnaire were taken by purposive sampling; 30 local residents respondents, 30 tourist respondents, and 30 municipal employees. Respondents are allowed to select more than 2 choices regarding the aspects that are considered important in the questionnaire. The topics include perception towards place identity, perception towards regeneration sustainability, and perception towards amphibious appropriate composition.

Intra-place Attachment in Kuin

Waterfront regeneration generally constructs places that are bound to local identity. The production of this place identity is packaged and marketed as a commodity (Goodwin, 1997). The identity of the waterfront area is linked to the effort to create a marketing brand for a place (Ward and Gold, 1994). Waterfront place identity development is closely related to the level of attachment and strength of the existing image. As a spatial and shape art, Architecture has to be able to integrate place construction and place marketing into a product that suits the needs of space use and users (Crilley 1993: 235). The

https://doi.org/10.1051/shsconf/20184107001

visual clarity of a place identity can be used as a basis for creating architectural images. The development of a waterfront area is a part of urban transformation on a wider scale. The themed, scripted and branded concept of place making is a result of the transformation of the waterfront region (Hannigan, 1998). The attachment value of place making in Kuin is largely determined by visual perceptions that are deemed important by citizens, tourists and municipal employees. The value of this perception should be based on a measurable planning and participation process (Hague, 2005, Nanzer, 2004). Hence the process of formulating perception is done in a participatory manner through a design workshop to validate the results of the questionnaire. The Kuin brand as 'The Thousand of Rivers' and 'The Venice of The East' is the identity of the city of Banjarmasin. However, the relationship between the city and its identity must be seen from the value of its space intelligibility (Proshansky, 1978). The attachment value of this place is largely determined by the connectivity and space integration values.

Therefore the assessment of place attachment should be judged both from place-based perception and space-based intelligibility. The diversity of perceptions towards place identity and space attachment is the embodiment of the symbols of the space fabric of the region (Hull, 1994; Patterson, 2005; Manzo, 2013).

Hybrid Customized Consolidation

Along with the decline of the space value there are also an increase in diversity of space use and self-organized users. The diversity of these activities is triggered by the diversity of existing livelihoods and tourism activities. A living space that is composite with the trade functions and tourism produces a hybrid form of customized consolidation. Hybrid customized activities can be seen from hybrid space, activities and governance aspects (Cho, 2016; Prayitno, 2017). The existence of this hybrid architecture adds value to the diversity of space and enhances the diversity of interactions. This process results in a much-found space structure in informal urban settlement areas in the form of co-existence space, co-working activities, and co-benefit governance (Prayitno, 2017). The architectural form of the existing spatial construction in Kuin has also undergone a customized hybrid process that combines the floating, pilling, embracing, spanning, and ascending forms that are composed in the amphibious settlement area.

Kuin Amphibious Kampong Regeneration

The attachment value of an amphibious territory determines the place identity of an area. The attachment of floating markets, which are place identities, is determined by the value of connectivity and space integration. The process of self-organized and self-customized kampong residents and tourist with local floating market assets is an interaction of a local social space. Local governments also greatly influence the quality of this place identity because they have the authority in the process of regulation and governance. Therefore the reinforcing of place identity has to be done by manner of sustainable governance approach by the municipal government.

The government's approaches generally are responses to negative impacts rather than positive outcomes of development (Karmona, 2010). Hence the concept of regeneration becomes very important in the effort to redefine, reconnect, revitalize, reconstruct, and re-imagine. To redefine is to reconstruct the remainder of the riverside settlement. To reconnect is to construct a link for spatial asset access that forms place identity. To revitalize is to create new hybrid functions to reinforce dwelling, working and access functions. To reconstruct is to create the floating, pilling, embracing, spanning, and ascending composition structures.

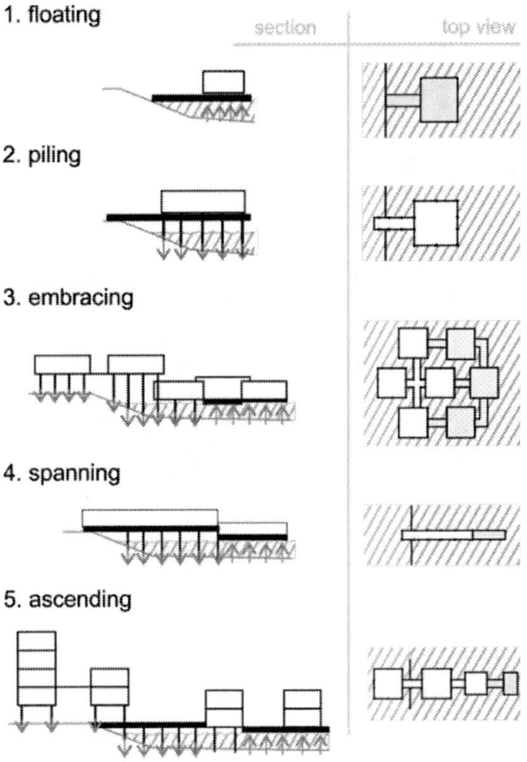

Fig. 1. Typology of Amphibious Space Composition.

To re-imagine is to create a contemporary architectural image for residential, commercial, and tourism functions in the context of amphibious composition identity.

Discussion

Based on the results of visual observation, it can be identified that the elements of Kuin riverside's place identity are the floating market, traditional boats boat, lanting traditional floating houses, traditional waterfront alleys, historic mosque, and traditional festivals.

https://doi.org/10.1051/shsconf/20184107001

Fig. 2. Floating Market.

Fig. 3. Lanting Traditional Floating houses.

Fig. 4. Historic Mosque .

Fig. 5. Traditional Waterfront Alleys.

Figure 6. Jukung Traditional Boats.

Fig. 7. Traditional Festivals.

This visual observation was carried out in a participatory manner either through field surveys, or through participatory design workshops. The assessment through questionnaires and validations from the workshop have identified that there exists different perceptions of place identity.

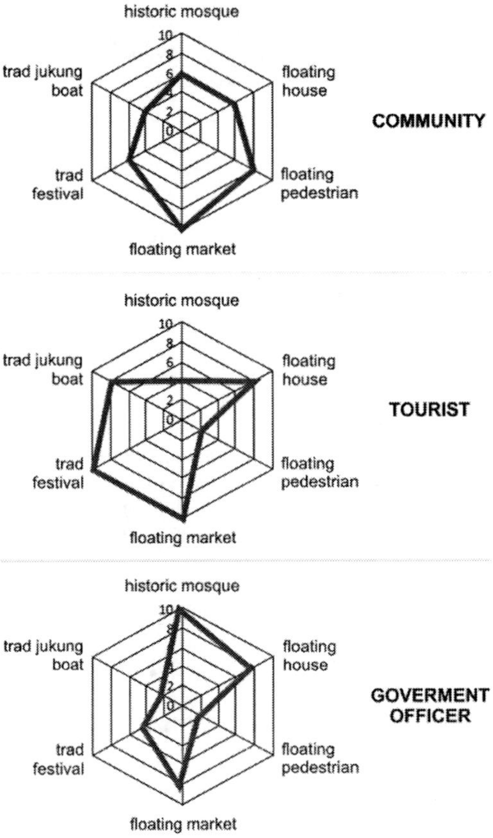

Fig. 8. Perception of Place Identity.

According to the tourists and the local community, floating markets is the main identity, while on the other hand the government officials viewed the historic mosque as the main identity. Tourists view that traditional festivals, floating houses, and jukung traditional boats are also an identity of the region (aside from floating markets), whereas the historic mosque is not. The local community sees waterfront pedestrian alleys, the historic mosque and traditional festivals are important identities while jukung traditional boat is not. Tourist also agreed that floating houses and traditional jukung are important identities and that the historic mosque and floating pedestrian are not. last but not least, government officials perceives floating houses and the floating market as an important identity while floating pedestrian, traditional boats and traditional festival are less important.

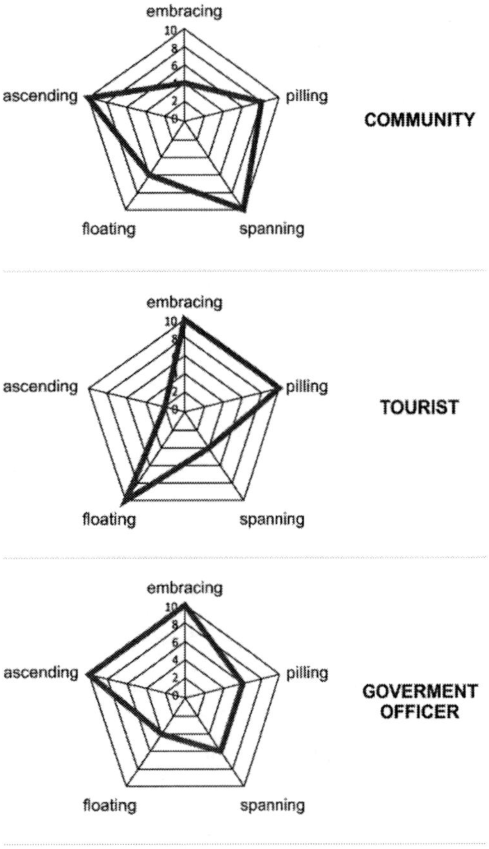

Fig. 9. Perception of Amphibious Appropriate Composition.

Based on the amphibious appropriate compotition aspect, the community considers spanning and ascending space constructions to be the most appropriate. Meanwhile, tourist views floating, pilling, and embracing space construction to be most appropriate. Government officials on the other hand sees ascending and embracing as the most suitable. Ascending is considered to be the most inappropriate method by tourists. Last, floating is considered to be the least appropriate method by government officials.

https://doi.org/10.1051/shsconf/20184107001

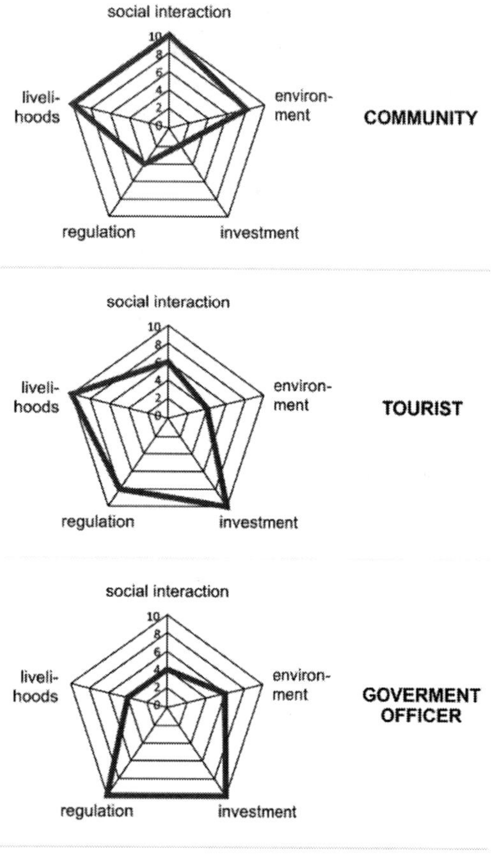

Fig. 10. Perception of Regeneration Sustainability.

The community views livelihoods and social interaction as the most important deciding factor from the perspectives of the regeneration and sustainability aspect. Tourists on the other hand put livelihoods and investment at the front, while government officials see regulation and investment as the most important deciding factors.

The differences in the perception of respondents (community, tourists and government employees) after validation through participatory design workshop lay in the basis of consideration in determining the value of each interest. Local community respondents consider aspects of livelihoods and economic affordability to be the basis of their main assessment, whereas tourists view attractiveness and market value the basis for their primary assessment. Government officials on the other hand view regulation, investment and ease of implementation at the forefront.

Based on the observation on the intensity of space use and simulation of space syntax it was identified that connectivity and space integration occur on a local level. This is due to the strong space attachment value of the values of local place identity. Attachment towards intra-based local territories cannot be up scaled and replicated on a global scale. The results of global connectivity and integration values are very low.

Fig. 11. Traditional Floating Market.

Fig. 12. Riverside Settlements.

Traditional floating market and riverside settlement integrated with historic areas (historic mosque and traditional festivals) have a very strong place attachment that cannot be scaled-up to citywide scale. Therefore, the process of consolidation and regeneration can only be sustained on a local scale. This is demonstrated by the diversity of spatial and building patterns that are responsive to the diversity of contemporary hybrid functions, which are done in adaptive, self-organized and, self-customized manner by the local community, tourist and municipal employees.

From the analysis of axial lines on global integration values, it was identified that the region has a low value. The relationship in the area is considered to be lacking, as seen in the scatter diagram value R2 = 0.088.

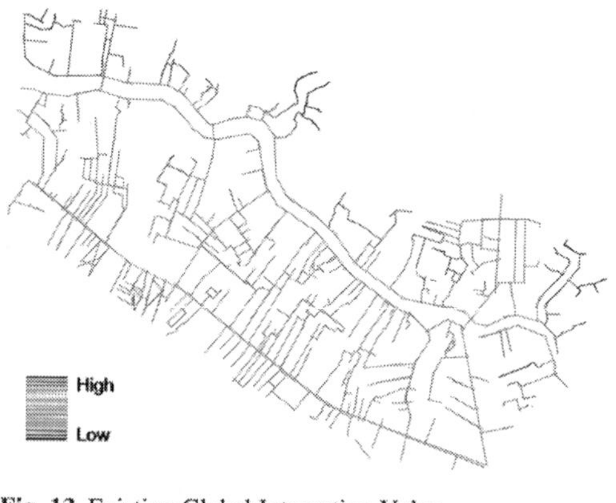

Fig. 13. Existing Global Integration Value

https://doi.org/10.1051/shsconf/20184107001

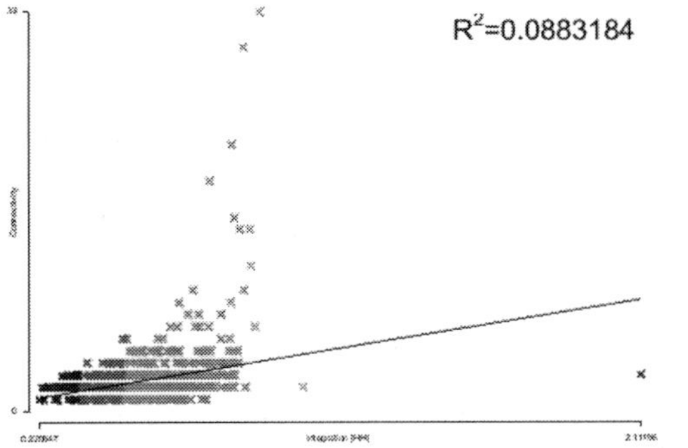

Fig. 14. Relationship Between Global Integration an Connectivity

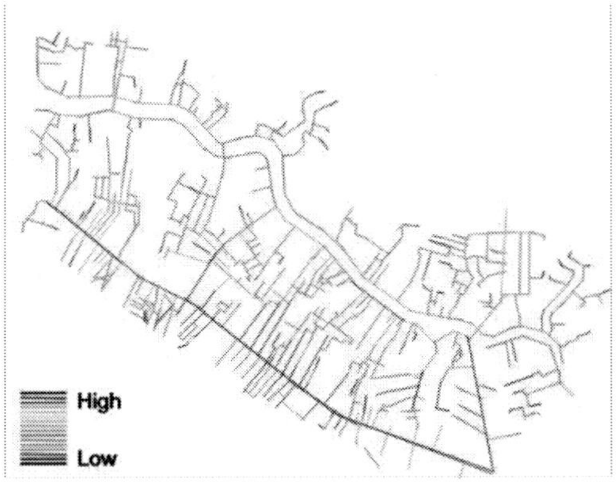

Fig. 15. Existing Local Integration Value

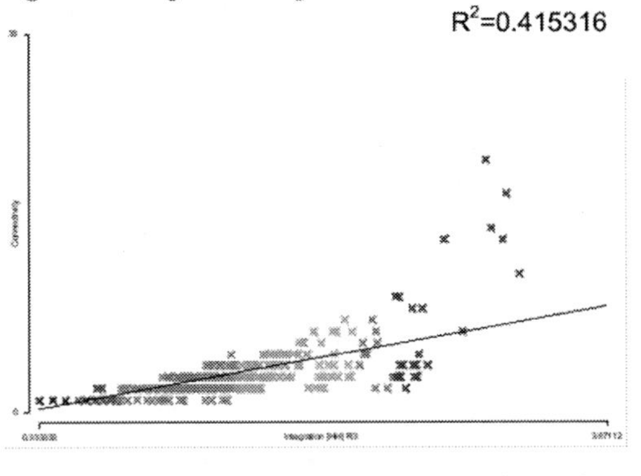

Fig. 16. Relationship Between Local Integration and Connectivity

It means that the accessibility value of the area with larger scale in Banjarmasin is low and difficult to reach from the city's network.

From the result of the analysis on space connectivity and integration of Kuin area through space syntax simulation, it is shown that the existing condition of Kuin area is much stronger locally (R2 =0,41) and globally integration (R2 =0,088) compared with its connectivity and integration on an urban scale (R2=0,017).

Fig. 17. Existing Citywide Value

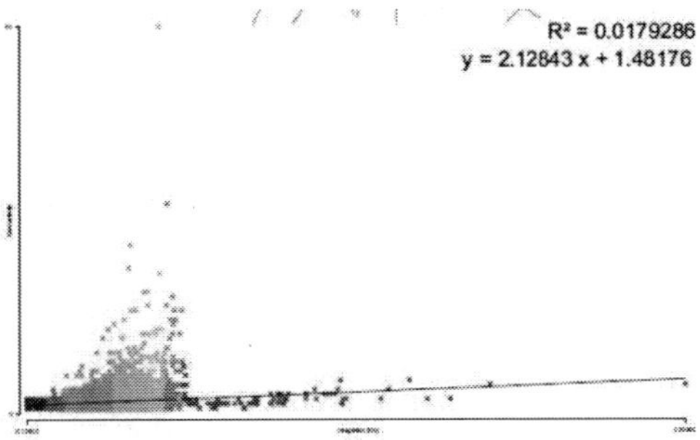

Fig. 18. Relationship Between Existing Citywide Global Integration and Connectivity

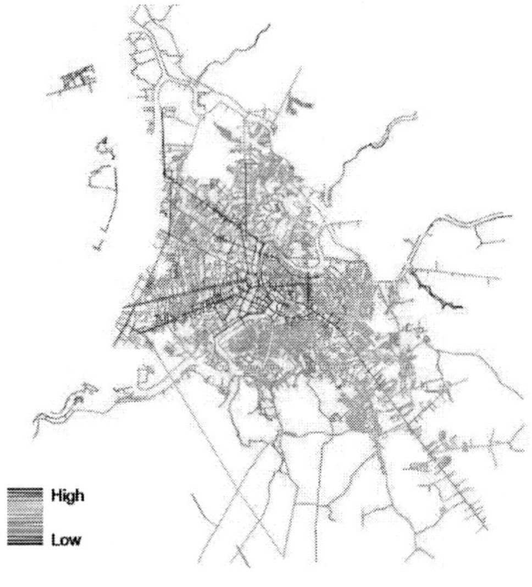

Fig. 19. Proposed Citywide Value

 https://doi.org/10.1051/shsconf/20184107001

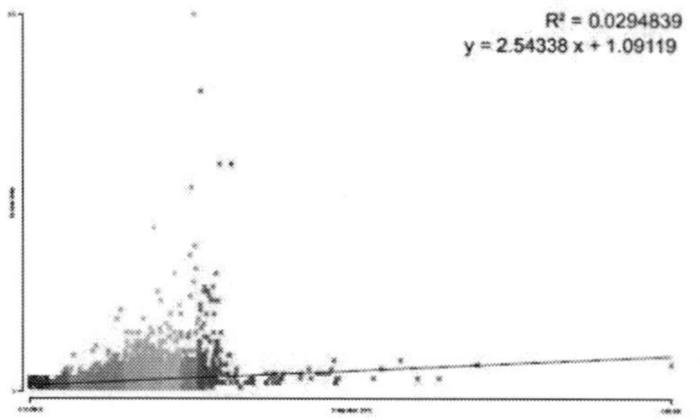

Fig. 20. Relationship between proposed citywide global integration an connectivity

Even when the connectivity is consolidated through trigger areas such as public open space, commercial, etc, its increase is not quite significant (R2 =0,029), which is only 0.012.

Judging from the comparison between local and global integration values, the current development of the river Kuin is still localized, meaning that its development has not been integrated into the city system and its development is only influenced by local values. By taking the comparison between high local integration value and global integration value as a basis, it can be interpreted that the area has a strong area value, seen from both social and land use aspects. However the weakness of a strong local integration value is that the region becomes separated from the city's developmental system, which lags the area's growth.

Conclusion

The study finds that changes in urban development trends and shifts in development paradigms in Kuin have led to hybrid diversity of riverside settlement villages. This diversity is formed in a sustainable customized consolidation. Consolidation represents the transformation of space use and perception of space-users and agencies. The existing spatial configuration can have place identity reinforced in an intra-network. This is because the forming elements of place attachment in Kuin cannot be up scaled or replicated elsewhere. The place attachment that occurred has changed its image due to the emergence of multiplicity activities and users who each conduct customized consolidation.

The existing amphibious architecture that is constructed through spanning, floating, embracing, pilling, and ascending are composed with local elements (titian, lanting, floating market, and jukung). The hybrid and mixed-use compositions that combined residential, occupational, and recreational functions are a form of residential regeneration village that was achieved through the redefine, reconnect, revitalized, reconstruct and re-imagine approaches.

References

Breen, A. & Rigby, D. (1994). *Waterfront Cities Reclaim Their Edge*. New York, NY: Mc Graw Hill.

Carmona, M., Heath, T., Oc,T. & Tiesdell.S. (2010). *Public Space, Urban Space*. London : Routledge.

Cho, I., Heng, K., & Trivic, Z. (2016). *Re-Framing Urban Space: Urban Design For Emerging Hybrid and High-Density Conditions*. New York: Routledge.

Crilley, D. (1993). 'Architecture as Metaphor'. *in : Kearns, & Philo, Selling Places*. 231-252.

Dovey, K. & Sandercock, L. (2002). Hype and Hope : Imagining Melbourne's Docklands, *City*, 6(1). 83-101.

Goodwin, M. (1993). The City as Commodity. in : Kearns,G. & Philo, C. (eds) Selling Places, Oxford : Pergamon. 145-162.

Hadinata, I. Y. (2010). *Tipomorfologi Kota banjarmasin.* Yogyakarta: Universitas Gadjah Mada.

Hague, C. and Jenkins,P. (Eds). (2005). *Place Identity, Planning and Participation*. New York: Routledge.

Hannigan,J. (1988). *Fantasy City*. London: Routledge.

Heldiansyah, J. (2010). *Kajian Peningkatan Kualitas Lingkungan Binaan Tepian Sungai Kota Banjarmasin.* Yogyakarta: Universitas Gadjah Mada.

Hull, B.R., Lam., & Vigo,G. (1994). Place Identity: Symbols of Self In The Urban Fabric. *Landscape and Urban Planning*, 28, 109-120

Lynch, K. (1969). *The Image of The City*. Cambridge. Massacht: MIT Press.

Malone, P. (ed) (1996). *City, Capital and Water*. London: Routledge.

Manzo,L.C. (2013). *Place Attachment: Advances in Theory, Methods and Application*. Routledge, London & Wright,P.D.

Mentayani, I. (2001). *Karakteristik Perumahan Tradisional di Tepi Sungai.* Makassar: Universitas Hasanuddin.

Mentayani, I. (2013). *Transformasi Adaptip Permukiman Tepi Sungai di Kota Banjarmasin; Kasus: Barito-Muara Kuin, Martapura dan Alalak.* Yogyakarta: Universitas Gadjah Mada.

Moore, C.W. (1994). *Water and Architecture*. London: Thomes and Hudson.

Nanzer,B. (2004). Measuring Sense of Place : A Scale for Michigan. *Administrative Theory and Praxis 26*(3), 362-382

Patterson,M.E., & Williams, D.R. (2005). Maintaining Research on Place : Diversity of thought and scientific progress. *Journal of Environmental Psychology*, 25, 361-380.

Philo, C. &. Kearns, G. (1993). 'Culture, History, Capital'. *in : Kearns,G. & Philo, C.* (eds) Selling Places, Oxford : Pergamon, 1-32.

Prayitno, B. (2013). An Analysis of Consolidation Patterns of Kampong Alley Living Space in Yogyakarta, Indonesia. *Journal of Habitat Engineering and Design*, 99-112.

Prayitno, B. (2017). Integrated Sustainable Kampong Hybrid in Code Riverside Settlement in Yogyakarta,

Indonesia. *The 3rd International Conference on Engineering of Tarumanegara.* Jakarta: Tarumanegara University.

Prayitno, Budi. (2012). A Morphological Analysis on Changing Patterns of the Banjarmasin Rivercity, Indonesia. *Journal of Habitat Engineering and Design*, 4(1). 23-32.

Prayitno, Budi. (2013). Analisis Permibilitas Keruangan dan Dinamika Fluida Angin dan Suhu Kawasan Permukiman Tropis Sungai di Banjarmasin. *Jurnal Manusia dan Lingkungan*, 20(2). 199-212.

Proshansky, H.M. (1978). The City and Self-identity, *Journal of Environment and Behaviour*, 10, 57-83.

Stedman, R.C. (2002). Toward A Psychology of Place: Predicting Behavior From Place-Based Cognition, Attitude, and Identity. *Environment and Behaviour*, 34(5), 561-581.

Torre, L.A. (1989). *Waterfront Development*. New York. NY: Van Nostrand Reinhold.

Uzzel,D., Pol,E. & Badenas,D. (2002). Place Identification, Social Cohesion, and Environmental Sustainability, *Environment and Behaviour,* 34(1), 26-53.

Ward, S. &. Gold,J. (eds). (1994). *Place Promotion*, Chichester:Wiley.

Yin, R. K. (2003). *Case Study Research: Design and Method Third Edition.* London: Sage Publication.

The Change of Space Use of Shared Space from Landed to High-Rise Settlement

Ayuko Setyo Adiyanti[1,*], *Ardhya* Nareswari[2], and *Alexander* Rani Suryandono[2]

[1]Student, Department Architecture and Planning, Faculty of Engineering Gadjah Mada University, Indonesia.
[2]Lecturer, Department Architecture and Planning, Faculty of Engineering Gadjah Mada University, Indonesia.

Abstract. Indonesia is entering an era of urban settlement transformation from horizontal landed living settlement to low-rise settlement, into the construction of vertical high-rise settlement. This resulted in the landed settlement residents that being moved to high-rise settlement, they have encountered a change of high-rise vertical living culture that different from their living culture before. This study aims to find out the use of shared space in landed living and high-rise settlement. The research method used in this research is qualitative descriptive study. The site of the research is in the landed settlement of *Kampung* Pulo and the high-rise settlement of Jatinegara Barat. *Kampung* Pulo settlement is a residence of Jatinegara Barat residents before they are moved. The result of the research shows the change of space use for shared space from landed to high-rise settlement; (1) landed settlement more accommodate the diversity of communal activities than high- rise settlement. (2) in the landed settlement there is a territory transition space that accommodates the needs of interaction, homebased business activities, and play, whereas high-rise settlement there is no more transition space, so the need for interaction and play is contained in the public space, while the trading activities needs are mostly found within the private area. (3) In landed settlements, the shared space used for communal activities is more multifunctional than high rise settlement.

Keywords: landed settlement, high-rise settlement, space use change.

Introduction

Indonesia is entering an era of urban settlement transformation from horizontal landed living settlement to low-rise settlement, into the construction of vertical high-rise settlement. This resulted in the landed settlement residents that being moved to high-rise settlement, they have encountered a change of high-rise vertical living culture that different from their living culture before. After being transferred to new settlement, some residents from Kampung Pulo have limited space to perform their daily activities. As a result, some activities such as selling, visiting, playing, and gathering, are carried in spaces that should not be used for such activities. this study will examine the change of space use of shared space from landed to high-rise settlement.

Literature Review

Space for Communal Activity

Shared space is a room function that always exist in Indonesian society. The existence of shared space is a symbol of the community, especially in a settlement, which has good relationships with others, marked by the communality. Shared space is a space that accommodates various communal activities of community (both positive and negative) from the economic,social , and cultural needs.

Kampong Pulo

Kampung Pulo is a very high density settlement like any other slums area in Indonesia. With an area of 8 Ha, the majority of the existing site is almost entirely utilize for building development. Residents of Kampung Pulo who live on the banks of the river and in the middle of the settlement almost entirely affected by the annual flood, so most of the housing in this area has 2 floors height.

Jatinegara Barat flat

Jatinegara barat Flats is located in Jatinegara Barat, about 300 meters from Kampung Melayu Terminal, intended to accommodate Kampung Pulo relocation which has been moved from Ciliwung River since June 2015. The Jatinegara Barat Flat consists of two towers, 16 floors, with housing capacity reaching 520 units.

Qualitative Methods

Research with qualitative methods that use qualitative approach aims to reveal events or facts, circumstances, phenomena, variables and circumstances that occur when

* Corresponding author: ayukoadiyanti@gmail.com

research takes place by presenting what really happened.

Sample Collecting Technique

Observation

Observation is a method of collecting data through direct observation in the field, the purpose of obtaining various concrete data in the research location.

Interview

The technique of interview according to Nazir (1988) is the process of obtaining data for research purposes by way of question and answer face to face between researchers and respondents.

Research Methods

This research uses qualitative method with descriptive approach. Research with qualitative methods that use qualitative approach aims to reveal the events, facts, circumstances, phenomena, variables and circumstances that occur when research takes place by presenting what really happened.

Data Collecting Methods

Primary Data

Primary data were collected by field observations, interviews with sources, as well as documentation in the field.

Secondary Data

Secondary data collection was conducted with the study of literature records, study results, and the results of previous research publications related to the title of this study.

Research Focus

The object of this study are landed settlement and high-rise flats. Based on variables used in some similar research, the variables that match the topic of space observation used for shared activities are; physical conditions of space, activities that occur, actors activities, frequency of activities, and space attributes.

Research Object

The research object is located in the landed settlement of Kampong Pulo and the Jatinegara Barat high-rise flat. The problem of frequent floods, resulting in residents inhabited in less decent housing. Floods not only harm the people of Kampung Pulo that have direct contact with the river Ciliwung, but also affected the entire village of Kampung Pulo is located far from the banks of the river Ciliwung. That is why, there is similar habit activities and culture between residents who live on the riverbank and on the village. Because of that, this research took place at landed settlement in Kampung Pulo that did not experience river normalization.

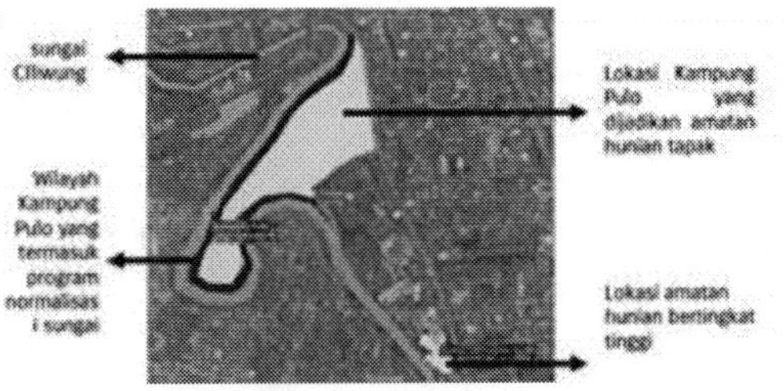

Fig. 1. Research Object.

1 Kampong Pulo Observation Results

Alleys at landed settlement in Kampong Pulo

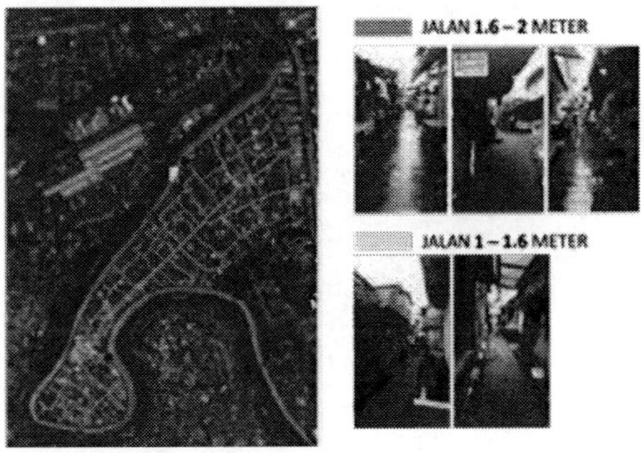

Fig. 2. Kampong Pulo Alleys.

The observations started from a few alley (blue circle) located on the banks of the Ciliwung river and on the edge of Jatinegara Barat street. There are two types of alleys in Kampung Pulo, the main alley (green) cuts the middle of Kampung Pulo area measuring 1.6-2 meters which can be passed by people and vehicles such as motorcycles and bicycles by passing each other. While the other alley branch (colored yellow) measuring 1-1.6 meters and can only be passed by a bike or people walking without passing..

Kampong Pulo's Characteristics

Node A

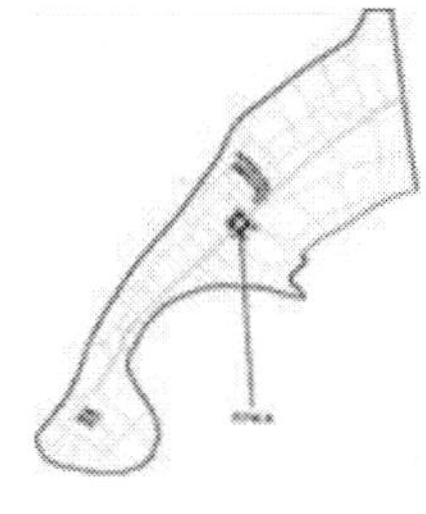

Fig. 3. Node A.

Node A is located in the middle of Kampung Pulo, at this point there is a buying and selling activity. In the hallway room measuring 1.8 meters with right-left alleyway there is a concrete covered drain, the placement of merchandise is placed in the right-left of the alley that is precisely above the concrete covered drains. When the buying and selling activity did not happen, there are trading tables that are left outside the stalls, but some are storing it back into the house.

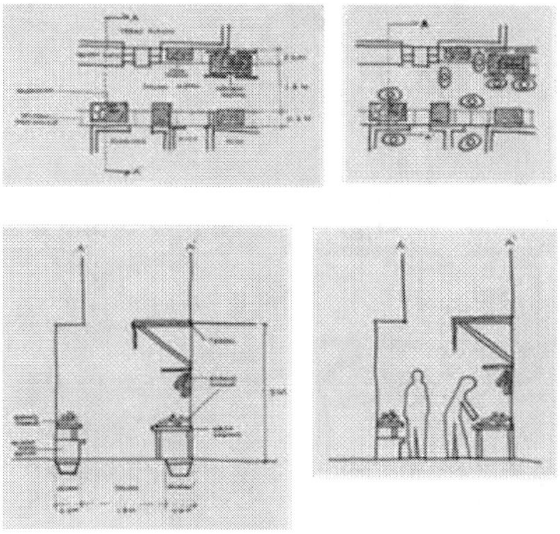

Fig. 4. Space Attributes.

Node B

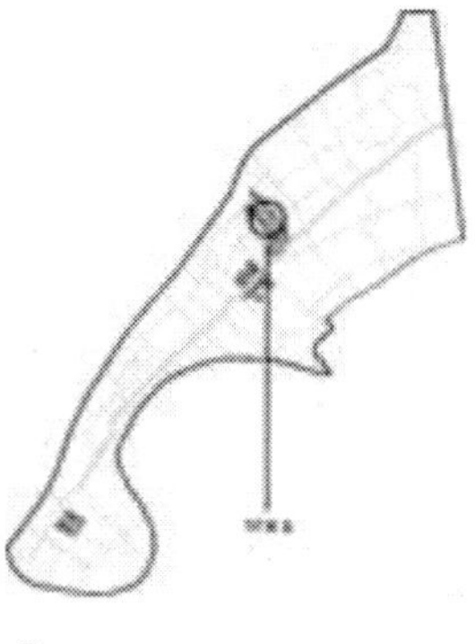

Fig. 5. Node B.

At node B, there are shared activities shown by the presence odong-odong traders who enlivened by the presence of children who plays and accompanied by parents who live in the house around. Odong-odong carts are parked in a rather wide location, usually they are looking for a vacant house that can be visited temporarily. Ceramic chairs in most homes are indeed made for relaxing chatting with neighbors.

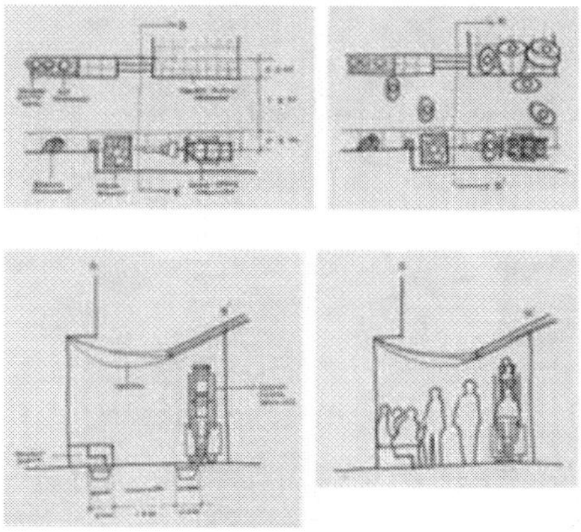

Fig. 5. Space Attributes at Node B.

Node C

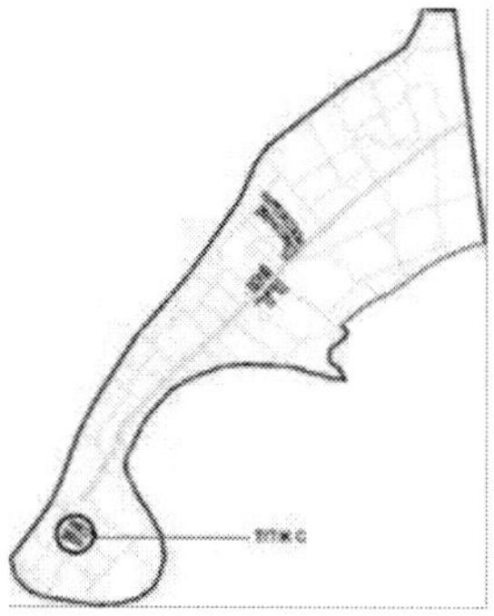

Fig. 7. Node C.

At node C there is activity such as communication with neighbors, some used to be sitting on chair, some other are usually just interact at front of houses without using any chair at all. Different home positions that go back and forth towards the alley are used as storage space, such as merchandise and parking of two-wheeled vehicles.

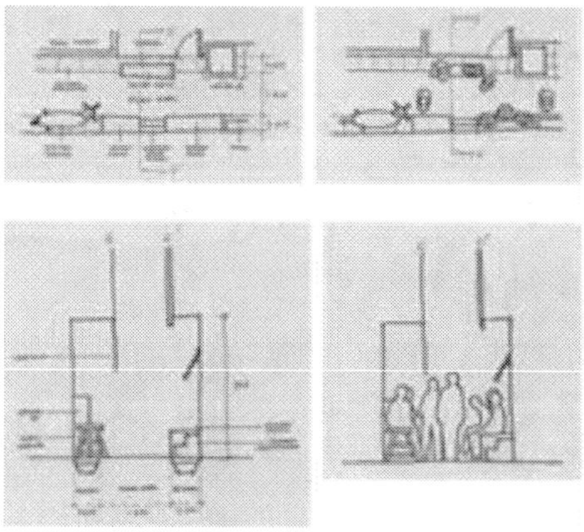

Fig. 8. Space Attributes at Node C.

https://doi.org/10.1051/shsconf/20184107002

Interview Result

Interviews and filling questionnaires were conducted on 17 respondents from Kampung Pulo residents. Of the 17 respondents, each with details of 9 people mothers, 5 fathers, 1 teenager and 2 children. The interview was conducted with the aim of asking some questions to the Kampung residents related to the profile of the residents of Kampung Pulo, the points of the tendency of people gathered, public space facilities available in Kampung Pulo. While charging questionnaires related to the verification of literature review data on the joint activities conducted by residents of Kampung Pulo. The results of the interviews show that some of the residents of Kampung Pulo are IDKs from DKI since birth, points of joint activity in alley, guard post, front of house and resident's house.

Observation Summary

Use of shared space for social activities

Some concrete seats are used as a place for casual conversation between the neighbors, the reason they choose to chat casually on the terrace because of the wide limitations of their homes. So entertaining guests and chatting with a comfortable neighbor is performed on the terrace.

Use of shared space for transactional activities

There are several types of selling in Kampung Pulo, the most common type is the type of selling on the terrace. The property trademark is dominated by the type of trade placed on the porch or front of the house. As for the type of selling in the house, which can only be performed by a residence that has large enough room to store the merchandise, and merchandise offered in the form of grocery stalls for everyday purposes.

Use of shared space for kids play

Child playroom takes place on the terrace of the house, street alley, vacant lot, field, game rental, and odong-odong game of traveling traders. Limitations of space and land resulted in children looking for land that they could use as a place to just talk about running, or playing bikes.

2 Jatinegara Barat Flat Observation Results

Ground Floor

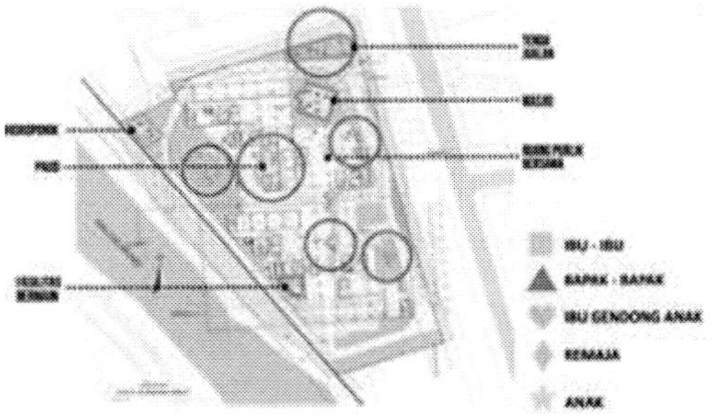

Fig. 9. Space Use at Ground Floor.

Shared activities on the 1st floor are scattered at some point, most of the residents choose to sit by the steps of entrance entrance. Groups of gangs consist of fathers, mothers to adults, some other mothers accompany their children in early childhood while talking to parents, there are also children playing in the back garden towers and mothers ktivitas together spread in some point, most of the residents choose to sit at the edge of the stairs entrance entrance.
Groups of gangs consist of fathers, mothers to adults, some other mothers accompany their children in early childhood while talking to parents, there are also children playing in the back garden towers and mothers who keep chatting while.

Second floor

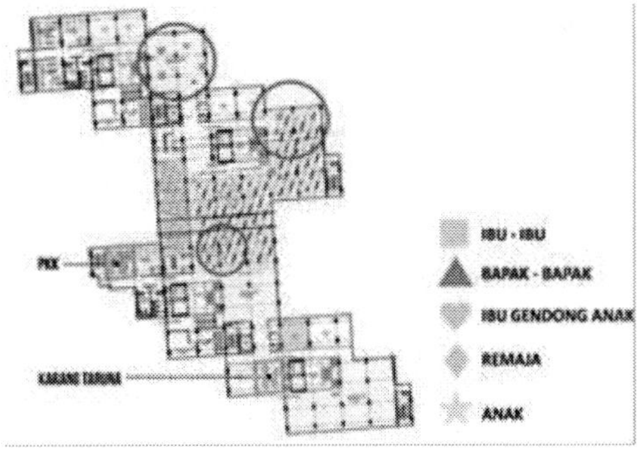

Fig. 10. Space Use at Second Floor.

The shared activity was done in several selling stalls, but because the kiosks were not so much, so the joint activity that happened was not so intense, while the activity was done by many children just running or playing badminton in the area together.

Typical floor

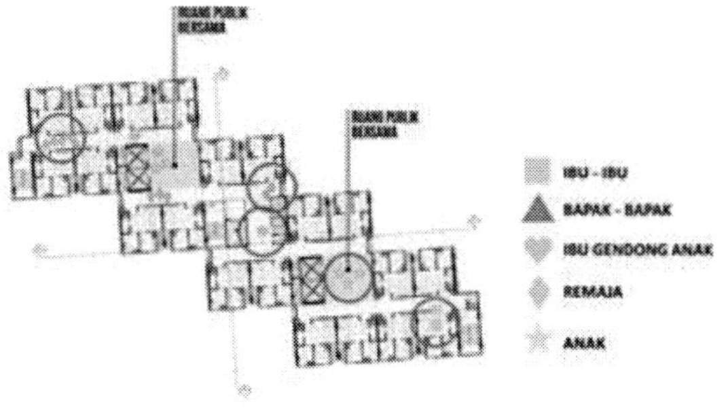

Fig. 11. Space Use at Typical Floor.

The typical floor of Jatinegara tower which is designated as residential floor, in the public area of the corridor and lobby lift is not every time used by many citizens. In everyday life, the elevator front lobby that serves as a public space together is not used. The space is used for the social gathering room of the mothers or the residents of the community every month, sometimes the children also cycling and running in the vestibule of the elevator. Actually has provided children's playroom and facilities on the 1st floor, but due to its location is far enough so that children choose to play near their residential units, and parental supervision is much more monitored through their residential units.

On a typical floor there are different rules. If on the 1st and 2nd floors is a floor intended for public activities, then a typical 14-storey is prohibited for public activities such as trade. Here's an overview of the rules that apply on typical floors:

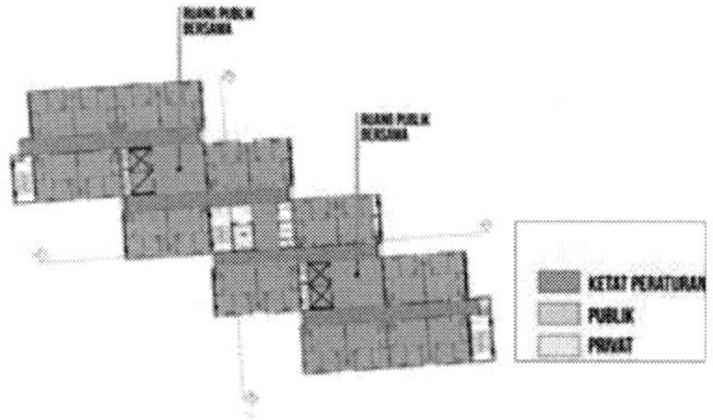

Fig. 12. Space Use Regulation at Typical Floor.

Interview Results

Jatinegara Barat Flats consists of 2 towers, namely tower A and tower B. From both towers are taken sample of respondents to be interviewed about the profile of residents of Jatinegara Barat Flats, the points tendency of people gathered in the tower area, public space facilities provided in flat , as well as additional questions about neighboorhood relationships, a market that becomes merchant buyers, and playground locations. Respondents from tower A counted to 27 people, while from tower B counted to 19 people. seller on the door of the North who made the respondents amounted to 4 people. In addition to several residents and sellers, one of the managers of the Jatinegara Barat Flats named Mr. Sarkim Sukarya as Head of Sub Division Housing Jatinegara Barat Flats was also interviewed related to the profile data of towers and shared space facilities provided.

The results of the interviews of this Jatinegara Barat Flat, among others; the work of most residents in the field of informal flat, co-existing facilities in the towers are on the 1st and 2nd floor, neighboorhood relationships are reduced compared to when living in Kampung Pulo, many people who work as traders and choose to sell in residential units and doors North on the 1st floor compared to the 2nd floor provided by the tower management as a public space for selling, and more children playing on typical floor corridors around their residential units.

Observation summary

Use of shared space for social activities

Shared activities gathered among the neighbors are more prevalent on the 1st floor in the public space area near the entrance stairs. In the area consists of groups of youth, fathers and even mothers who huddle to sit back together. Elsewhere, PAUD's front room and children's playground behind the towers are also used as chat rooms for mothers who are waiting for their children to play. North door selling arena is also quite crowded buyers from offices around the towers. While on the 2nd floor, communication between neighbors is not so happening because at least the stalls selling on the 2nd floor. Some people go by to buy things they need but not linger. Some children are seen playing in the 2nd floor public area. Similar to the 2nd floor, a typical floor also appears to be few people who gather together. Some children play together or play bicycles in front of the elevator, but for the fathers and mothers not many are seen clustering together..

Use of shared space for transactional activities

There are 3 places that use to become selling area provided for jatinegara barat residents. Such as selling in residential units, on the 2nd floor and at the north door. Here are the details:

Table 1. Flats Residents that Work at Informal Sector.

Trade location	total
Residential units	53
Second floor	12
North gate	31
Total sellers	96
Source: author	

Use of shared space for kids play

The Jatinegara barat flats have provided a children's playground on the 1st floor on the West side of the tower, but unfortunately most users are from the floor near the ground floor. While the children who live upstairs, more likely to play in the front elevator area and corridor at each floor.

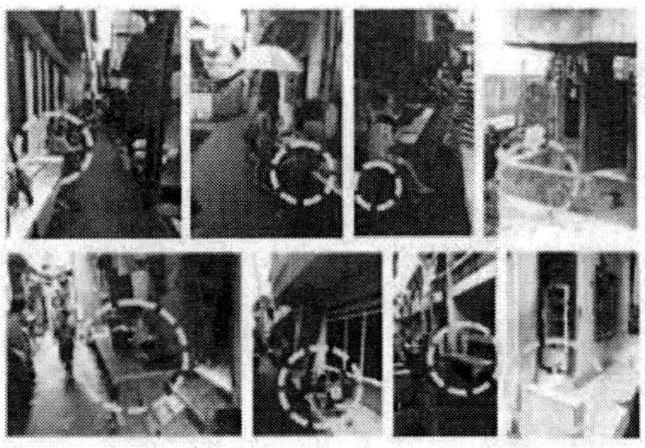

Fig. 13. Seating Provided at Landed Housing Kampong Pulo.

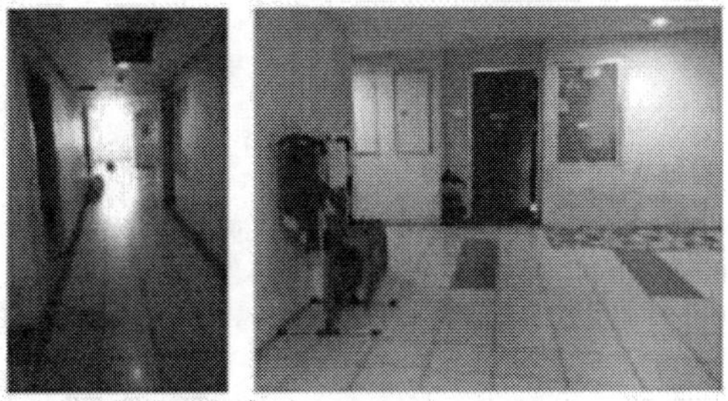

Fig. 14. Seating did not Provided at Flats Corridors.

Fig. 15. Entrance Area for Sitting and Sosializing.

Fig. 16. Seating at landed housing kampong Pulo.

Fig. 17. Activity Supporting Element at Kampong Pulo.

Fig. 13. Some Traders at North Gate Jatinegara Barat Flats.

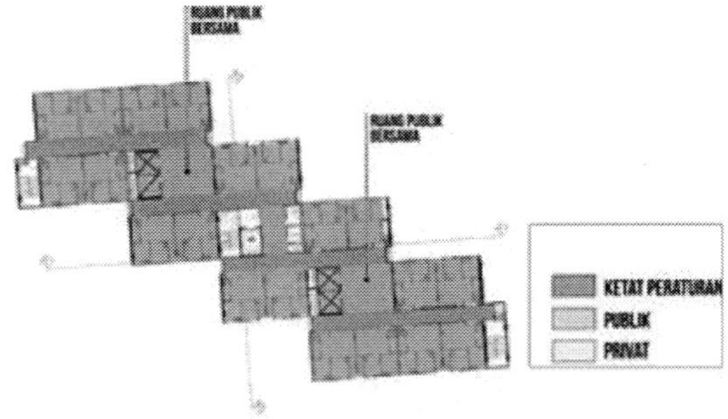

Fig. 13. Space regulation at Jatinegara Barat Flats.

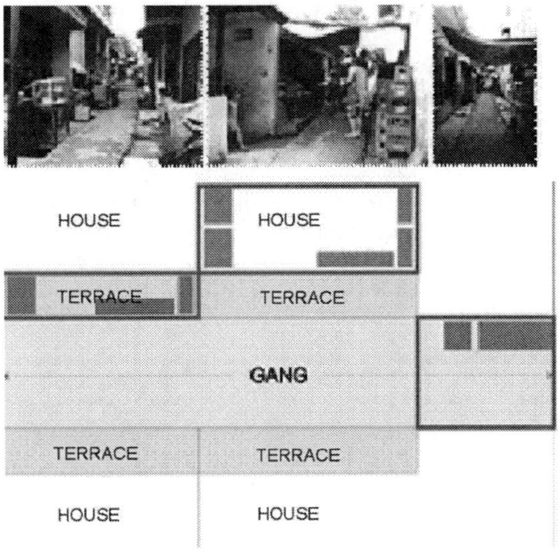

Fig. 13. Shared space use illustration at kampong pulo.

In contrast to the Jatinegara Barat Flats, most dominated by the form of 'trade-in' area type. Apparently this relates to enforcement of existing regulations, it causes pressure on the occupants who want to trade and expanding into the dwelling whose purpose is to avoid enforcement of regulations by the manager / management of towers.

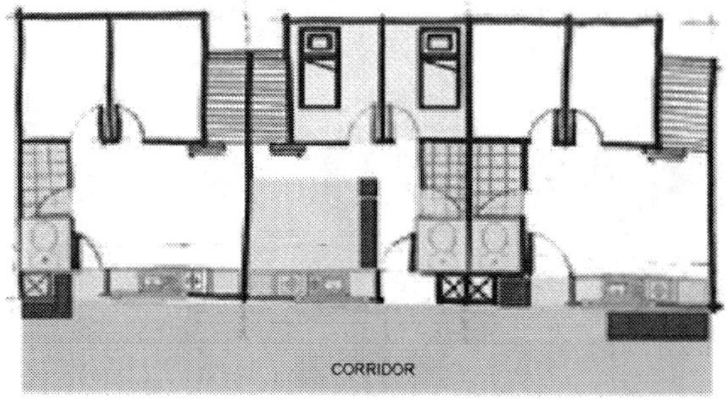

Fig. 13. Type of Trading Area at Residential Unit in Jatinegara Barat Flats.

Conclusion

Change of space use for shared space from landed to high-rise settlement

1) Landed settlement accommodate more diversity of shared activities rather than high-rise settlement. in Kampung Pulo, various kinds of shared activities occur in many common spaces, whereas in Jatinegara Barat flat, some variety of shared activities only occurs on the 1st floor.
2) In the landed house, there is a transition area such as terrace that accommodates the needs of interaction, trade, or play. In contrast to high-rise occupancy, there is no longer found transition space because of space limitation, so the need for interaction is only occur in the public space, while the needs of trading are mostly found in private areas.
3) Spaces used for shared activity in landed settlement are more multifunctional, it is different from high-rise settlement because every space is only used for 1 activity function.

References

Altman, I. (1975) *The Environment and Social Behavior.* Monterey, CA: Wadsworth.

Ananto, Sofian D. (2015). Adaptasi Teritorialitas pada Permukiman Horisontal ke dalam Permukiman Vertikal. *Temu Ilmiah IPLBI 2015.*

Darmiwati, Ratna. (2000). Studi Ruang Bersama dalam Rumah Susun Bagi Penghuni Berpenghasilan Rendah. *Dimensi Teknik Arsitektur* 28(2). 114-122.

Dwiputra, Citra. (2009). Konsep Perencanaan dan Perancangan Rumah Susun Sederhana Sewa di Surakarta. *Skripsi S1 Universitas Sebelas Maret.*

Fitrie Aisyah dan Evawani Ellisa. (2013). Pengaruh Elemen Lungkungan Terhadap Interaski Social Di Ruang Public Hunian Kampung Adat Penduduk. *FT UI. 2013*

Haryadi & Setiawan, Bakti. (1995). *Arsitektur Lingkungan dan Perilaku : Teori, Metodologi dan Aplikasi.* Proyek Pengembangan Pusat Studi Lingkungan, Direktorat Jenderal Pendidikan Tinggi, Departemen Pendidikan dan Kebudayaan Republik Indonesia.

Madanipour, A. (1996). Urban design and dilemmas of space. Environment and Planning D: *Society and Space* 14, 331–355.

Najib, Mochammad., Ramdlani, Subhan., dan Asikin, Damayanti. (2015). Ruang Bersama Kampung Temenggungan Ledok Malang. *Jurnal Mahasiswa Jurusan Arsitektur* 3(1)

Nasution, Ivan K. dan Mediatrich Triani. (2012). Kecerdasan Ruang Marjinal dan Potensi Ruang Publik. *Proceding Temu Ilmiah IPLBI, 21-24.*

Ningsi, Sari Widya. (2014). Perluasan Bangunan Perumahan Swadaya sebagai Ruang Fungsi Hunian dan Usaha Kota Makassar. *Tesis Pascasarjana Universitas Hasanuddin Makassar.*

Novitika Sari, Hendria. (2010). Hubungan Antara Persepsi terhadap Pemanfaatna Ruang Publik dengan Penyesuaian Sosial Penghuni Rumah Susun. *Skripsi S1 Universitas Muhammadiyah Surakarta.*

Purwanto, Edi. dan Wijayanti. (2012), Pola Ruang Komunal di Rumah Susun Bandaharjo Semarang. 39(1). 23-30.

Rakodi, C. (2002). *Urban Livelihoods A People-Centred Approach To Reducing Poverty.* Earthscan, London.

Rifqi Ihsan, Faris. (2014). Pola Penggunaan Ruang Komunal di Kampung Deret, Tanah Tinggi, Kecamatan Jogar Baru, Jakarta Pusat. *Tesis Pascasarjana Universitas Indonesia.*

Shalih, Osmar. (2012). Adaptasi Penduduk Kampung Melayu Jakarta terhadap Banjir Tahunan. *Skripsi, Fakultas MIPA, Departemen Geografi. Universitas Indonesia.*

Siema, Jesieca And Michael Tedja, M.T., Ir. And Indartoyo, M.T., Ir. (2013). Peremajaan Pemukiman Kampung Pulo Dengan Pendekatan Perilaku Urban Kampung. *Undergraduate Thesis, Binus.*

Sitorus, Freddy Masito. (2015) Fleksibilitas Pemanfatan Ruang pada Permukiman Rawa Banjir Kampung Pulo, Jakarta Timur. *Tesis Pascasarjana Universitas Gadjah Mada.*

Suminar, El Yanno. (2016). Kampung Vertikal Kalianyar dengan Pendekatan Arsitektur Perilaku. *Skripsi S1 Universitas Sebelas Maret.*

Sunaryo, R.G (2004) Penataan Ruang Publik Yang Memadukan Pola Aktivitas Dengan Perubahan Fisik Kawasan, Kasus Kawasan Tambak Bayan - Babarsari, Yogyakarta, *Seminar & Lokakarya Nasional Ikatan Arsitek Indonesia*, Jakarta.

Sunaryo,Rony Gunawan and Soewarno,Nindyo and Ikaputra,and Setiawan, Bakti. (2010). Posisi Ruang Publik dalam Transformasi Konsepsi Urbanitas Kota Indonesia. In: *Seminar Nasional Bidang Ilmu Arsitektur dan Perkotaan:Morfologi Transformasi dalam Ruang Perkotaan yang Berkelanjutan,* Universitas Diponegoro, Semarang, Indonesia.

Yasmin. (2009). Pola Pemanfaatan Ruang pada Perumahan Massal Vertikal sebagai Refleksi Gaya Pemanfaatan

Ruang Bersama. *Disertasi S3* Universitas Pendidikan Indonesia.

Zega, Nifka Ulrico Giovanni. (2015). Rumah Susun Sederhana Sewa di Denpasar. *Skirpsi S1 Universitas Udayana.*

The Relationship Between Spatial Configuration and Social Interaction in High-Rise Flats: A Case Study On The Jatinegara Barat in Jakarta.

Rifan Ridwana[1,*], *Budi* Prayitno[2], and *Adi* Utomo Hatmoko[2]

[1]Student, Department Architecture and Planning, Faculty of Engineering Gadjah Mada University, Indonesia.
[2]Lecturer, Department Architecture and Planning, Faculty of Engineering Gadjah Mada University, Indonesia.

Abstract. The construction of high-rise flats to minimize urban slum areas in Indonesia still create space use behavior problems for its residents due to the changes of space configuration between high-rise landed housing. Conventional high-rise flats cannot well accommodate the needs of social interactions happened in landed housing because of its spatial limitation that leads to uncertainty of space use behavior settings. This study aims to understand the relationship between spatial configuration on high-rise flats and social interaction levels of its residents. The object of study is Jatinegara Barat high-rise flats built to relocate slums community from the landed housing in Kampong Pulo. This study applies mix method research using space syntax method to analyze spatial configuration by looking at connectivity and space integration values, and then comparing it with social interaction data from place centered observation to find out the level of interaction and spaces tendency used interaction place.

The results of this study indicate that: (1) the relationship of spatial configuration to social interaction level in Jatinegara Barat flats can be positive or negative. (2) Positive relationships are found on the1st and 2nd-floor areas. High configuration values with high interaction levels are found in shared spaces on the 1st and 2nd floors with characteristics such as open space, large space, and availability of interaction supporting elements, while low configuration values with low interaction levels are found in more confined spaces such as private spaces and narrow corridors. (3) Negative relationships are found in the corridor and shared space in front of the elevator on each typical floors. Shared space in front of the elevator that has high spatial configuration value with large area show a low level of social interaction. While corridor with lower configuration value with the narrow area but have supporting elements such as chairs, mats, and shops have a higher level of social interaction. (4) This study shows that in the case of the relationship between spatial configuration and social interaction, availability of interaction supporting elements has greater influence rather than any other spatial factors.

Introduction

The issue of slum settlement which is not a new problem in Indonesia until now has not been completely resolved. In order to eradicate slums in Indonesia, since the 1970s the government has introduced a village improvement program to improve informal settlements. Flats become one of the programs being intensified by the government at this time. But the construction of flats raises other problems. Social problems often appear in the flats related to the issue of cultural behavior of residents. They face obstacles in interacting with each other, especially because of the limitations of the space available to the flats. In an effort to support the development of a more suitable and appropriate vertical shelter to the character of eastern society, especially Indonesia, a study of the relationship between spatial configuration and interaction levels in the towers is required. This study will examine the relationship between the configuration of flat spaces to the social interactions of its inhabitants based on space syntax theory in the Jatinegara Barat, East Jakarta.

Research Questions

1. How is the relationship between spatial configuration and social interaction in Jatinegara Barat flat?
2. What are the factors that play a major role in the relationship between spatial configuration and social interaction in Jatinegara Barat flat?

Research Purposes

1. Understanding the relationship between spatial configuration and interaction level of residents of Jatinegara Barat flat.
2. Understanding Factors that play a major role in

* Corresponding author: rifanridwana@gmail.com

the relationship of spatial configuration to the interaction of residents Jatinegara Barat flat..

Literature Review

Flats

According to regulation no. 16 year 1985 about flats, flats are high rise buildings built in an environment, divided into structured parts functionally in horizontal and vertical directions that each units can be owned and used separately, especially for shelter, equipped with shared parts, shared objects and shared ground.

Social Interaction

Social interaction is a mutual relationship between people in a group or community / society. The power of existence in a social interaction is very meaningful in relation to the formation of the physical environment space. (jan Gehl, 1996). some aspects that play a role in the process of social interaction:

1. Personal space : Hall (1966) argued that the personal space is a distance to communicate, where the distance between individuals is a distance communicate. This distance can also be called proxemic distance or proximity distance.
2. Territory : Territory is a tendency in control of an area. (Altman, 1984)
3. Space Elements : (Rapoport, A. 1982) The interaction between the behavior and the environment as a nonverbal communication has the meaning of being divided into three elements: Fixed-feature element, Semifixed-element feature, Nonfixed-feature element.

Space Syntax

Configuration can be interpreted as a set of relationships where there are interdependent objects in one structure (Hillier: 2007). The technique of configuration analysis is called space syntax (Darjosanjoto, 2005, extracted from Hillier and Hanson, 1984; and Hillier, 1996). by analyzing the spatial pattern, can be known how the relationship relation in an object either micro scale up to macro. (Hanson, 1998; Dawson, 2002). There are 3 variables in space syntax analysis:

1. Connectivity. Connectivity is a dimension that measures the local property by counting the amount of space directly connected to each of the adjacent spaces in a spatial configuration (Hillier et al :1993 dan Hillier et al: 1987).
2. Integrity. Integrity is a dimension that measures the global property of the relative positions of each space over the other spaces in a spatial configuration (Hillier et al: 1987 dan Hillier et al: 1993).
3. Intelligibility. Intelligibility is the highest stage of measurement in space syntax. The value of intelligibility indicates the level of correlation between the measurement of local scale (connectivity) with the measurement of global scale (integrity)..

Research Methods

This research use a method of combining quantitative with qualitative methods called Mixed Methods Research. This research uses Sequential explanatory approach. This method is characterized by data collection and quantitative data analysis in the first stage, and followed by the collection and analysis of qualitative data in the second stage. space syntax simulation is used for quantitatif method using Depthmap software, where place centered mapping observation and interview are use for qualitative method.

Research Object

The location of this research is Jatinegara Barat flats in West Jakarta. This flat consists of 2 16-storey towers with 14 floors of residential and 2 public floors. The Jatinegara Barat Flat consists of 2 towers with 2 floors which basically functioned as a public area for two towers. Most of the people's livelihoods work in the informal sector as a seller of grocery, shop, open sidewalks, porters, SPG markets, laundry workers.

Fig. 1. Jatinegara Barat Flat.
Source: UPT Jatinegara Barat's Document.

This research is focused on studying the space used for interaction activities. this interaction space is an informal space that accommodate any interaction activities. unit observation to be studied are all spaces used by the residents of Jatinegara Barat flat for interaction activities on residential units, hallways corridors, public spaces along with the type of interaction activities that occur.

Variables

Table 1. Space Configuration Variables.

	Variable	Data	Parameter	Relationship
1	Space Integration (value of space relation with all other room in 1 floor)	• Amount of space • Space Configuration • Space layout • the location of the room furniture	Using Convex map Analysis. Depth. Range of color spectrum from blue to red. Blue: low integration level Red: high integration level	The higher the value of integration of a space, the greater the tendency for social interaction
2	Space connectivity (the amount of connected space with the space around it)	• The location of the Door • Circulation layout • The amount of circulation • Visibility of space	*Depth. Range of color spectrum from blue to red.* *Blue: low level of connectedness. low depth value* *Red: high connectedness, high depth value*	The higher the value of depth, the higher the value of space connections so as to be directly proportional to the tendency of meeting and the movement of people.

Table 2. Social Interaction Variables.

	variable	Data	Parameter	relationship
1	*Personal space*	• the amount of space • Number of people in one room • Distance between people	Distance between people. Standard Social distance (2.1-3.6 meters)	Comfort in interaction is influenced by personal space, personal space optimal in conducting social relations is about 2.1-3.6 meters.
2	territory	• Space boundaries • Ownership of space (private-public) • Space elements	Level of ownership: • Public • Semi-public • Semi-private • • private	The more public the nature of the territory of a space, the higher the tendency of social relations in the space.
3	Space elements	• the amount of space • space openings • circulation layout • furniture elements	Availability of furniture	The more facilities provided make it more comfortable for people to stay and increase the likelihood of social activities

Research Stages

The research phase of this mixed method research starts from (1) the quantitative research phase using space syntax simulation to find out how the configuration and connectivity of the flat and identify the spaces tendency used for social interaction. followed by (2) qualitative research stage using place centered mapping observation and direct interview. this second phase is used as a verification step of the simulation result. whether it is appropriate or different from the actual data that occurred in the field. (3) data comparation from phase 1 and stage 2.

Result and Discussion

1 Result and Discussion Correlation Between Space Connecitivity and Social Interaction Level

Ground Floor

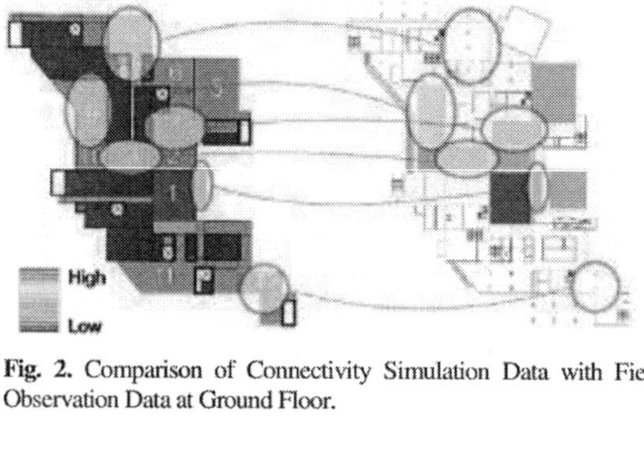

Fig. 2. Comparison of Connectivity Simulation Data with Field Observation Data at Ground Floor.

Positive Correlation

Judging from the comparation of simulation data and field observation, spaces that have high connectivity and also have high interaction levels are room 1 (hall), 2 (hall), and 5 (shared open space). Meanwhile, spaces with low connectivity as well as low interaction levels include closed spaces such as classrooms, offices, utility rooms, and other private spaces, as well as some open spaces such as rooms 6, 11 and 12.

Negative Correlation

Spaces on the 1st floor that have high connectivity but have low interaction levels are 3, 4, and 8, while spaces that have low connectivity but have high interaction levels are open space 9, and 10.

Second Floor

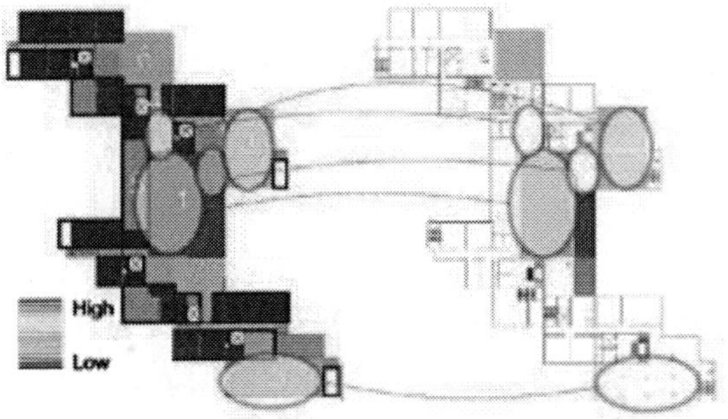

Fig. 3. Comparison of Connectivity Simulation Data with Field Observation Data at Second Floor.

Negative Correlation

Positive Correlation

Space that has connectivity and also has a level of interaction is almost the same is in space 3, 7, and some areas in space 1. This space is open space that is provided as a place of economy residents Jatinegara Barat tower. Meanwhile, spaces with low connectivity and low interaction rates include closed spaces such as closed kiosks and other private spaces, as well as some open spaces such as spaces 2, 6, 8.

Negative Correlation

Spaces that have high connectivity but have low interaction levels are 1, 2, and south corridors. While spaces that have low connectivity but have a high level of interaction is space 3.

Typical Floor

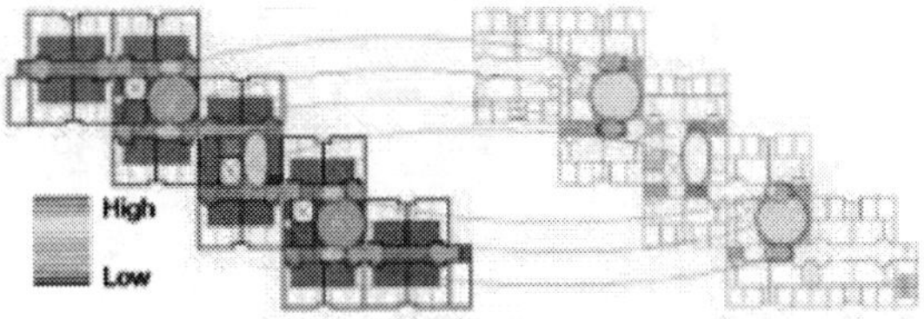

Fig. 4. Comparison of Connectivity Simulation Data with Field Observation Data at Typical Floor.

Positive Correlation

Spaces that have connectivity and also have a comparable level of interaction include corridor spaces in residential areas in general with low connectivity and interaction rates.

Negative Correlation

Spaces with high connectivity but low interaction rates are 2, 4 and 6 spaces (elevator front room), whereas low-connectivity spaces with high interaction levels are spaces on the north and south sides of the front room of the elevator.

2 Correlation Between Space Integration and Social Interaction Level

Ground Floor

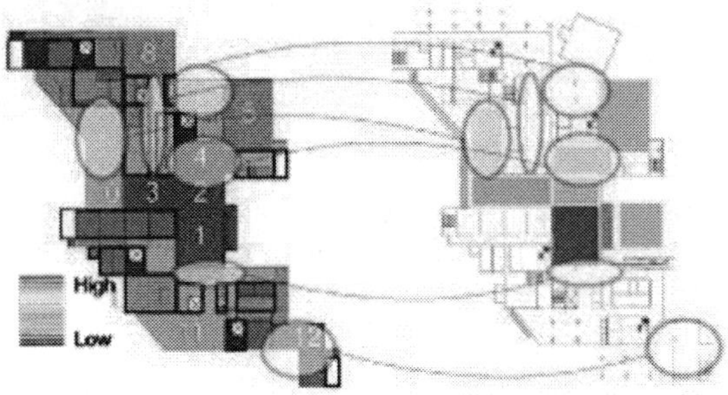

Fig. 5. Comparison of Integration Simulation Data with Field Observation Data at Ground Floor.

Fig. 6. Social Activities at Ground Floor.

Positive Correlation

Judging from the comparation of simulation data and field observation, spaces that have high integration and also have high interaction level are room 1 (hall), 2 (hall), 5 (joint open space), and 9 (park paud). These spaces are a common open space that is provided as a gathering place residents Jatinegara Baratern. Meanwhile, low-connectivity spaces as well as low interaction rates include closed spaces such as classrooms, offices, utility rooms, and other private spaces, as well as some open spaces such as spaces 8, 11 and 12.

Negative Correlation

Judging from the comparison of simulation data and field observation, the chambers that have high integration but have low interaction level are the northern corridor room, space 3, 4, and 6.

Second Floor

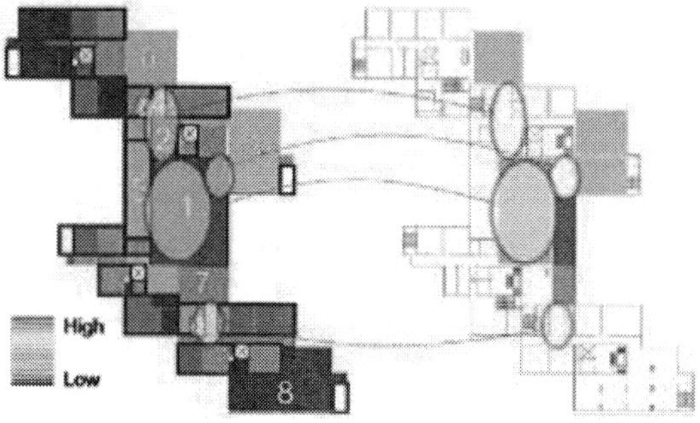

Fig. 7. Comparison of Integration Simulation Data with Field Observation Data at Second Floor.

Fig. 8. Social Activities at Second Floor.

Positive Correlation

Spaces that have integration and also have a level of interaction that is almost the same on the second floor is in space 3, 7, and some areas in space 1. These spaces are open space shared as a place of economic transactions residents Jatinegara Barat . Meanwhile, low-connectivity spaces as well as low interaction rates include closed spaces such as closed kiosks and private spaces such as offices and utilities, as well as some open spaces located in the north and south of such buildings open space number 6 and 8.

Negative Correlation

Spaces that have high connectivity but have low interaction levels are open space 1, 2, and south corridor. While spaces with low connectivity but high interaction rates are open space alongside number 3 on the north west side of the building.

Typical Floor

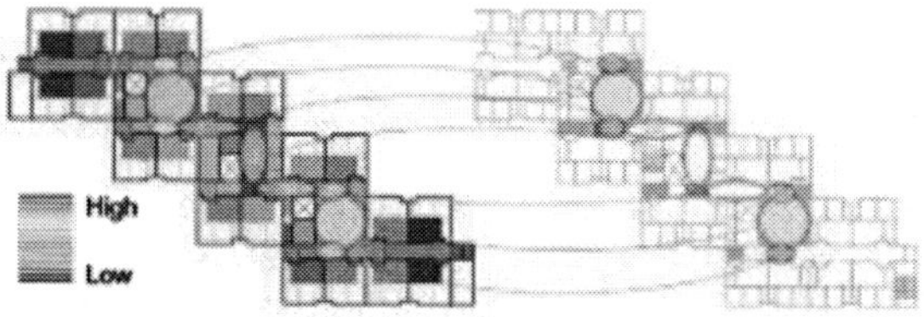

Fig. 9. Comparison of Integration Simulation Data with Field Observation Data at Typical Floor.

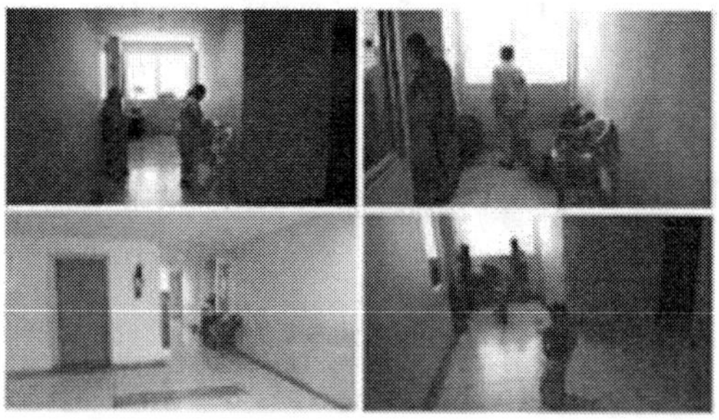

Fig. 10. Social Activities at Typical Floor.

Fig. 11. Comparison of Integration Simulation Data with Field Observation Data at Ground Floor.

Positive Correlation

Judging from the comparison of simulated data and field observations, spaces that have integration and also have a comparable level of interaction on a typical floor tower include corridor spaces in residential areas in general with low connectivity and interaction values, and some areas on the north side of south room 4.

Negative Correlation

Spaces that have high connectivity but have low interaction levels are 2, 4 and 6 spaces (elevator front room), while spaces that have low connectivity but have a high level of interaction is the space on the north and south side of the front room elevator.

Literature Review

Conclusion

1. Configuration of space (integration and space connectivity) affect the level of interaction residents in Jatinegara Barat flat. The correlation of spatial configuration to the level of interaction can be positive or negative.
2. Positive correlations with high interaction rates are found in several common spaces on the first and second floors with open space characteristics, broad, and high integration and connectivity values. While a positive correlation with low interaction rates was found in spaces with narrow private territories enclosed with low integration and connectivity values.
3. Negative correlations found in spaces that have interaction support elements such as the presence of chairs, mats, stalls, although the configuration value of space is small. These conditions are found in typical corridor areas on the floor, as well as some common open spaces. While positive correlations are found in some common room on the first floor, the middle area of the hall hall of the second floor, as well as the front area of the elevator on a typical floor that does not have the interaction supporting elements as found in other spaces in the towers with high interaction levels.
4. The main factors that play a role in the relationship of space configuration to the level of interaction of citizens in the rusun Jatinegara Barat is the existence of spatial elements supporting interactions such as furniture and chair mats stalls selling citizens, compared with the size dimensions of space.
5. Other factors that have an influence on the level of interaction among residents of west jatinegara towers include: dimensions and characteristics of space, territorial space, and the regulation factor towers Jatinegara Barat.

Suggestion

1. the provision of interaction supporting elements is needed in spaces that have high potential for social interaction so that space can function optimally to accommodate interaction activities.
2. In addition to the space configuration settings, the design of flats should be balanced with consideration of physical, psychological and social environmental factors into the design.
3. to develope a better and more innovative high-rise housing in the future, it is necessary for architects, developers, communities, and governments to play a role in the realization of flats that adopt the social life of this urban village in the midst of society.

References

Alexander. C, dkk. (1977). *A Pattern Language*. NY : Oxford university Press.

Alitajer, dkk. (2016). Privacy At Home: Analysis Of Behavioral Patterns In The Spatial Configuration Of Traditional And Modern Houses In The City Of Hamedan Based On The Notion Of Space Syntax. *Frontiers Of Architectural Research.*

Altman, dkk. (1984). *Environmental Psychology.* London: Holt, Rinehart and Wiston, Inc.

Ananto, S.D. (2015). Adaptasi Teritorialitas Pada Permukiman Horisontal Ke Dalam Permukiman Vertikal. *Prosiding Temu Ilimiah IPLBI 2015.*

Andrade, Pamela, dkk. (2011). *Experiment In Public Housing Vol. 1 - A Reexamination Of Subsidized Housing Policy & Design.* Massachusetts: Northeastern University.

Aziz, A.A, dkk. (2012). Relationships Of Spatial Properties In Low-Cost Flats Configurations And Residents Local Social Contact. *ASIA Pacific International Conference On Environment-Behaviour Studies.*

Barada, W.P & Mutlari, Dhani. (2013). Analisis Space Syntax Rumah Susun Berbasis Gang Kampung. *Simposium Nasional RAPI XII FT UMS* Semarang.

Hillier, B. And J. Hanson. (1984). *The Social Logic Of Space.* Cambridge: Cambridge University Press.

Hillier, Bill. (2007). *Space Is The Machine.* London: University College London.

Peponis J, Wineman J. (2001). *Spatial Structure Of Environment And Behavior, Handbook Of Environmental Psychology.* Canada.

Prayitno, Budi. (2013). An Analysis Of Consolidation Patterns Of Kampung Alley Living Space In Yogyakarta, Indonesia. *Journal Of Habitat Engineering And Design 2013*, 5(1).

Siregar, Johannes P. (2014). *Metodologi Dasar Space Syntax Dalam Analisis Konfigurasi Ruang*. Malang: Universitas Brawijaya.

Sofian, D. Ananto. (2015). *Adaptasi Teritorialitas pada Permukiman Horisontal ke dalam Permukiman Vertikal*. Bandung: Institut Teknologi Bandung.

Turner, Alasdair. (2001). Depthmap: A Program To Perform Visibility Graph Analysis. Atlanta: *Proceedings 3rd International Space Syntax Symposium*.

Storyboard as a Representation of Urban Architectural Settings

Arif Rahman Wahid[1,*] and *Paramita* Atmodiwirjo[1]

[1]Lecturer, University of Indonesia, Indonesia.

Abstract. Narrative, as one of the closest aspect to human experience, is sometimes overlooked because of its subtlety with everyday life. The study of its representation, narratology, discusses the manner of selling space and time as a series of interconnected events instead of independent occurrences. It offers an interesting view in rethinking architecture, in particular, its representations. In architectural education, dialogue between the two fields will change the way the next generation architecture graduates tell their ideas. Current architecture schools should not be looked as factories that produce only future architects. It is proven by the fact that some, if not most, graduates will have jobs other than professional architects. They may work as set designers, filmmakers, artists, or invent their own field. Architecture schools, at their best, are more about how to ensure the students can perform spatial thinking and expressing their ideas through representation. All of these circumstances lead to the importance of multidisciplinary discourse in architecture education.

This paper aims to explore the potential of storyboarding practice in Basic Design 2 studio as part of architectural education at University of Indonesia. Adopting a narrative element, storyboard in this studio is used to read urban architectural settings and retell everyday life events; scene by scene, unfold in space and time, through different kinds of creative representations. By doing this exercise, the students 'sense of spatial arrangement is developed by their understanding of position and orientation of objects settings. They also learned about how the time works; both in compressed or expanded ways. Decision-making in choosing the key events within the storyboard plays a role in making engaging visuals. In conclusion, storyboarding exercise to represent urban architectural settings will enhance the students 'sensitivity of space, time, and how their ideas are being told by making a rich, multi-layers of narrative.

Keywords: Narrative, Representation, Storyboard, Multidisciplinary, Architecture education

The Intriguing World of Narrative Learning in Architecture Education

"Once upon a time, not too far far away, there is a boy in an architecture studio, a student, sitting in the corner. He reached his smartphone, then start catching monsters with Augmented Reality on the screen, while at the same time prattling with some of his friends about last week's TV serial's season finale. On the table, there are abandoned detail drawings task, and a CAD application opened on the computer. That day is just one of those boring times in the studio for him. Not that he hates it, the boy just does not want to be a professional architect; he planned to make a film after he finishes the course for good."

Let's admit it. Not all architecture school graduates will become architects for their life. They may want to work as filmmakers, artist, set designers, or something else that are entirely different for their career path. Architectural education should not be threatened by this. On the contrary, it means that architectural education can infiltrate many alternative routes other than solely the route of being professional architects. In return, they have to know several skills related to it, like the utilisation of 3D printing, modelling, CNC cutter, etc. "Even though many skills have to be learnt by an architecture student nowadays, there is a constant need for a future architect to be able to communicate through representation" (Clear, 2014, p.97).

An architecture school, at their best, is more about how to ensure that the students can perform spatial thinking and expressing their ideas through various media that will be delivered to particular audiences. Seeing today's advancement in technology, information overload, and a plethora of media channels used to spread it, now is the time, if not too late, for architectural education to rethink its approach in edifying the students to represent their ideas. Are models, plans, sections, and elevations drawings being all we need to deliver a project?

The infinite possibilities that lie ahead in the future of the students lead to the press of multidisciplinary discourse in architecture education. As the boundaries between disciplines are blurred, multidisciplinary approach provides a way to question traditional architectural education's approach and creating architecture as an integrated spatial design (Clear, 2014). Among many of disciplines, there is narratology, the study of narrative representation, which discusses the manner of telling space and time as a series of interconnected events

* Corresponding author: hello@arifrwahid.com

instead of independent occurrences. "But the narrative is seen not simply a sequence of events, but renders such sequences, and, through it, 'narrative' itself, as an integral part of human experience" (Koeck, 2013, p.19). "It is something we all engage in, artists and non-artists alike (Abbott, 2002, p.1)". It is in the vein of our everyday life.

> *"Here are two views on this question. The first comes from the eminent psychologist Jerome Bruner: "A good story and a well-formed argument are different natural kinds. Both can be used as means for convincing another. Yet what they convince of is fundamentally different: arguments convince one of their truth, stories of their lifelikeness. The one verifies by eventual appeal to procedures for establishing formal and empirical proof. The other establishes not truth but verisimilitude."*
>
> Abbott, 2002, p.162

By diving into a narrative area, the knowledge of storytelling will enhance communication skills. In the dialogue within architectural education, for a student, it can be used to engage the audience, whether other students, facilitator, or prospective client to convey his ideas. More importantly, narrative can be used as a new way of rereading a design. "It also might integrate itself with human and cultural memory, and it might be reflexive and performative – in real time or retrospectively" (Spiller, 2014, p.84).

Encounter with the Good Witch of the North

> *"No, but I am their friend, although I live in the land of the North. When they saw the Witch of the East was dead the Munchkins sent a swift messenger to me, and I came at once. I am the Witch of the North."*
>
> *"Oh, gracious!" cried Dorothy. "Are you a real witch?"*
>
> *"Yes, indeed," answered the little woman. "But I am a good witch, and the people love me. I am not as powerful as the Wicked Witch was who ruled here, or I should have set the people free myself."*
>
> (Baum, 2000, p.9)

Just like when Dorothy first met Glinda in the Land of Oz, this is the time for architectural education meets its mentor. Universitas Indonesia gave us the opportunities to explore this belief. Feeling the urge in studying architecture with a multidisciplinary approach, we looked a few steps back and reflected on the purpose of architectural education in Department of Architecture, Faculty of Engineering, Universitas Indonesia. According to its curriculum, the expected learning outcome for their graduates; both in Architecture and Interior Architecture Programme are:

> *"Able to create architectural design by integrating basic architectural knowledge, applying design and communication skills, applying ability for imagination, creative thinking, innovation and three-dimensional thinking."*
>
> *"Able to analyse context in which architecture is designed and integrate it through design that responds appropriately to the context."*
>
> (Department of Architecture, Faculty of Engineering Universitas Indonesia, 2017 p.55)

These learning outcomes should be integrated since the beginning of the course to make sure that every student has basic knowledge in architectural design and how to communicate their ideas. In particular, in Basic Design 2, in second-semester studio which we taught, its learning objectives include the ability to produce spatial works that respond to a context through appropriate media and exploration of representation techniques (Faculty of Engineering Universitas Indonesia, 2017).

With one of us has a background in the narrative field, we introduced the students with the practice of storyboarding to help them tell their story. Storyboard is a plan for the visualisation of a story; usually for a film, originates from Disney's artists who pin their individual panels up on the wall to make it easier to be viewed and change things (Glebas, 2009). "By doing one's own storyboards carefully and thoroughly, you know exactly what is going to be done before the actual filming begins" (Hart, 2008, p.3). However, a storyboard is a different form of narrative compared to its planned output media (film, theatre play, etc.). "It is telling a story through a series of pictures" (Glebas, 2009, p.6); even though in the following cases, pictures are not conceived only as a two-dimensional surface that contains images or texts but also as another form of representation.

The storyboarding practice was exercised in a task entitled Urban Walk Project. This brief two-week project involves a group journey to several traditional markets in Jakarta: Pasar Citayam, Kemiri Muka, Lenteng Agung, Pasar Minggu, Cikini Ampiun, Gondangdia, and Petak Sembilan. The selection of those specific places is based on the fact that most of the students had never been to the particular traditional markets. Therefore, the experience will be a novel, unbiased with their preliminary knowledge or memory. In this project, the students are expected to learn how to communicate events and human activities unfold in space and time through different kinds of creative representations of the architectural setting. They also asked to demonstrate appropriate ability in documenting, gathering, and filtering information from urban spaces.

> *"...the cinematics of a city can be understood without the linguistic knowledge of units, syntax and grammar. It regards the city as a layered, multifaceted system that is open to interpretation; a poststructuralist sphere that is shaped and understood by more than just functional programme, economic constraints, environmental concerns and social impacts (the measures that normally lie at the fingertips of architects and planners when considering urban interventions). Instead of trying to define a linguistic understanding of urban space, it builds on an existing and habitual system of film language through which we might be able to view, and interpret, urban spaces."*
>
> (Koeck, 2013, p.74)

By doing a storyboard, the students perceive the city through a cinematic vision, allowing it to be interpreted in human scale, senses, and feelings. After the preliminary trip to the market, the students are required to develop diverse approaches in reading the market as an urban architectural setting. They have to tell their action plan for the second visit with a purpose of buying some ingredients from the market vendors. Working in pair; the students are maintaining a collaboration process while creating a storyboard which consists of 9-12 frames to depict the field observation. They were asked to engage with the environment and observe in detail the events and settings of the trading process between their partner and vendor. Since they are Architecture or Interior Architecture students, three-dimensional creative storyboards are encouraged to be created instead of mere story, on boards.

Glebas (2009) argues that in a live action movie, a storyboard is a visual guide for the director instead of the actors. Since in this project the students take role as both director and actor, the storyboard becomes their guide to undergo the journey. Picking up the Disney's artists idea that the frames in a storyboard could be (re)-arranged easily, it indicates that the journey's sequences could happen in so many ways. The role of the storyboard is to maintain this logical narrative sequence, so it does make sense for its audience (Hart, 2008). Hence, in this paradigm, a storyboard is seen as a planning tool, a representation of one possibility of event that will happen. "[It] is the bridge that allows the tight synchronisation involved in creating live-action actors interacting with virtual monsters, supernatural forces, or moving through complex architectural spaces" (Glebas, 2009, p.48). Treating a storyboard as visual plan could give an adequate direction on what to do or what kind of escapade is there going to be experienced while giving the character enough room for improvisation in the upcoming events.

The Red Shoes and the Winged Monkeys

> *"Now the charm began to work. The sky was darkened, and a low rumbling sound was heard in the air. There was a rushing of many wings, a great chattering and laughing, and the sun came out of the dark sky to show the Wicked Witch surrounded by a crowd of monkeys, each with a pair of immense and powerful wings on his shoulders."*
>
> (Baum, 1900, p.83)

While working with the storyboards, the students are experimenting with many techniques. Since they play the role as author, h, and actor simultaneously, they tend to tell the story of the journey to the market with their corporeal body and senses as the centre of the experience. It runs through their point of view as the walker in the urban space, not as if the God looking from above. The students are reminded to be aware of the minute drama that might happen in the market and involve a bit of exaggeration. This awareness to see ordinary events and brings out qualities to show how significant they are (Glebas, 2009) is genuinely needed in tell a compelling story and architecture as well.

A storyboard often consists of character, setting, and event inside a set of panels. The positioning of the panels will affect the reading experience as they act as an indicator that time or space is being divided (McCloud, 1994). Regarding storyboarding practice in the studio, the shape of the panels is infinite. Because of the three-dimensional aspect of it, the panels could be appeared as rectangular frames, spatial borders between models, or even looked barely panel-less. In accordance to Dorothy's journey which progresses with the appearance of many characters, in this exercise the students created various elements that enrich their storyboard.

Figure 1 shows a relatively basic storyboard form by one of the student pairs. With three panels made more prominent than the others, this storyboard easily emphasise its key events, which is the trading of goods process, marked with the existence of the stall keepers. It also illustrates the insignificant of the characters in that story which can be seen from the colourless figure that represents the protagonist (the student). Whereas the other actors (stall owners) are painted in red, and the goods are kept in their default vibrant colours. This storyboard relies much on photographs to keep its context in the market. However, the setting pictures of the place is also made desaturated. At a glance, the diminished details of the actor might have aimed at highlighting the commodities that were traded, but actually, the act exposes the actors even more. The paradox in the absence of details occurred because of the play in the layer of representations. Other than the main panels consisting of characters, highlighted objects, and backgrounds, there are also texts that work as information of the precise stand location and time of the event.

Fig. 1. More Conventional Storyboard with Emphasis on Some Key Panels

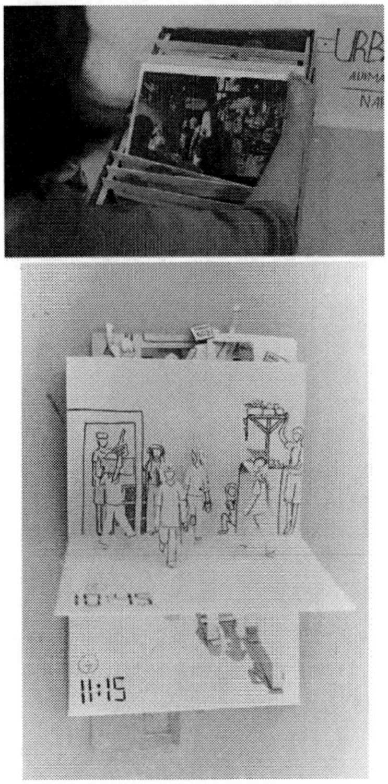

Fig. 2. Layered and Pop Up Book-Style Storyboards.

Other students made the storyboard with the essence of a book in mind. To comprehend the whole story, the reader is forced to see the pages one by one. In this kind of representations, all scenes are always experienced as the present. When the reader opens a page, the previous page will be the past, and the next one is the future event. As illustrated in Figure 2, the example on the bottom picture made in a pop-up book style to resemble a three-dimensional space. However, here the urban space is read as a combination of people and loose objects in the market such as stalls, openings, and other objects that have a direct interaction with the character. The students leave all ground and far setting in white, so they do not distract the description of the occurred story. The time, in this example, appears explicitly again in every corner of the page with the clock time.

In Figure 3, traditional architectural representation techniques are blended in storyboards in the manner of map and plan connected with panels as spatial references. The interrelation showed by a set of lines or colour codes.

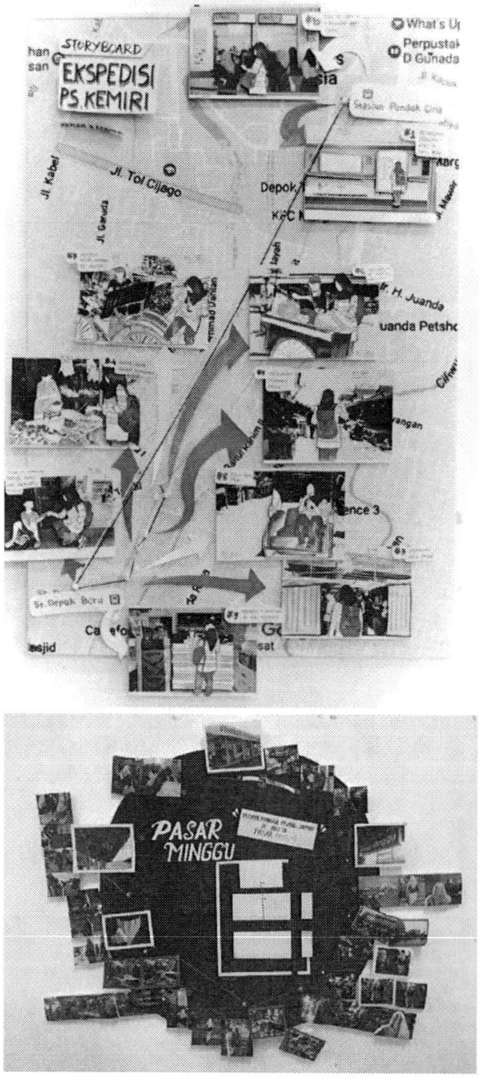

Fig. 3. Storyboards Embedded with Plans

Quite distinctive, the storyboards in Figure 4 do not use fixed borders between panels. Amazingly, the audience still can read the narrative without much hassle. It is our consciousness of narrative that explains how the transition between panels happened. Pictures and the intervals between them create the illusion of time through closure (McCloud, 1994). It has shown cleverly in the use of the train sides. By seeing the surrounding of each front of the train, the audience understands that the train carriage had moved from one station to another. Here, the train acts as the panel and part of the setting as well. The expression of characters in Figure 4 denotes that these students value their character as the main protagonist. The protagonist, printed neatly with a full face and gestures shown to the audience is outstanding from the rest of the setting or the cameos.

Fig. 4. Blurred Frames, Three-Dimensional Storyboards

Students who worked on the model in Figure 5 represent the market using black colour to distinguish the stalls, actors from the actual space. Characters are shown with red and blue figures; the red indicated the consumer and the blue one marked the shop owner. Solid coloured figures scored the characters base position, while more transparent figures are showing the character's movement. In this model, the story told in a linear way-by the number put on several places-but could be read in a non-linear direction. It is also a noticeable instance of time squeezed in a storyboard. From one sequence, many figures do not represent many people in the market. Instead, they represent the movement of the people which means each figure has its own time.

When the Two Worlds Collide

"Hush, my dear," he said. "Don't speak so loud, or you will be overheard--and I should be ruined. I'm supposed to be a Great Wizard."
"And aren't you?" she asked.
"Not a bit of it, my dear; I'm just a common man."
(Baum, 2000, p.105)

From what students had accomplished, there are three key points at least that can be deducted to support the importance of this exploration method. First, the students' sense of spatiality is developed by their discernment of position and orientation of objects in settings. In recreating an urban setting, they consciously observe the surrounding and take note of what placed where and when. The existence of many human beings in the market also become their reference to measure scale, proportion, and rough dimension that make that setting works as a place. This practice challenged the students to record and retell what Koeck (2013, p.73) claimed that "we actually have learned to mentally select and process the flood of moving images to which we are exposed every day".

They also learnt about how the time operates; not only within architecture, but also its representation. By walking in the market, the students are aware of how nimble everything is changing. Specifically, as suggested by McCloud (1994), in a comic-or storyboard in this case-we perceive the time spatially. In other words, space and time are merged. From several cases mentioned above, each panel of the storyboard displayed how the time can be compressed or expanded. For instance, in Figure 1, in which the protagonist buys vegetables, and the vendor prepares to pack them as well, or the double actors and vendors that exist in Figure 5. In a real world, the actual sequence would be somebody asks the shop owner to buy vegetables, only then the seller prepares it. And the duplication of characters in Figure 5 obviously shows their movement throughout the time. Nevertheless, our mind knows the order of the event. As Abbott (2002) suggested, the storyboards as a narrative form, functioned to be a mechanism for expressing our awareness of time.

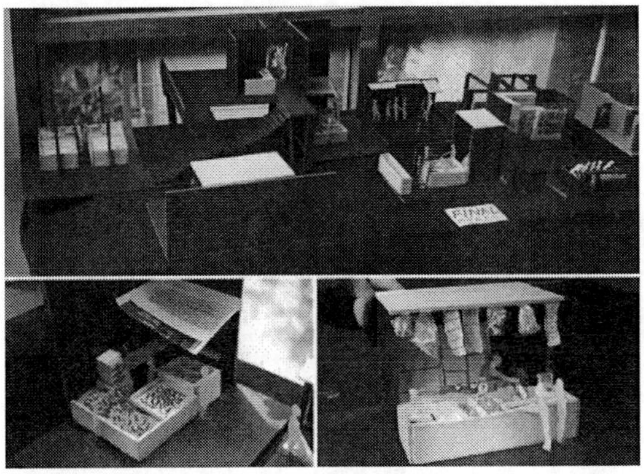

Fig. 5. Storyboarding with Detailed Architecture Model

Creative thinking and decision making are the third important aspect addressed in this exercise. It includes the planning of the structure, composition, and overall look of the storyboard. Although there is flexibility to arrange the scenes back and forth in a storyboard, at the same time it forces the students to have the fixed order of the panels. Chosen events that appeared in the panels must indicate the critical stage of the story. Also, they should be able to hook the audience by making engaging visuals through various methods of representations. Creative thinking plays a role in deciding what format the storyboard should be, depending on who the audience is. In a real movie pitch, a storyboard is widely used as a tool to explain the whole story visually and convince the stakeholders. Likewise, in architecture crit, the students need to articulate their ideas and designs, in a logical narrative using their choice of media (models, drawings, diagrams, etc.) to the reviewer as the audience. Hence, the storyboarding practice could help the students to make captivating presentations via its narrative structure.

The Aftermath

"My greatest wish now," she added, "is to get back to Kansas, for Aunt Em will surely think something dreadful has happened to me, and that will make her put on mourning; and unless the crops are better this year than they were last, I am sure Uncle Henry cannot afford it."
Glinda leaned forward and kissed the sweet, upturned face of the loving little girl.
"Bless your dear heart," she said, "I am sure I can tell you of a way to get back to Kansas."
(Baum, 2000, p.144)

Now we should return to the root of this discussion, the architectural education. We would argue that multidisciplinary learning in architectural education indeed broaden the perspective on architecture. Storyboard, borrowed from narratology and filmmaking provides an exciting view of seeing architectural representation. By using that practice, the students are given the task to plan their journey to an urban architectural setting. Along the project, the practice of storyboarding advanced the students' sensitivity of space, time, and how their ideas are being told by making a rich, multi-layers narrative. Moreover, in the subsequent studios, the mixture between well-formed argument and a vast amount of lifelikeness storytelling through various methods of representations can help the students convince their audience and create an engaging design.

As the curtains are going down, we may conclude that the use of storyboard does affect the way representation is made in architectural education. Probably not all sets of architectural drawings are required for all types of tasks. There is no doubt that models, plans, sections, elevations, or perspectives drawings are needed at some point in the process. But at certain times, using a storyboard is more relevant in showing students' creative thinking with the point of view of the protagonist, the walker in the urban spaces. However, the practice of storyboarding cannot be taken as the elixir that brought by the hero in Campbell's hero's journey structure. Rather, we can put it as one of the thousands of the hero's faces. Storyboard is only one approach that offers an alternative representation technique which will help the learning process in architecture education. Finally, there is no happily ever after as the real world is a never-ending field of conflicts. It all depends on the students' shoulder as the future architects, which will be the right tools for the right jobs at the right time. Fin.

Acknowledgement

The photos included in this paper are student works of Urban Walk Project in Basic Design 2 Studio, class of 2016 from Architecture and Interior Architecture programme at Universitas Indonesia.

References

Abbott, H. P. (2002). *The Cambridge introduction to narrative*. Cambridge: Cambridge University Press.

Baum, L. F. (2000). *The Wonderful Wizard of Oz*. Retrieved from www.e-booksdirectory.com/details.php?ebook=1254.

Clear, N. (2014). Convergence: Architecture as intregrated spatial design in Spiller, N., & Clear, N. (Eds.). (2014). *Educating architects: how tomorrow's practitioners will learn today*. New York: Thames & Hudson.

Faculty of Engineering, Universitas Indonesia. (2017). *Universitas Indonesia, Faculty of Engineering Academic Guidebook: Architecure Department* (2017 Edition). Retrieved from http://eng.ui.ac.id/wp-content/uploads/BUKU-ARSITEK-INGGRIS.pdf

Glebas, F. (2009). *Directing the story: professional storytelling and storyboarding techniques for live action and animation*. China: Elsevier/Focal Press.

Hart, J. (2008). *The art of the storyboard: a filmmaker's introduction*. Boston: Elsevier/Focal Press.

Koeck, R. (2013). *Cine-scapes: cinematic spaces in architecture and cities*. New York: Routledge.

McCloud, S. (1994). *Understanding comics: the invisible art*. New York: HarperPerennial.

Spiller, N. The architectural mixing desk of the surrealist city in Spiller, N., & Clear, N. (Eds.). (2014). *Educating architects: how tomorrow's practitioners will learn today*. New York: Thames & Hudson.

Urban Design Dimension Of Informality At The Perimeter Of Brawijaya University And UIN Maliki Malang

Binar Tyaghita Cesarin[1,*], *Himasari* Hanan[2], and *Agus* Suharjono Ekomadyo[2]

[1]Magister Student of Urban Design Department of Architecture, School of Architecture, Planning and Policy Development; Bandung Institute of Technology, Indonesia
[2]Lecturer, Department of Architecture, School of Architecture, Planning and Policy Development, Bandung Institute of Technology, Indonesia

Abstract. Informality is one of the commonly emerged issues in urban design which rarely explored, especially informality within university's perimeter. Brawijaya as one of the biggest and oldest University in Malang over time has boosted the development of several of its perimeter, provided several hotspots for students and youth. These rapid hotspots growth is related to the growth of informal practices. For cities that developed by its universities, it is necessary to understand both of the formal and informal practices within its perimeter. Through this study I would like to know the characteristic of informality within university perimeter, which formed by Brawijaya, UIN Maliki, and ITN; and to frame it within Carmona's Urban Design Dimension. Mapping is utilized as a primary method to analyze both formal and informality within site. The formal aspect consists of formal activities, function and site user. While informality mapper consists of street vendors and street art. The research found that while urban informality within Brawijaya and UIN Maliki are related to the character and morphology, its formal structure has; its relation is reciprocal. The morphological, visual and functional dimension of university perimeter is driven by the social, perceptual and temporal dimension formed by its user and showed through informality.

Keywords: University Perimeter, Informality, Street Vendor, Street Art

Introduction

Informality is one of the commonly emerged issues in urban design which rarely explored, especially informality within university perimeter. University perimeter in Indonesian cities usually grows to support the student life activity which interrelated with the life of its local residents. As informal practice character intermeshed with its formal structure (Dovey, 2012), it is interesting to explore the characteristics of informality within university perimeter.

The complexity and severity of informality cases often made it difficult to understand and manage which explained why the government, planners and urban designer's efforts to manage informality are rarely succeeded. Urban design is integrated design field concerning the urban setting, which is both process and product. With that framework, Carmona et al (2003) proposed six dimension or urban design that are overlapping and interrelated within the everyday urban activity. Framing informality within urban design dimension might provide a new way to understand and manage the urban informality. Which to control and to tolerate; and which to ignore and recognize.

Malang is the second largest city in East Java, Indonesia and currently known as the 'education city' as it has at least 55 higher education institutions including universities. One of its university perimeters is located around the Sumbersari – Gajayana Street. Wedge by three Universities: Brawijaya, UIN Maliki and ITN and consists of two kampongs (Ketawanggede and Sumbersari); this site has the unique characteristic of Indonesia's university perimeter which has a diverse user. Through this study, I would like to understand the characteristics of informality within university perimeter and how to frame it within urban design dimension so that it can be used as consideration for government, planner and urban designer to understand and manage informality, especially within University perimeter.

Understanding Urban Informality

Informality is actually part of human oldest civilization which organically developed from informal settlement and trading. After 1950 the process is reversed, as informality process started after the formal structure was built. (Dovey, 2016). Informality actually first mentioned by Juan Pablo Pe'rez Sa'inz in 1989, but arguments and concept about informality started to developed in the 1950s. During that period, informality within urban space often associated with immigrants and poverty. Informality refers to certain practices done by groups of people which consist of petty traders, street vendors, coolies and porters, small artisans, messengers, barbers, shoe-shines boys and personal servants at alleyway or street side (Al-Sayyad, 2004 as cited in Dovey,

* Corresponding author: tyaghita@gmail.com

https://doi.org/10.1051/shsconf/20184107005

2016).

Generally during that period urban informality defined as urban practice outside formal planning and regulation which usually done by certain groups of people with limited working and economy capability. However after the 2000s more positive understanding of urban informality started to arise. Al – Sayyad (as cited in Dovey, 2016) defined urban informality as a new way of life, which done to overcome poverty, availability of jobs and settlement which cannot be supplied and/or achieved formally. Informal trading also showed a form of entrepreneurial flexibility, adaptability skills and creativity (Neuwirth, 2012).

Urban informality, especially in form of street vendor, is a widespread phenomenon in Indonesian's city. It is a part of every day's life and can be found where there is a high concentration of population. The existence of street vendor at urban spaces often triggers problems such as traffic congestion, contribute negatively to the physical appearance of urban space, contribute to lack health issues and reduce the feeling of safety amongst the street user. It also considers a threat to local business and often triggers conflict. Although there is also a positive aspect of urban informality which is potential to enrich the character of urban space, enhance the lively atmosphere, animate the life of the city; generally urban informality was considered out of place (Yatmo, 2008).

Street vendors are often characterized by mobility and flexibility. Those aspects manifest into the physical setting of their activity which in Indonesia it mostly consists of stationary with a permanent structure such as warung or kiosk, stationary with non permanent structure that is easily assembled and disassembled such as *gelaran* or mat and tent, ambulatory by using pushcart, *pikulan* and *bakul* (Yatmo, 2008). Mobility and flexibility are the key survival tactic when the informal trader is subject to prosecution and confiscation of goods (Dovey, 2016).

Fig. 1. Types of Street Trading in Indonesia: 1) Kiosk, 2) Mat, 3) Tent.
Source: Yatmo, 2008.

The types of the street vendors are also geared to the urban morphology in subtly different ways. In general terms, the more fixed types are more likely to be tolerated in relatively informal parts of the city, while the more mobile vendors are able to infiltrate the more formal city. The fixed kiosk requires a high level of protection, while mobile traders can more easily escape from street cleaning' operation. The fixed kiosk is essentially a form of informal settlement since a temporal stall can become a permanent building overtime as wheels become dysfunctional and a roof is added. There are also daily and weekly cycle of informal trading, where several enterprises using the same space at different times of the day. Sometimes, the uses of place also show collaboration between traders with adjacent shopkeeper/ shop owner.

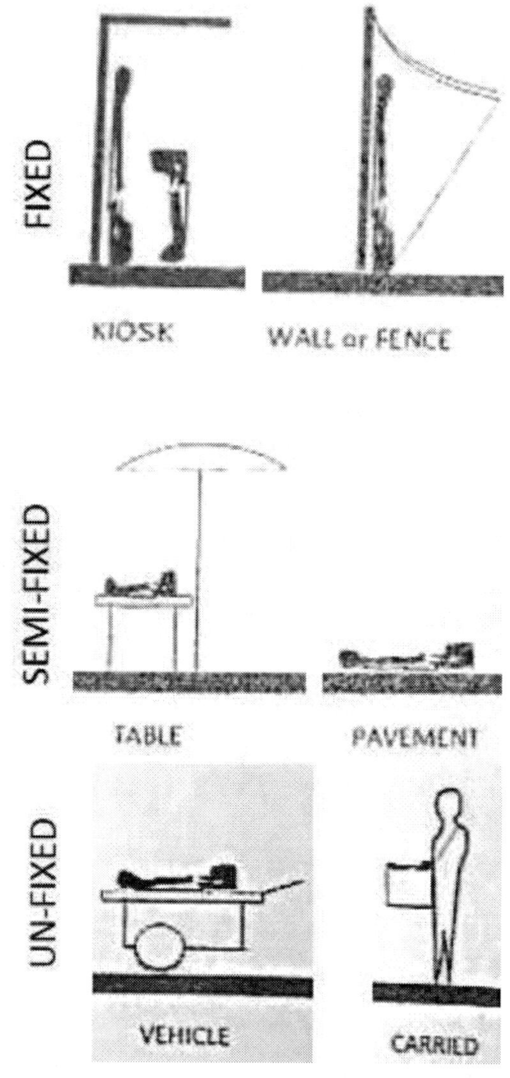

Fig. 2. Spatial Typologies of Street Trading.
Source: Dovey, 2016.

Graffiti is also considered an 'out of place' elements of urban design (Cresswell, 1996 in Yatmo, 2008). While it actually the oldest art form with its roots in cave paintings, currently it is a practice that slips between crime and arts

(Dovey, 2016). The varieties of graffiti writing range from the quick tagging of a signature (in felt pen or spray paint) through various forms of 'paste-ups', slogans, 'throw-ups' and stencils to larger pieces or extended wall.

Urban graffiti is mediated by urban morphology and depends upon the production of blank and relatively smooth public walls that are not strongly identified with the occupants. Grafitti emerges within the interstices of a streetscape, or side or rear walls, laneways, derelict buildings, retaining walls, bridges, and railway cuttings. While laneways are relatively saturated, some main streets are entirely unaffected; where graffiti transgresses an existing identity, then it is much less likely to be produced and much more likely to be erased.

Fig. 3. Types of Grafitti.
Source: Dovey, 2012.

In locations where dynamism, change and creativity are seen as integral to the emergence of place identity, graffiti is often an important contributor to the production of an edgy and trasgressive urban character (Dovey, Woodcock and Wood 2012). When it's out of place it will considered vandalism, but when it enhance the characteristic of the urban space it considered a street art. It symbolic capital allows a slippage into place branding and advertising, adding economic value to a street art.Street art at its best engages with the street and draws its potency from the urban situation in which it located.

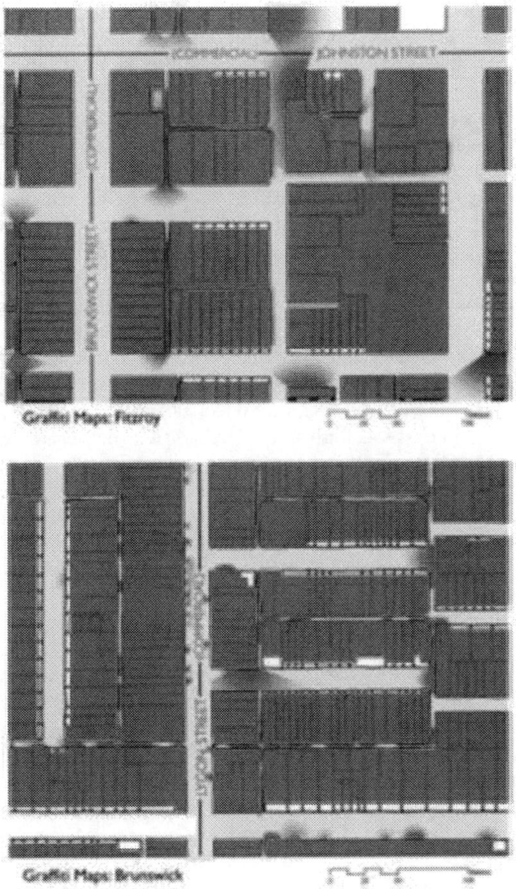

Fig. 4. Graffiti and morphology in Fitzroy and Brunswick.
Source: Dovey, 2012.

Urban Design Dimension

Urban design is integrated design field which main concern is public space within urban settings. It is a shared interest between at least city planning, landscape architecture, civil engineering, and architecture field (Carmona et al, 2003; Lang,2005). The concern in Urban Design developed over time, from the visual aesthetic tradition which emphasising the visual qualities of buildings and space during the 1890s – 1960s; the social usage tradition which primarily concerned with the social qualities of people, places and activities during 1960s – 1980s; which then fused and developed into the 'making places' tradition.

Through the 'making place' tradition, contemporary urban design simultaneously concerned with both the aesthetic entity and behavioral setting, which shifts the focus of urban design on diversity and activity. With this concept comes the notion of urban design as the design and management of the 'public realm' - defined as the public face of buildings, the spaces between frontages, ·the activities taking place in and between these spaces, and the managing of these activities, all of which are affected by the uses (Carmona et al, 2003).

Within the 'making place' tradition, there have been a number of attempts to identify the desirable qualities of successful urban places and/or 'good' urban form by Lynch (1981), Jacobs and Appleyard (1987), Bentley (1985), Tibbalds (1989) and Congress or New Urbanism (1993). Each of those has a different degree of prescription and created with a different context. Urban design dealt with a rapidly changing environment which driven by diverse user and activity, and involve diverse process and field. Thus it is necessary to see urban design as both design process and product (Carmona, et al, 2003; Lang,2005). With that framework, Carmona et al (2003) proposed six dimension or urban design that are overlapping and interrelated within everyday urban activity which consists of morphological, perceptual, social, visual, functional and temporal.

The morphological dimension of urban design mostly focusing on issues of urban form and urban layout regarding permeability, integration, and segregation; and vehicular/pedestrian access. The urban form and layout are formed by land uses, building the structure, plot pattern, street pattern and public space network. Within the urban form and layout, there are parts which do not change or change slowly and the part which changes over a much shorter period of time.

The perceptual dimension of urban design emphasize on people and how they perceive, value, draw meaning from and add meaning to the urban environment. Perception is more than just seeing or sensing the urban environment. It refers to the more complex processing or understanding of stimuli found in the urban environment. The consequences of these process on urban design are the construction of place, which involves the design of land use and built from considering the meaning/image it wants to achieve. One of the current issues found in the urban environment is placelessness which caused by globalization, mass culture and loss of attachment to the territory. The other is invented places which involves superficiality, 'other-directedness' and lacking authenticity.

The social dimension raises issues concerning values and difficult choices with regard to the effect of a design decision on individuals and groups in society. Within social dimension, the boundary of the public realm is often difficult to identify and define precisely. The public realm has 'physical' dimension which is the space and setting that support or facilitates public life; and 'social' dimensions which are the activities and events occurring in those spaces and settings. Because the physical setting is often difficult to identify, it is suggested that urban designer focuses on the social dimension which is the public life. Public life can be broadly grouped into two interrelated types: 'formal' and 'informal'. Of most interest in urban design is informal public life, which occurs beyond the realm of formal institutions and entails choice. The role of design is delivering particular social goals which are the provision of an accessible, safe and secure, equitable public realm for all. For that urban designer will need to work with a wide range of other public and private stakeholders to effect significant sound benefits. Rather than determining human actions or behavior, urban design can be seen as a means of manipulating the probabilities of certain actions or behaviors occurring. While urban designers cannot 'make' places, they can create more 'place potential'.

The visual dimension emphasized the need for the urban designer to consider the whole context in which they approach the visual aspect of additions to the urban environment. The Urban designer must be careful to avoid equating consideration of the visual dimension of urban design with the consideration of architectural design. Buildings, streets, and space; hard and soft landscaping and street furniture should be considered together, to create drama and visual interest and to reinforce or enhance the sense of place.

The functional dimension of urban design focusing on how places work and how an urban design can make 'better' places; which in detail concerning the use of public spaces and the public / private interfaces; mixed uses and density consideration; environmental design and capital web. The design of public space and public/private interface should consider how people use it which related to social use of space, movement,it shape and edge, visual and aural privacy; and what they need from it which are comfort (which supported by environmental design), relaxation, passive engagement, active engagement, and discovery. Mixed used are important for creating a lively environment, however, it should be done with economic consideration as it will be more difficult to develop, manage and might lessen the value of the investment. With density, the concern is how to protect amenity and provide a livable environment as sometimes it equated with the poor quality environment. While the major capital web consideration in urban design is: the provision of public open space; road and footpath design; parking and servicing; and other infrastructure.

The temporal dimension concerns the temporal or 'time' dimension of urban design which includes three key aspects which are time cycle, continuity and stability; and the management of change. Activities are fluid in space and time, environments are used differently at different times. Facilitating and encouraging the use of urban spaces requires an understanding of the effects of the cycles of day and night, the seasons, and related cycles of activity. At different times of the day and night, the urban environment is perceived and used differently. Although environments relentlessly change over time, a high value is often placed on some degree of continuity and stability. Urban designers need to understand how environments change, what stays the same and what changes over time. They also need to be able to design and manage environments that can accommodate the inevitability of time's passage. Urban design is a massive project which needed quite a lot of time to be finished, it also involves diverse field which might also change over time. Thus it is necessary to consider how to manage the change through three key processes of urban design which are development, control, and communication.

Formal Structure of University's Perimeter at Sumbersari - Gajayana

The method used to understand the urban informality phenomenon and how it related to its urban design dimension is mapping. In this study, the mapping is not only done as an illustration or to shows territory but as thinking tools to investigate (Dovey, 2014). The information collected through observation which partially utilizing the google maps and street view. The component of formal

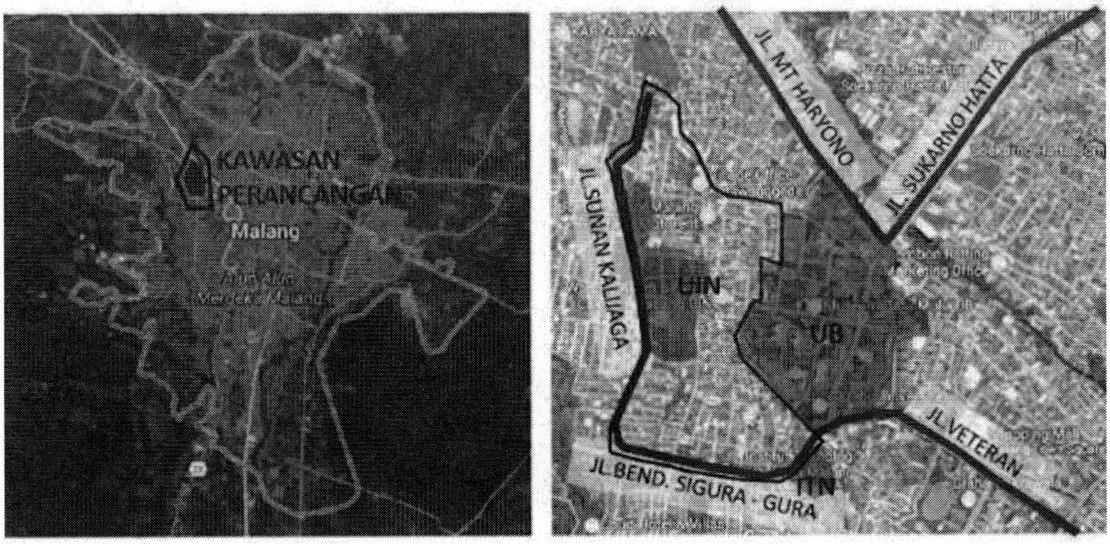

Fig. 5. Site Context.

practices observed and mapped consists of site users, activity and function. While the component of informal practice consists of street vendors and street arts.

Sumbersari – Gajayana street is the main street corridors which connecting the university's perimeter area formed by Brawijaya University at the east, UIN Maliki at the west and ITN at the south. It also connects two local kampongs which are Ketawanggede (the university perimeter of Brawijaya) at the east and Sumbersari (the university perimeter of UIN Maliki and ITN) at the west. The study area is limited to areas affected and formed by the university's activity which is the main street and local street located within the kampong (as shown in figure 2).

The site is part of North Malang which is the center of higher education in Malang and it's highly accessible through public transportation. The main road went through the site also connecting the site with other University's perimeter that becomes student's hotspots which are Bendungan Sigura – Gura, Veteran and MT Haryono Street. Through its local street network, the site also connected through other hotspots which are Sunan Kalijaga and Watugong. Compared to other hotspots, the site actually has a more diverse mix of user and activity, because of the stronger presence of local kampung resident. However, it's currently the most unattractive hotspots for students.

Fig. 6. Study Areas.

The main use for this area is local resident; which also the owner of several formal and informal retails in the area, and the university student consists of mostly Brawijaya and UIN Maliki's students. As Brawijaya grew its students is coming from the wider radius, which includes a student from bigger cities such as Jakarta, Bandung and from cities outside Java. This student demographics change allegedly affected the social dynamic in the site, which then rendered the site unattractive. The functional mix consists of housing, public facilities, retails, consumption and small-scale production. The housing consist of kampong resident's house (which sometimes also provide room for students), boarding house (cost) and boarding house which mixed with a retail function on the first level. The strong presence intense use of public facilities which consist of civic center Kantor kelurahan), kindergarten and elementary school, mosque and health facilities (posyandu). The unique types of retails found within the areas, which are not usually found outside the university perimeter area, are those which provide the student's academic needs such as stationery, photocopy, printing, translation, and data analyzing services and additional study courses. While the most dominant

Main Street: Sumbersari Entrance, Right Side Is Entrance To Brawijaya University, Left Side Connected To Bendungan Sigura – Gura Street And ITN.
Source: Google Street View, 2016.

Main Street: Sumbersari Street 1). The Ruko's Development, 2). The Old And New Retails, 3). The University owned retail and privately own retails.
Source: Google Street View, 2016.

Local Street: 1). Academic related services anc consumption at Sumbersari Gg 3 Street, 2). The Sunan Drajat Alleyway, 3). Academic related Retails at Kertoleksono Street, 4). Local housing and boarding house at Kertosentono Street.
Source: Google Street View, 2016.

Fig. 7. The main and local street within site.
Source: Google Street View, 2016.

function found on the areas are retailed which supplies both students and local residents daily needs such as Toko kelontong and small-scale-home based consumption function such as warung.The small-scale production found on site is consists of food production, tailor services, printing, merchandise production, shoe artisan, and florist. The most dominant production are food production and printing.

Fig. 8. The Greening Competition Event at Ketawanggede. (Source: http://kelketawanggede.malangkota.go.id, 2016)

While the mix of function showed the entanglement of the students and local residents, the activities do not. There is no shared activity within the site other than brief interaction which only happened between students and the local resident which acted as its landowner. The student's routine activity is showed through the academic's retails available throughout the area, but their other activities such as performances and hangouts are not found within this area. There are also potential university's events and programs, such as the recycle and agripreneur programs form Brawijaya's University and student's events form UIN, which rarely or never done within the site. Most of the student activities are either done in their living space or in university.

The local resident's activity is mostly done utilizing the space of local street and an empty lot. The interaction between residents often found at the local streets, which they consider the extension of their home's terrace. They also conducted several events such as greening competition and historical parade and bazaar. However, most of these activities and events are still not accommodate spatially and mostly done within Local Street. The main road mostly consists of retails, further covering the unique characteristic of the kampong behind it.

Fig. 9. The Historical Parade.
Source: https://tawangsarikampoengsedjarah.wordpress.com, 2016.

Informal Practices within University's Perimeter at Sumbersari - Gajayana

Complementing those formal practices, the informal practices found within the site consists of a street vendor and street art. Street vendor found at the site mostly sell food and drink, although there are also small numbers of vendors which sell groceries in the morning, gasoline and newspaper. The types of food sold are mixed of local home-cooked food (such as *'warung nasi'*, *pecel*, *Soto*, *sop ayam*), specialty food (such as *restoran padang*, *chinese food*,*rica – rica*, *siomay bandung*, *pempek*) and cheap student food (*nasi lalapan*, *nasi goreng*, *nasi ayam*). The type of snack sold is consist of desserts such as ice cream, *martabak* (pancake) and light snack such as *'cilok'*. The product is sold and its prices showed the customer it's targeted, which are kids of local resident and university's student. It also shows the cultural diversities of the student and how it exists along with local food.

The dominant typologies found are kiosk and vehicle, although there are also found a table and pavement typologies. The table typologies are not technically using a table to sell but using the shelf. The wall/fence typologies are not found at the site. The fixed type of street vendors found at site mostly attached to local resident's house, utilizing their terrace to sell food and drinks, some developed into part of the housing interface as warung. While the semi-fixed and un-fixed type found at the local street mostly sell vegetables, groceries, and snacks which are the most of products which the fixed type didn't sell. This show how street vendor as informal practices can be developed into formal practices and how both of formal and informal practice can actually can goes along harmonically through some sets of 'informal rule and organization'.

Street art found in the site, based on how it drew consist of two types which are graffiti and mural. The graffiti consisted of those which only to show identity through tagging on storefronts an those which convey protest

Fig. 10. Typologies of Street Vendors.

https://doi.org/10.1051/shsconf/20184107005

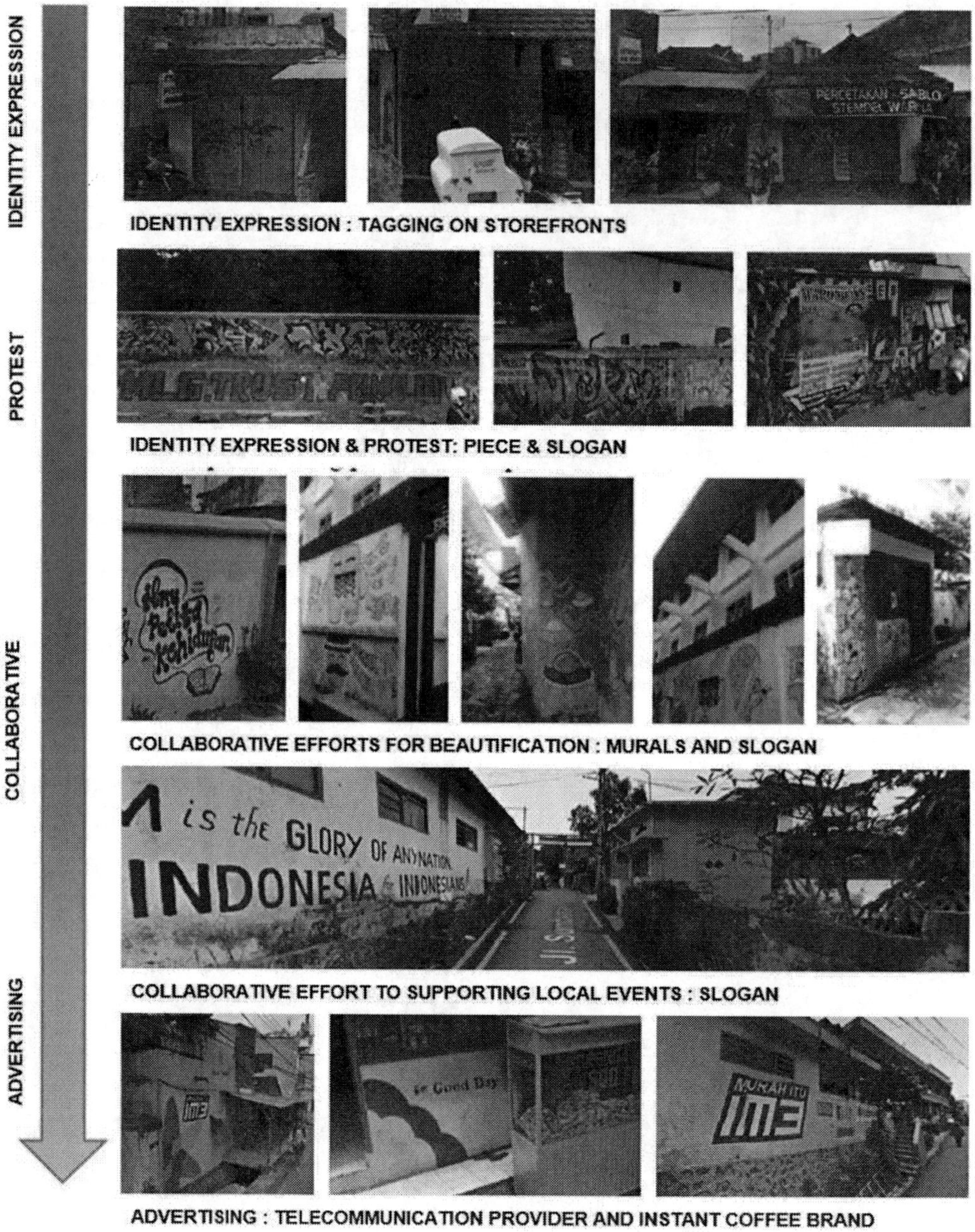

Fig. 11. Typologies of Street Arts.

message.

One of the message identified was '*Malang sumuk*' or roughly translated Malang's daily temperature is becoming hotter. Malang was known for its airy and cool weather, the development of its urban area is causing the lost of urban interaction space, lack of identity and environment issues such as the rise of its temperature. That message was written by one of the biggest street art community in Malang which called themselves Joy O Klan as a form of protest to city's development which degrading the urban identity through the rapid and massive growth Ruko and left them no place to socially interact. As a form of their protest they usually only tagging their community's name to the Ruko wall, but to convey a message which most people can understand they started to write 'Malang Sumuk' (Ferianto, 2015).

Fig. 12. 'Malang Sumuk' Graffiti by Joy O Klan.
Source: Google Street View, 2016.

While the murals found on the site consist of three types based on its motivation and how it made: the collaborative for neighborhood beautification, collaborative to support local events and commissioned murals made to sell and promote products. UIN's architecture faculty had a program called 'Grebeg Arsitektur' which tried to provide an architectural solution for kampong (Tri, 2015). That program resulted in murals along the alleyway connecting UIN entrance with Sumbersari Gg3 Street. This alleyway is where the activity of the UIN's students more intense as it provides academic, daily and tertiary needs from students. The messages they tried to convey through the murals are mostly normative and more directly such as don't do drugs, '*hati-hati di jalan'* (be careful on the way) and 'hijaukan kampung kita' (make our kampong green). However there is one message which conveys the assemblage of the neighborhood user and the interrelationship between them which say '*yang lebih tua menghargai yang muda, yang lebih muda menghormati yang lebih tua*' (the older generation appreciate the younger generation while the younger generation respect the older generation).

Fig. 13. 'Greek Arsitektur' Murals by Student of Architectural Faculty's UIN Maliki.

The other collaborative murals found are those which support the local events at Sumbersari which are the historical re-enactment parade and bazaar. Most of them consist of slogans written on white painted wall quoting the existing nationalism slogan by influential figure during Indonesia's Independence. They mostly located along the pathway towards the Kantor Kelurahan Sumbersari where the bazaar and historical war theme re-enactment took place. As seen in fig. 7 it showed how the murals complementing the bazaar and war enactment settings.

Fig. 14. 'Naskah Proklamasi' along with other slogans and trace of local historical war-themed bazaar.
Source: Google Street View, 2016.

The advertising murals found on site mostly promoting telecommunication vendors and instant coffee brand. Those are a form of wall painting advertising strategies usually done to promote telecommunication vendors, coffee, machine lubricant, lamp, and cigarettes, however, what's found at the site are telecommunication vendor and coffee product. Advertising is done considering its consumer characteristic, thus the type of product advertised on site showed that the site is considered where their potential consumer, which are students, are mostly gathered and doing their activity. While these advertising murals have good visual quality and might contribute to the positive visual quality of its environment, their existence might be associated with the neighborhood characteristics. For example, people might associate some part of the neighbourhood as the 'indosat' street than Kertosentono Street.

All the types of street art found on the site showed how street art as an informal practice is representing its user and complementing its formal practices. To understand more about how formal and informal practice related to its urban area's character, productivity and vitality I will analyze further through its clustering.

Clustering, Alliance, and Association

Mapping analysis of function within the site found clustering of retail function which is academic and daily needs; consumption and services; and daily and tertiary needs. Academic and daily needs clustering formed by the intensive activity of Brawijaya University's student and Ketawanggede's residents. Consumption and services clustering formed mostly by local retails and UIN's Maliki students. While daily and tertiary needs clustering formed in the parts of the site which located near the intersection which connects the site to another university's hotspots and Malang's retail corridor.

 https://doi.org/10.1051/shsconf/20184107005

Within those function clustering, I have also found clustering of activity formed by local residents and university's programs. The clustering of local resident's activity are usually located near or within existing public facility which is the Kantor Kelurahan and it seemed like it related to the function clustering. Greening competition activity is found within academic and daily needs clustering, while historical parade and bazaar were located within consumption and services retail clustering.

university's student, by trying to show their roots through historical reenactment.

Fig. 15. Clustering of Retail Function.

The greening competition is a beautification activity the local resident done to their house and public spaces. Clustering of academic retails showed the intensities of the student activity within local kampong, which creates issues such as cleanliness and safety. Thus, the greening activity showed their sense of belonging of their public space which might be done to show the attempt to reclaim their space, kept it clean and safe. While the historical parade and bazaar were probably done to counter the effect of commercialization within and surrounding Sumbersari's kampong, which mostly driven by the existence of

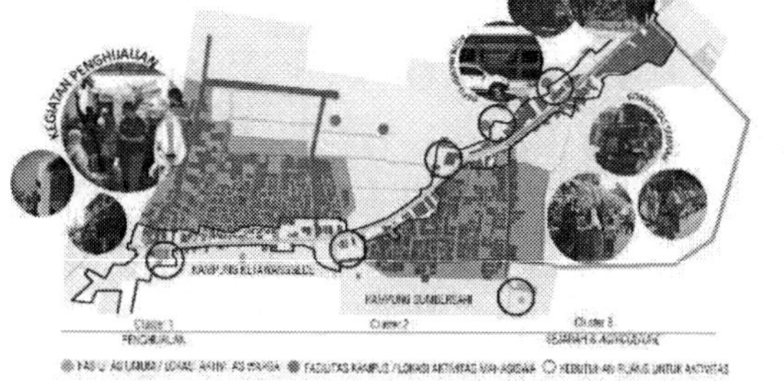

Fig. 16. Clustering of Activities.

Other than local activities, there are also university's programs which located near its perimeter which is recycling center and agripreneur, which runs by agriculture faculty of Brawijaya University. Mapping analysis found that along the university boundaries which include the recycling and agripreneur center, also found vandalism and graffiti. Murals were mostly found in the alleyway towards UIN Maliki and along the local streets towards the Sumbersari Civic Center (Kantor kelurahan). While advertising was found scattered along the main road and one local road connecting the main road to Brawijaya University's entrance.

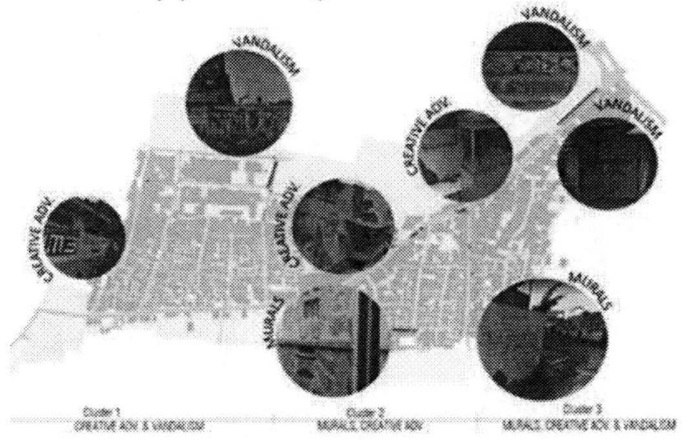

Fig. 17. Clustering of Street Art.

Graffiti or street art quite literally 'takes place' in the sense that it appropriates the street and claims a 'right' to the blank walls of the city and has a captive audience (Dovey, 2016). The typologies and placement of street art in the site are related to the sense of ownership of the place which related to the user in each cluster. While vandalism which found on street front located in cluster 3 can be considered 'bad' and disrupting, it's still unclear whether graffiti a 'good' or a 'bad' thing. Graffiti finds a place in those parts of our cities where identities and practices are open and unfinished. Thus the clustering of graffiti showed that those clusters were still underdeveloped, lack of identities and urban practices.

The clustering of murals showed the resident efforts to reclaim their space by showing their identities through images. Murals found in the alleyway connecting Sumbersari's kampong with UIN Maliki showed the efforts of the students to contribute to their surroundings. It also shows that the boundary between university and its surroundings can be blurred. While murals located near the historical bazaar location showed how street art can be utilized to supports formal activity, which in turn will create a new positive identity for the urban space.

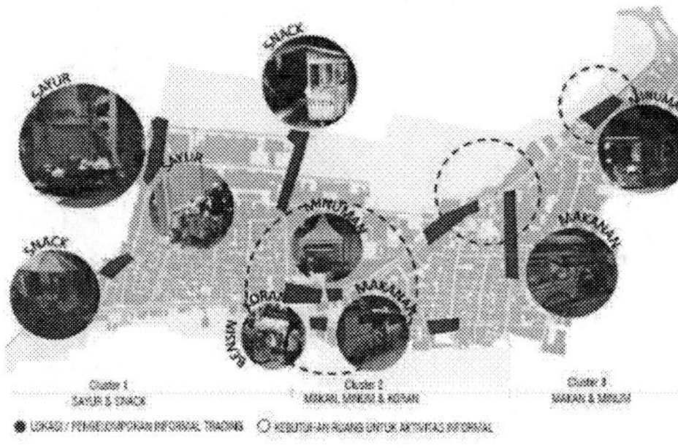

Fig. 18. Clustering of Street Vendor.

The clustering of street vendor also related to its formal clustering which is its user and the functional clustering. Within academic and daily needs clustering and greening activity, the typologies of street vendor found are a kiosk, vehicle, and pavement. The more fixed type of vendor is slowly transitioning into a more formal form of retail selling mostly food and drink, the semi-fix mostly selling vegetables in the morning and the unfixed selling snacks such as ice cream and meatball. These informal organization of street vendors is not found in cluster 2, where both street vendor and local retail are both selling the similar product of food. Although on the main street, where the retails mostly selling clothes, accessories and offering services for motorcycle; the design of the kiosk/vehicle are more elaborate and the product sold was more diverse, including gasoline and newspaper.

Parallel with street art, street vendor clustering also related to cluster identity. Within cluster 3 the street vendor consists of a mostly kiosk with the type of food it sold varied from local / known food such as bakso or nasi goreng up to the newer type of food and/or drink such as kebabs and blended drink. The spread of more modern newer style of food within cluster 2 while also related to its connection to other more developed student hotspots (ITN and Veteran Street) also related to lack of identity the cluster had. Food and drink are the type of product which most street vendors sold because its profits are more certain as people might always need food and/drink.

Urban Dimension of University Perimeter

Findings on urban informality within Brawijaya and UIN Maliki confirmed that informality is related to the characteristic of its formal structure. While street vendor type might relate to the morphological dimension, its product selection is the response to the identity it's formal structure providing which are the existing retail, activity, and user. However urban informality within settlement in urban perimeter was not just a byproduct of the formal structures, findings confirmed that street vendors can evolve into a more formal form, a kiosk which extended from local resident's house. The strong correlation to a characteristic of the formal structures also presents in the cases of street art. Although urban morphology was considered for placement, the types and message it conveyed is a response to the character of its place which formed by the assemblage of the user and existing activity. It shows how formal and informal practices are both influenced and influencing its formal structure. These findings confirm that informality indeed is part of public life which apparently not occurs beyond the realm of the formal institution as Carmona has stated (2003), but intermeshed with the formal structure (Dovey, 2012). It is a mode of production of urban space that is also its product as the emergent place assemblage (Dovey, 2012).

Informality within university perimeter showed the diversity of its user and how differences can be bridged over. This showed through informal rule and organization of formal and informal retails at cluster 1, also through the collaborative efforts on the alleyway murals. The diversity of user creates a diverse perceptual dimension of the urban environment. These manifest through a range of street art found within the site which conveys a wide range of message from showing identity, protest to normative slogans. This street art also was done as a way to counter the 'placelessness' of the urban environment, which caused by the loss of attachment of the university's student to this area. To some parts of the site, the street art is able to reinforce the sense of the place, but still unable to contribute to the visual interest of the whole environment.

University perimeter is formed through the interaction of local resident which use the area for long period of time and students which change periodically. That temporal dimension affected the informal practice more than the formal practice. The function of the university stayed relatively similar within 10 year period, while the informality changes more dynamically. The changes of students will also affect functional dimension of the environment, especially the density. Higher density resulted in the poor quality environment within university perimeter which showed through the lack of comfort for pedestrian and public space on the site.

Urban space is a place of exchange and traffic in people, ideas, and goods (Dovey, 2016). Within university perimeter, there are potential for exchange and spread of ideas and goods through its formal and informal practices. Informal practices might consider urban problems, but this research showed that it actually intermeshed with the formal practice and has the potential for 'making the place' when collaborative efforts are present. The morphological, visual and functional dimension of university perimeter is driven by the social, perceptual and temporal dimension formed by its user and showed through informality. These not only explained why the current efforts in managing informalities in Indonesia were rarely succeeded but also shows how it might works which is as stated by Dovey (2016) through tolerance and recognition. It is necessary to recognize the social, perceptual and temporal dimension of informality; to consider which morphological, visual and functional dimension of it tolerable concerning public good.

References

Arifrahara, G., Haswanto N, & Wahjudi D. (2013). *Analisa Karakteristik Visual dan Strategi Visual Iklan Wall Painting Studi Kasus Iklan Wall Painting Empat Operator Telekomunikasi Nasional, Wimba Jurnal Komunikasi Visual & Multimedia*. Vol. 5 No.1 Tahun 2003. Retrieved from http://jurnalwimba.com

Bappeda Kota Malang. (2016, 10 11). *Ringkasan Eksekutif Rencana Induk Jaringan Jalan Kota Malang*. Retrieved from Bappeda Kota Malang: http://bappeda.malangkota.go.id

Carmona, M and Tiesdell S. (2003), Public Spaces, Urban Spaces.Oxford : Architectural Press.

Carmona, M and Tiesdell S. (2007), Urban Design Reader.Oxford: Architectural Press.

Dovey, K. (2010). Informal urbanism and complex adaptive Assemblage, International Development Planning Review, 34 (4) 2012 pp 349 - 367. doi:10.3828/idpr.2012.23

Dovey, K. (2016). Urban Design Thinking: A Conceptual Toolkit. New York: Bloomsbury Academics.

Dovey, K., & S., &. W. (2015). Creative Multiplicities: Urban Morphologies of Creative Clustering. Journal of Urban Design Vol 20 no 1, 52 - 74. DOI: 10.1080/13574809.2014.972346

Dovey, K., Wollan, S. & Woodcock I. (2012) Placing Graffiti: Creating and Contesting Character in Inner-city Melbourne, Journal of Urban Design, 17:1, 21-41, DOI: 10.1080/13574809.2011.646248

Ferianto, G.E. (2015). *Street Art: Representasi Identitas Dan Kritik Social (Studi Kasus pada Generasi Baru Street Art Joy O Klan Kota Malang)*. Retrieved from Jurnal Mahasiswa Sosiologi http://jmsos.studentjournal.ub.ac.id/index.php/jmsos/article/view/76

Garis Putih (2016, October, 20). *Graffitude - 10 Years Ngaco Fams* [Video file]. Retrieved from https://www.youtube.com/watch?v=HdzErhSai50

Hamid, A. [Abdul Hamid]. (2016, April 25). *Festival Mural #Earth Day Uin Maliki Malang* [Video file]. Retrieved from https://www.youtube.com/watch?v=-Kqch-OdMrI

Homes and Communities Agency the UK. (2015, 09 12). Guidance Urban Design Compendium. Retrieved from Gov.UK: https://www.gov.uk/government/publications/urban-design-compendium

Lang, J. (2005). Urban Design: A Typology Of Procedures And Products. Oxford: Architectural Press

Porter. L et all.(2011) Informality, the Commons and the Paradoxes for Planning: Concepts and Debates for Informality and Planning.Planning Theory & Practice, 12:1, 115-153, . DOI: 10.1080/14649357.2011.545626

Syrian, Y. [Syidan Yusuf]. (2016, August 28). *Karnaval sumbersari malang* [Video file]. Retrieved from https://www.youtube.com/watch?v=leD8rvAYtLg

Tri, A. [Agus Tri]. (2015, August 22). *Grebek Arsitektur 2015 UIN Maliki Malang Sumbersari bermural dan berkebun (review)* [Video file]. Retrieved from https://www.youtube.com/watch?v=C4iB9nGEYBQ

Yatmo, Y.A & Atmodiwirjo, P. (2011) "Understanding the Metaspaces of Street Vendor in the Cities: Temporality, Strategies, and Tactics." Originally published in the Proceedings of 10th International Conference on Quality in Research (QIR), Faculty of Engineering, University of Indonesia, 4-6 December 2007. Retrieved from http://arsitektur.net/doctorwho/wp-content/uploads/2011_vol_05_03-04_Understanding-the-Metaspaces.pdf

Yatmo, Y.A. (2008) *Street Vendors as 'Out of Place' Urban Elements*, Journal of Urban Design, Vol. 13. No. 3, pp 387–402. DOI: 10.1080/13574800802320889

Reassembling Traditionality and Creativity: A Study of Place as Assemblage in Pasar Cihapit Bandung

Agus Surharjono Ekomadyo[1,*], *Tyas* Santri[2], and *Andika* Riyadi[1]

[1]Institut Teknologi Bandung, Indonesia
[2]Langlangbuana University Bandung, Indonesia

Abstract. There are many efforts to revitalize traditional market as urban economic and socio-cultural place. Several creative communities in Bandung have been trying to pep up the vitality of Pasar Cihapit Bandung through socio-cultural events. Through place as assemblage approach, this paper traces the actors involved and their relation in the place-making, and also to seek how the creative movement contribute to the traditional economic activities which have been taken place before. This creative movement has identified to be able to attract public attention to the market, signified in the media publication. The emergence of new actors also generated the contested space in Cihapit market, influenced by actors' relation to another group outside the market. The market publicities is also attracted political actors concern, stimulated by their popularity desires, that impact to space contestation intensity. It is found that creative movement can increase the physical image of the market by socio-cultural events, but it is not found its significant contribution to improve the economic role of the traditional market. This paper contributes knowledge to understand how everyday practice of power works in designing and creating built environment

Keywords: creative community, public market, Pasar Cihapit, assemblage, place-making, contestation, controversies, translation

Introduction

Discussion about assemblage in urban studies and architectural design emerges to unveil the real-practice of power in everyday life of built environment production. Two French Philosophers, Deleuze (1980) and Latour (2005) use the term of "assemblage" to explain "social practice" and "the social". In architectural discourse, Deleuzian thinking is used by Dovey (2010) to explore placea in becoming and Latourian by Yeneva (2012) to map the architectural controversies. The study of assemblage is important to answer how design is put to guide the better future in real practice of everyday life of built environment production.

This paper exercises the public market as urban place in assemblage perspective. Public market is important urban artifact, where everyday economic activities is taken place along the urban history. Here, information, knowledge, and values are exchanged by society through goods and services trading, so public market is recognized as place for economic and socio-culture. By this reason, the public market revitalization program has been run by Indonesian government since 2014s and generate any national and local political concerns.

Pasar Cihapit in Bandung is urban artifact that is examined by place as assemblage approach. This market has specific charracter because it serves the middle-up consumers who resides in city center area and well-known as a place to sell high qualified dailiy commodities. The market exoticism attracts people to visit this place as recreational destination in small, temporary, and informal activities. Since the end of 2014s, two "creative" communities try to pep-up this market by organizing several creative events that be able to attract public attention and political concerns. But this movement also generate controversy: they also utilize the exotic public market as their groups identity for social existence or can be regarded as social movement to revitalize public market as economic and socio-cultural place?

By assemblage approach, the controversies in Pasar Cihapit is hoped to be mapped to be learnt explicitly. Public market as like as another urban spaces is a place where the people struggles their desire in the city. By learning the assemblage in Pasar Cihapit Bandung, it can be reflected the revitalization effort of public market in Indonesia, and is related with the emerging of creative-class, who mostly come form urban middle-class, to contribute in traditional economic development.

Method

This research is conducted by assemblage approach for place-making study. Analysis of place making is framed by combining Dovey's Deleuzian framework of place as assemblage and Actor-Network Theory to map the actors-artifacts relation in built environment production. Data is

* Corresponding author: aekomadyo00@gmail.com

collected by secondary data, historical study, observation, and depth interview with the actors

Public Marketplace as Urban Assemblage

The concept of assembly introduced as a concept of engineering, referring to an act of putting something together to achieve desired results. Understandings of this concept are broadening with emergence of various technology studies. Assembly understood as an interaction of various entities, and conceiving a new function for a purpose. Some philosopher use the concept of assembly to refer a collection of interacting contingent entities, such as Deleuze and Guattari in their Thousand Plateaus (Deleuze: 1980).

Deleuze concept of assemblage differentiates organic entities functioning as a whole and interaction of many contingent parts that made to fulfill certain function or purposes. In this process, he argues that three sequences are needed to make an assemblage: Coding, Stratification, and Territorialization (Phillips: 2006). Deleuze argues that assemblage is articulated by means of imaginations within myriad heterogeneous elements, such as constellations (Wise 2014). This process of articulation through organizing matter around body, called Coding (Smith 2008). A constellation, of course, consists of heavenly bodies and we select a few, organize it around to make a connection. In order to become an assemblage, we select, compose and complete a territory; in this case we call it a constellation of stars. When we complete a territory, there are needs to set a hierarchical body to organize priority; this process is called stratification (Smith 2008). Assemblages are forged through connecting and disconnecting entities, this process called territorialization. When an entity included in an assemblage, it is a territory. Deterritorialization refers to disconnected entity and likewise the term Reterritorialization, which is regarded to reconnecting an entity into assemblage (Wise: 2004). One of the unique feature using assemblage as an approach is that this approach do not strictly rely on human as primary entity that has to be accounted. Deleuze also said that assemblage is heterogeneous object rather than homogenous. That is to say the 'entity' refers to human and non-human bodies, actions and reactions (Wise: 2004). Heterogeneous approach is what we will need when conceiving complex structures like cities or marketplaces.

Dovey (2010) use an assemblage approach directed from works of DeLanda (2006) to understand how places can be made. Point that differs from Deleuze-Guattari works is that DeLanda regards assemblages is a 'state of affairs' instead of a 'thing' or collection of parts. DeLanda views on assemblage is that every entity inside an assemblage is mutually contingent, a machine can be removed from an assemblage and used for another assemblage. Dovey gave an example of street that is not a street on its own, but it become a street when it is connected to the buildings, cars, sidewalks, trees, goods, people, signs, etc.. He emphasizes that use of Assemblage Theory is to avoid reductionism trap and incorporates sense of meaning into analysis of the cities, rather than limited to material urban or simply adding into objects.

Dovey's thinking about spatiality and sociality is influenced by concept of representational space by Lefebvre (1991:26) in production of space. He argues, while space is socially constructed, the social is spatially constructed. In space production, power is used in everyday manner to share the understanding. There are desires that generates power to gain the ends and power over to control action of others. The power is mediated in architecture and urban design.

Interesting practical use of this approach is Dovey's works on conceptualization of Functional Mix (Dovey: 2017). In this work, Dovey argues that every one of three cities have three distinct function. The functions are Live, Work and Visit and three cities are New York, Bogota and Barcelona. Interestingly, Dovey accommodate dynamics of assemblage into three functions. He argues that there are no solely distinct functions are being used in those three cities. Relevant question that being addressed in Dovey's work will also relevant for this research: What mix are relevant and in what proportion?

Public market is a good example as an urban assemblage. This place contains functional mix: although its main function is place for economic purpose, but any other non-economic activities is taken place here. Ekomadyo (2012) exercised public market by place as assemblage approach, and uncover the various and richness social-relation in traditional market place construction.

But what is the market itself? In simple understanding, market is understood as a place for goods and services trading, where buyers and sellers meet each others. Sabiani (2012:6-16) explored many perspective to understand market. As a place, market is understood as physical placa or geographic location where supply and demand meet and are exchanged. As a segment, market refers to certain customer groups. As exchange mechanisme, market is classified into various kind of exchange mode. As a process, market is understood in social-behaviour perspective of market actors. In institutional perspective, market sistem is a social coordination method with collective adaptation among members without coordinator. As a network, market is heterogenous institution assembled with human and non-human elements that interrelated by a control.

Callon works propose an idea of a viewpoint about market. A market is, in fact, a collection of interrelated calculative devices (Callon: 2005). This made effectiveness of a market is defined by its ability to do complex calculations and produce a solutions that are impossible by merely theoretical reflections. Basis of this idea is a process of calculation, which is explained in Callon's earlier works. Calculation, or Qualculation in Callon's words, is an act of decision making, that is split into three distinct steps (Callon: 2003). First is framing, it is a way to account who is involved and who is not relevant to the issue being addressed. Second is display, those who relevant will be able to view position of the issue being addressed in a medium, it can be a paper, board, digital devices, screen, or even drawing on sand. Finally those who accounted put a decision based on their judgments on the issue, this is called totalization.

In Callon's view, market is a collection of objects that helped people calculate and decide what they are going to do. In a market, objects that became a medium to help people decide what are they buying are designed, spread and optimized by its placement. Various discount labels repeatedly shown in the ceiling, halls, or even product's display to inform the visitors about reduced prices of particular products. Product packages that not only efficiently designed to contain what is inside, but also eye catching and invite us to grab and see. These things are referred to calculative devices in Callon's works. Every object that aid relevant entity of their calculations is calculative devices. The more effective of assemblage of calculative devices then the market will be more effective to calculate, display and decide.

Since this paper will specifically target marketplace that have an extra activities when creative community joins to develop the market, Functional Mix concept became relevant when talk about place making. This research differs from Dovey's since using a qualitative approach to market that guided by Michel Callon's work in building assemblage. This research will review on how calculative devices work on making places as a part of contingent entities in the assemblage, yet it has pivotal role on how market assemblage does its primary function. By case of Pasar Cihapit Bandung, this paper exercises to combine Deleuze-Dovey's and Actor-Network Theory (Latour-Callon-Yeneva) concepts of assemblage to unveil the desire and translation on everyday practice of place-making.

Pasar Cihapit Bandung as Urban Assemblage

Pasar Cihapit is one of historical public market in central area of Bandung city. Cihapit was one of neighborhood cluster designed by Dutch Indies colonial government (Wiryawan, 2012). Economic activities had been starting while some Chinese traders opened shops to serve surrounding demands. It attracted local traders to occupy empty space in the back of the shops and start to build informal market. In 1985, government of Bandung build the market with permanent shelter, and it was renovated in 2006 (Dinas Pengelolaan Pasar Kota Bandung, 2007:26).

Pasar Cihapit is well known as public market provides high-quality daily goods (vegetables, fruits) and serves the middle-up consumers who resides in central area of Bandung. Cihapit traders are traditional traders, who continue their family business in this market since 1940s. Some commodities have own specialties and well known abroad, such Warung Mak Eha, Surabi Cihapit, and Kue Balok Cihapit. This market exoticism attracts some people to visit this market as tourist in small scale, informal, and temporal.

In the end of 2014,0 there was a "creative" movement to pep-up Pasar Cihapit as urban identity in Bandung. Here, the creative movement refers to concept of Florida's creative class to capture the phenomenon of creative people who influences the economic growth related to specific location that endorse innovation, diversity, and tolerance. The creative class is distinguished as people who work with creating new forms: they can be scientists and engineers, university professors, poets and novelists, artists, entertainers, actors, designers, architects, as non-fiction writers, editors, cultural figures, think-tank researchers, analysts, and other opinion-makers (Florida, 2005:34).

The movement started with sporadic and spontaneous mode, mostly initiated separately by two communities: Apit Tjihapit and Los Tjihapit. They also open cafés inside the market as their daily business. As "creative" communities, both groups has strong informal relationship with creative people generated by universities not so far from the market. This community initiative got good responses from market management, the spontaneous movement become accumulated, and successfully creates many creative events and installs artistic elements to make the market more attractive. This movement has been able to attract public attention signified on many media publication about Pasar Cihapit revitalization (Helmy, 2016:11). The recent popularity of Pasar Cihapit also attracted some political actors to visit, contribute their power, and publish their activities.

The presence of newcomers in Pasar Cihapit impacts the contestation among actors in utilizing the space. There are new installed artifacts and reproduced spaces that represents images and symbols of involved actor groups involved. Contestation generates controversies about the role and relation among new actors in public market. Essentially, the role of public market is a place for economic and socio-cultural activities for the society. When controversies is happen, the socio-technical assemblage is interesting to be traced to learn how representational space in public market is reproduced related to its role as economic and socio-cultural place.

Tracing the Assemblage of Place-Making in Pasar Cihapit Bandung

Place-making is a term used in architectural design domain to describe how built environment can be designed and created to deliver purposeful meanings to the users. Morgan (2010:11) recognizes three broad approaches to place theory: phenomenological and humanistic approaches explore the deeper significance of place to human existence and the subjective and emotional quality of people's relationship to places; psychometrics, explores the relationship between the physical environment and the human psyche by attributing numeric measures to psychosocial phenomena, and social constructivism, sees it as a socially constructed phenomenon. Assemblage approach of place-making sees mutual relation among society and the technical objects: it is socio-technical approach rather than social constrictivism, to keep see the relation is mutual and flat.

Assemblage in place making is traced through identify the actors involved specifically the new actors who stimulates unstabilized controversies. It is identified there are three main actor groups involved: traditional traders, market management, and creative groups. Traditional traders refer to the existing traders who traded in this market for a long time continuing their familiy business. The term "tradition" is used refers to something is handed down as an actual transmission from one generation to the next (AlSayyad, 2004:7). Market management consists the officers who are delegated by government city of Bandung to manage the marketplace. The creative groups, the new comers, are refer to the community who installing various creative events and artistic technical object in this market, and these groups have strong informal relationship with creative movement in Bandung in common.

All involved actors create the technical objects As artifacts, the technical objects is like an inscription that contains messages and meanings. Tracing socio-technical place-making is to make a description, an antonym of inscription, how actors and artifacts is related each others as an assemblage.

Actors Desire/ Power

Power/desire is used by Dovey (2010: 14-15) as the starting point to trace place as assemblage. Desire is the main source of human strugle of life, flowing through various activities, and becomimg identities, organisms, things, and assemblages. Human use power to achieve the purpose of their desires. They also use their capacity to create power over to control the others actions. The "power over" appears in force, coercion, manipulation, domination, seduction and authority.

Traditional traders have desire to earn of live through continuing familiy business in the market. Traders have capacity to achieve the desires by selling and promoting commodities through displaying them to the consumers. Traders control commodities supply and quality through relationship with suppliers, and control displayed commpdities through arranging the shop lay-out.

The market management has task to make the market well-managed: the officers work as city apparatus. The officers work motivated by professional career, it is indicated in promotion of former Head of Cihapit market based on his achievement to improve the market creatively. The market management has authority to open and closed the market gates to control the beginning and end of market activities, to draw retribution from the traders, and control common facilities and utilities.

The creative groups have desire to run their new business in the market: it also represents their existence among their communities. Public market is chosen because the affordable rent-price and its socio-cultural contents. To run their business sustain, they create events to attract and gather communities who will consume their commodities. As like as common traders, they control the supply and quality of commodities by develop sustain relationship with suppliers. Apit Tjihapit leader delegate employees to run their cafe daily, while Los Tjihapit leaders attend themselves daily to serve the consumers and play as the node for community informal discussion.

Representational Artifacts

This term comes by combining Lefebvre's concept of epresentational space and Latour's concept actor and artifact network. Representational artifacts refers to technical objects created by actors representing his/her desires, power to, and power over in the place-making. Technical objects represents groups identities, so the assemblage can be read and traced.

Traditional traders represents their artifact located on their each trading spaces. To promote and sell the commodities, they design their trading spaces through commodities lay-outing and inserting additional element in front side of the trading spaces (lettering, signage, color). Trading space lay-out represents trader control of commodities and consumers. The boundary among trader and consumers can be rigid or fluid. Market management plays as government agency, so market boundary physical elements such as gates, walls, and roofs are territorial signifier for management authority. Public facilities such sinks, toilets, or nursery room also represents there are management officers who maintain the market. When creative movement emerged, market government contributed by creating decorative elements such as artistic umbrellas or organizing ethnical events in every Wednesday. Political concerns stimulates construction of new permanent management offices, where previously located temporarily using empty kiosk inside the market.

Creative groups install new technical object inside the market. The new forms are more creative, supported by resources outside the market. Apit Tjihapit community supported by artists, designers, architects, and university students creates trading tables for informal alleys traders, education facilities, and mural in market wall. This community has also strong relationship with political actors, that contribute the permanent building for public market library. Lost Tjihapit community designs a small café based the leaders interior design background, fulfilled by many posters of cultural discussion show the strong relation with the artists, journalists, and cultural communities in this city.

The actors also create publication on social media to show their activities. Apit Tjihapit gets many publication based on their social entrepreneurship events. Los Tjihapit create posters spread through social media to broadcast their discussion events. Bandung governments also broadcast their achievement claim to revitalize this market. Political actors visit this market and then post it by popularity motives. The social media construct another space contestation: actors' desire can be read.

Sense of Consumed Space

Simply, a marketplace is a place where traders meet buyers, where production meets consumption. When assemblage perspective concerns how spaces are produced, it is important to concern how spaces are consumed. It is related to the concept of sense of place about how spaces are used and give meaning to the users, on the other words, how spaces are consumed by users. Dovey (2010: 25-26) describe, place as assemblage implies in concerning to place intensity such as activities, character or another degree of some quality, not merely on the materiality of physical form. On a marketplace, intensity is mostly determined by lively trading activities. There are local proverb that said the market lively activities depends on the blessing of the spirit who own the market. It is the market's Genius Loci (Ekomadyo, 2012:1). Who consume the space? Again, they are the actors. But for consumption perspective, another actors classification is needed: traders: who use space to sell, and consumers: who use space to buy. Traders consumes their each trading spaces. The space dimension is commonly just enough to put commodities and traders only. Consumers mostly use circulation area to interact with the traders, bounded by commodities area. This border is commonly rigid, but it blurred in special commodities such as food stall or in informal trading space.

In general, Pasar Cihapit has two traders territory: formal territory inside the building, siginified by permanent big-span roof structure; and informal territory signified by the row of

 https://doi.org/10.1051/shsconf/20184107006

trading tables in the entrance alley to the building. In the building, there are two kind of trading space: kiosk and open platform, which occupied 70% of all formal trading spaces.

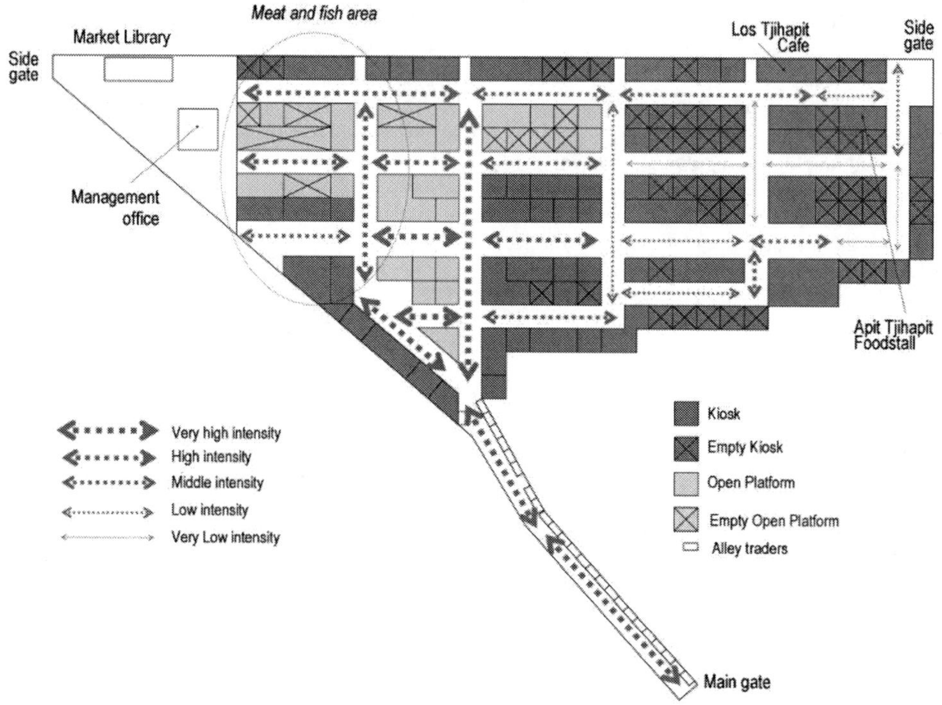

Fig. 1. Map of Pasar Cihapit

The empty kiosks are located in uneasy accessible area. Meanwhile, all of alleys space is fulfilled by the traders. It is indicated that intensity of space consumption by traders is determined by accessibility (fig.1).

The intensity of consumed space by market consumers is identified in circulation space. Different with trading space, circulation space intensity, beside is determined by accessibility, is also influenced by specific commodities as attractor. Meat and fish commodities, and also cafe of Los Tjihapit can attract people to come although the kiosks is far from the main entrance (fig.1).

Controversies Mapping

Many actors in Pasar Cihapit generated many controversies, and the controversies also involve the artifacts creation. In Actor-Network Theory perspective, controversies are seen as actors process to gain knowledge. Mapping the controversies is a way to understand how actors build their collective existences. The controversies is a 'hybrid forum',: a space of conflict and negotiation between actors (Latour, 2005: 23-25, Yeneva, 2012:4,60). By this standpoint, a public market is good example to interpret as "hybrid forum", because historicaly the urbanized human civilization begin on a place named "Agora", a greek name of city center containing public market activities, when transform in Roman period as a "Forum" (Tangiers, 2008:9-10).

Controversies in Pasar Cihapit actually is happen among traditional traders, but the controveries have been stabilized. Controversy among formal traders inside the market building, for example, can be seen in fairness of zoning for commodities and trading space types. Properly by functional consideration kiosks should be used for dry commodities, where open platforms for wet commodities. But in the field, there are wet commodities sold in the kiosk and vice versa: not by function but rather by rent-price consideration. There are also controversies between formal and informal traders who occupy the side of entrance alley. They alley traders got more good accessibility to be reach by consumers, although they are not administratively admitted by the market management.

Two creative groups, Apit Tjihapit and Los Tjihapit, are also generate controversy. Apit Tjihapit community works and is organized by capitalism mode through building formal relation with bussiness school of an university, that can be attract political actors and big enterprises to gain economic capital to fund their social activities. On the other side, Los Tjihapit community is more socialist, works and organizes through informal relation with grassroot activists. Artifacts produced by Apit Tjihapit tends to be monumental and able to attract the public massively, meanwhile Los Tjihapit created small events with deep of social struggle contents.

Market management is most controversial actors. In general, the city and municipal government who have authority to manage the public market, was accused the declination of public market performance that caused public market slum, dirty, and lack of competitiveness as shopping facilities (Poesoro, 2007, p.7). Comparing with modern shopping centre, the management are responsible to create excellent services, price discount, or attractive events to guarantee the high performance of the facilities. In Indonesia, the market management just play as administrative agent to draw retribution from the traders. They do not have capabilities and authority to innovate the market and give incentives to attract more consumers. Although Pasar Cihapit management involves in many creative events, but this effort does not increase the market economic performance significantly. There are many empty kiosks still empty, no many traders joining to the the market.

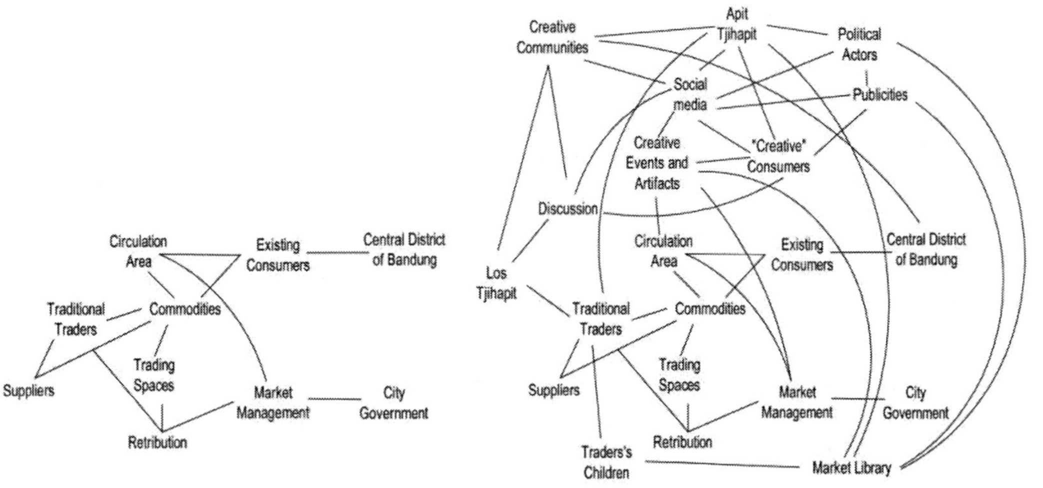

Fig. 2. Actors-artifacts network before (left) and after (right) creative movement

Traditionality and Creativity Translation

Translation is a term of Actor-Networ Theory used to decribe the process how network is made. Callon (1986, 19–34) describe four stages in actors or actor groups translation: problematization, interessement, enrolment, mobilization. Socio-technical assemblage among actors and artifactscan be illustrated on a sociotechnogram. Actor-artifact relation map is able to disentangle the controversies. Before the creative movement, Pasar Cihapit was a common market: its network are constructed by relation among consumers, traders, and suppliers of daily goods. This market has special consumers who came from middle-up class. The market management rented the trading spaces and collect retribution from traders (fig.2 left).

After creative movement in the end of 2014, more actors are involved, network enlarges. The existing actors network is still kept. Creative groups create creative events, and the social media publish them to attract other actors come to the market. Public attention improve the market image as Bandung cultural identity.

Although be able to attract public attention, there are no mobilized translation between the existing traditional consumers and new "creative" consumers. Two consumer groups come from different group, and they do not have significant mutual relationship that can accumulated market consumers significantly. Because there are no consument accumulation, there are significant economic increase in this market: the empty trading spaces is still empty.

Conclusion

This research get early finding, that creative movement in Pasar Cihapit Bandung can contribute public market role as socio-cultural place, but have not already yet as economic place. This movement is able to create public attention to increase urban identity of the public market. But there are no significant translation to accumulate "creative" consumers and "traditional" consumers that potentially increase economic activities in this market.

By exercising assemblage approach, this research shows that a small place such as public market is constructed by urban assemblage. A social movement will involve many actors who generate controversies and contestation, because each groups will strugle and represent their desires into produced artifacts. Research about urban assemblage in public market directs to the concept of "urban learning assemblages", to see an urban place is representation of the actors' feeling of way of finding to their world through combined action of human and nonhuman agencies' (Macfarlane, 2011:1)

As a working paper, this research have already exercized Deleuzian assemblage to unveil desire and power represented in place-making, and Latourian actor-network to map the controversies and translation among actors and artifacts. Furthermore, this research will identify the collective desires for better future of the public market as economic and socio-cultural place, and what translation should be concerned to create the collective desires. Urban assemblage approach is hoped can help to read the public market revitalization, especially to trace how creative activities from urban intellectual middle class can contribute to draw traditional economic. In long term, this research is directed to answer how built environment design can guide everyday practice of power to reach better future as collective desires.

References

AlSayyad, N. (ed.). (2004). *The end of tradition?* London: Routledge

Callon, M., Law, J. & Rip, A. (eds.) (1986). Map-ing the Dynamics of Science and Technology. London: Macmillan, p.

Callon, M. & Law, J (2003), On Qualculation, Agency And Otherness. http://www.lancaster.ac.uk/fass/resources/sociology-online-papers/papers/callon-law-qualculation-agency-otherness.pdf (retrieved Dec 20, 2016)

Callon, M. & Muniesa, F. (2005). Economic Markets as Calculative Collective Devices. *Organization Studies,* Vol. 26, 8, 2005. Pp.1229–1250

Dinas Pengelolaan Pasar Kota Bandung (2007). *Data Potensi Pasar Kota Bandung 2007.*

DeLanda, M. (2006) *A New Philosophy of Society.* New York, NY: Continuum.

Deleuze, G. & Guattari, F. (1980) *Thousand Plateaus: Capitalism and Schizophrenia.* London, UK: Continuum.

Dovey, K. (2010) *Becoming Places: Urbanism/Architecture/Identity/Power.* London and New York, UK & US: Routledge

Dovey K., & Pafka, E. (2017): What is functional mix? An assemblage approach: *Planning Theory & Practice,* DOI: 10.1080/14649357.2017.1281996

Ekomadyo, A.S. (2012). Menelusuri Genius Loci Pasar Tradisional sebagai Ruang Sosial Nusantara. *Seminar Nasional Semesta Arsitektur Nusantara 1 (SAN1)* Universitas Brawijaya Malang,

Ekomadyo, A.S., Zahra, A., dan Najmi, I. (2012). Tracing Social-Economic Forces In Construction Of Urban Place; Case: Pasar Pamoyanan Bandung. Paper presented in *International Seminar on Place Making and Identity: Rethinking Urban Ap.roaches to Built Environment* . partment of Architecture, Universitas Pembangunan Jaya,

Florida, R. (2005). *Cities and the creative class.* New York: Routledge,

Helmy C. (2016). Semangat Perubahan di Jantung Kota. Article in *Kompas.* June, 5, 2016.

Latour, B. (2005) *Reassembling the Social An Introduction to Actor-Network-Theory.* Oxford: Oxford University Press

Lefebvre, H. (1991) *The Production of Space,* Oxford: Blackwell.

McFarlane, C. 2011 *Learning the City Knowledge and Translocal Assemblage.* West Sussex, UK: Blackwell Publishing - John Wiley & Sons John Wiley & Sons, Ltd.,

Morgan, P. (2010) Towards a developmental theory of place attachment. *Journal of Environmental Psychology,* Vol. 30, pp. 11-22.

Philips, J. (2006). Agencement/ Assemblages. *Theory, Culture & Society,* Vol 23, Issue 2-3, 2006. Pp. 108-109.

Poesoro, A. (2007). Measuring the Impact of Supermarkets of Traditional Markets. *SMERU Newsletter,* 22, 3-10

Sabianti, D.S. (2012). *Pasar Bahan Bakar Nabati (BBN) Nyamplung Di Jawa Tengah dalam Perspektif Teori Jaringan – Aktor.* Theses. Develompet Studies Master Program. Institut Teknologi Bandung

Smith, D., & Protevi, J. (2008). Gilles Deleuze. plato.stanford.edu. Retrieved 29 Sept 2017, from http://plato.stanford.edu/entries/deleuze/#ThoPla

Wiryawan M. R. (2012). Cihapit: Surga dan Masa yang Kelam Bandung. https://komunitasaleut.com/2012/03/01/cihapit-surga-dan-masa-yang-kelam-bandung/

Wise, J., & Slack, J. (2014). *Culture and Technology.* New York, NY: Lang, Peter New York.

Yaneva, A. (2012). *Mapping Controversies in Architecture.* Surrey, England: Ashgate Publishing Limited,

The Techniques of Participatory Design for Inclusive Public Space Provision in Kampung Kota of Surakarta

Kusumaningdyah N.H[1,*], and *Lia* Sparingga Purnamasari[2]

[1]Architecture Department, Faculty of Engineering, Universitas Sebelas Maret Surakarta, Indonesia
[2]Rural-Urban Design and Conservation Laboratory, Achitecture Department, Faculty of Engineering, Universitas Sebelas Maret Surakarta, Indonesia

Abstract. The technique of participatory design is essential for the future development knowledge of architecture to inform the improvement of architectural education and practice in Asia. Therefore to provide of inclusive public spaces for the community, this research article presents a field study of the usage of community participation design method in design and planning to provide inclusive public spaces within the kampung (settlements) in Surakarta, Indonesia. Surakarta has a problem to provide inclusive public space for its dense organic-informal *kampung's* for the vulnerable populations such as children. Multi dimensional complexities of the *kampung*'s often challenge the notion of public space provision. Community participation approach arises as one of the preferred alternative methods to solve the problems with its basic community-based foundation. This research chose Sangkrah and Gandekan sub-districts as research locations both of the areas represent *kampung*'s characteristics and inhabited by many children population. Steps toward the goal of the study are; (1) field observation of the implementation of a community participation design through participatory place making project, and (2) examining typology assessment perspective, and (3) clarify participation techniques. This research used: (1) descriptive analysis for the project implementation, and (2) qualitative analysis for examining community's perspectives. Through this multi-perspective analysis, this research presents the an alternative framework for implementing techniques community participation methods in design and planning to provide inclusive public spaces especially for children in Indonesia and other developing countries.

Keywords: Community Participation Methods in Design, Inclusive Public Space, *Kampung* Kota

Introduction

In 1996, UNICEF established the Child Friendly Cities initiatives as the embodiment from The 1989 Convention on the Rights of the Child that was synthesized systematically from 1924 Children Rights' Declaration (UNICEF, 2009). Indonesia adopted child-friendly concept into central-government policy since 2006 and equipped with jurisdictional standing in 2009. Six institutional indicators of child-friendly city were formulated and categorized into five clusters; (1) civil rights and freedom, (2) family and alternative care, (3) basic health and wellbeing, (4) education, leisure, and cultural activities, and (5) special protection (Indonesian Ministry for Women and Child Protection, 2010). Surakarta is one of the cities in Indonesia that had adopted the child-friendly concept in 2013 and had been declared as Indonesian' best practice ever since (Antara, 2017).

Along with this issue, UNICEF (1989) stated at UN Convention on the Rights of the child that playing is a children's right. This is because playing is beneficial for children's growth, as it allows children to use their creativity while developing imagination, agility, and physical, cognitive, and emotional strength (Ginsburg, 2007; Burton, 2011). This is also adjacent with Indonesian's child friendly city's education, leisure, and cultural activities indicator. The provision of the amenities that support this right has become a necessity. Public spaces that help to support this activity is critical (Committee on Environmental Health, 2009; Vargas, 2015). Surakarta at this context has also put an effort towards the issue by starting to build *Taman Cerdas* ("Smart Park") citywide that adopted the child-friendly concept (Surakarta Development Planning Agency, 2017). In 2014, there were 6 *Taman Cerdas* (public park) and gains into 10 parks citywide in 2017 (Utami, 2014; Surakarta Development Planning Agency, 2017). But the efforts arrive into the question about the provision of public space in slum settlement and dense *kampungs* that spread across the city, as Surakarta is still entangled with slum problems (Astuti, 2014; Surakarta Department of Public Works, 2015).

In 2014, 10% of Surakarta's urban area is categorized as slum settlements (Surakarta's Mayor Decree, 2014). The slum settlements spread across five districts and

* Corresponding author: kusumaningdyah_nh@staff.uns.ac.id

covers up to 465 Hectares of land. The main concentration of slum settlements in Surakarta are located alongside the Kali Pepe riverbanks. *Kampung* Sangkrah and Gandekan are two of Surakarta's prominent slum area. These *kampungs* are characterized as densely populated area, with high building-density and lack of adequate infrastructures. Therefore, there are lacks of space left for a new public space. Besides, the child-friendly parks built by the municipality are built on the district-level in which they're not reachable by all of the children everyday comfortably and safely (Putri, et. al, 2014). This then results in children using street and narrow vacant space to play.

The *Kampung Layak Anak* Project with The Participatory Approach Design

Departing from concerns over the lack of adequate public spaces that supports children activities within slum settlements (*kampung*) in Surakarta and the tendency of continuous top-down approach in child's supporting infrastructure provision, Urban-Rural Design and Conservation (URDC) Laboratory at Department Architecture, Faculty Engineering, Universitas Sebelas Maret of Surakarta initiated a project called "*Kampung Layak Anak*" (child-inclusive *kampung*). This project aims to gather and visualize data about the existing public spaces that the children use within the dense *kampungs* in Surakarta that could turn into a tool to persuade government's interventions towards the issues. Along the process, the main activity has been extending to not only gathering data from the children, but also implementing design collaboration for making a public space provision through community participation methods in design and planning, included in this method is place-making method.

The Research's Aims

Since the project in both *Kampung* Sangkrah and *Kampung* Gandekan had finished, it is important to evaluate the **techniques** of design collaboration to outline lessons from the implementation of the project. The design collaboration in this research is being observed and evaluated from the (1) ladder of participation, (2) typology assessment perspectives, and (3) participation techniques. Steps towards the goal would be preceded by describing the process and the actors involved in the project extensively.

Methods

This research could be categorized as an evaluation research that aims to evaluate *Kampung Layak Anak* project from the participation level perspectives. Neuman (2013) described evaluation research as an applied research in which one tries to determine how well a program or policy is working or reaching its goals and objectives. The quantitative data gathered for this research is collected through continuous field observations during the implementation process of the project that includes observation in each stage or activities of the project. This research used descriptive method within its analysis. There are three analysis conducted;

1. Participatory Ladder Analysis
   This analysis is respectively based on the work of Arnstein (1969) that since has been adopted into youth and children context by Hart (1992; 1997; 2013) extensively. This works becomes the baseline to understand in which level children participate in both mapping and designing process.
2. Designer-Children Typology Analysis
   The second analysis is based on the work of Lee (2007; 2008) that had been developed from the philosophy of Lefebvre (1972). This analysis is intended to understand the relationship of designer and the community (children) within the process of design implementation of Kampung Layak Anak project and the result (product) of each stages of the process.
3. Participatory Techniques Analysis
   The third analysis is based on the theory of participatory techniques developed by Sanoff (2000). This analysis is intended to understand techniques / methods used in each stage of the process and how the results it created differs from each of the project.

The Context of *Kampung Layak Anak*, Place -Making Project Profile

Kampung Layak Anak Stage Process

Supported by a collaboration with *Kampung Kota* (Urban *Kampung*) 2017 class at the Department of Architecture, Universitas Sebelas Maret, this project was enabled to deployed 14 students to do collaborative design with children. The students were from the class of *Kampung Kota* as the facilitators in this project. The facilitators will be responsible in assisting in managed focus group discussion, data collection and analysis data. *Kampung* (sub-district) Sangkrah and Gandekan are selected based on its slum characteristics (See Figure 1). The process spans for five months, from March to June 2017 that took place in one neighborhood (*Rukun Warga- RW*) in each *kampung*; *Kampung* Sangkrah (RW 13 – RT 03) and *Kampung* Gandekan (RW 6 – RT 2).

The processes of community participation design divide in to four stages in both *kampungs* as case studies (See Figure 2). **Stage I** is mapping and focus group discussion (FGD). Facilitators with children are doing participatory mapping existing public space location and condition. Facilitators were identified perception and collecting aspirations about public space in both *kampungs*. This process was set up an information of existing condition and collected the perception and aspirations of children about public space. **Stage II** is the process of Design Collaboration. In this process the stu dents and children were verifying the potential spot for public space through place making methods activities. This process is picturing inspiration, ideas, and creativity through the children's sketch. **Stage III** is technical collaboration (design action). In this process the facilitators were translating the idea from children's sketch in to applicable design.

Fig. 1. Study Area of Child Inclusive Kampung.

Design action for building inclusive public space for children in both *kampungs*. **Stage IV** is the monitoring process. The aim is to evaluate and monitoring the design action of inclusive public space in both *kampungs*.

During the implementation, there are several actors from different background (stakeholders type) involved in the project, and the configuration varies between both *kampungs* and in each stage of the program. This is caused by the difference of children participant's profile; *Kampung* Sangkrah gathered children through elementary school, while *Kampung* Gandekan gathered through existing child-based local community (*Taman Pendidikan Al-Quran*). Therefore, school authority appears only in *Kampung* Sangkrah and takes a role to provide permission for the project. Local community appears in *Kampung* Gandekan with a role to help gathering children for the project. Government representatives appears in both *kampung* as they also provide permits and help monitoring the implementation. Children are involved in focus group discussions and design collaborations as the main participants, but their ideas are being counted as a base for technical collaboration. Architecture student in this projects poses for two different roles; as facilitators in Stage I and Stage II, and as designers in Stage III. At the Stage III, they collaborated with local artists and handy-man.

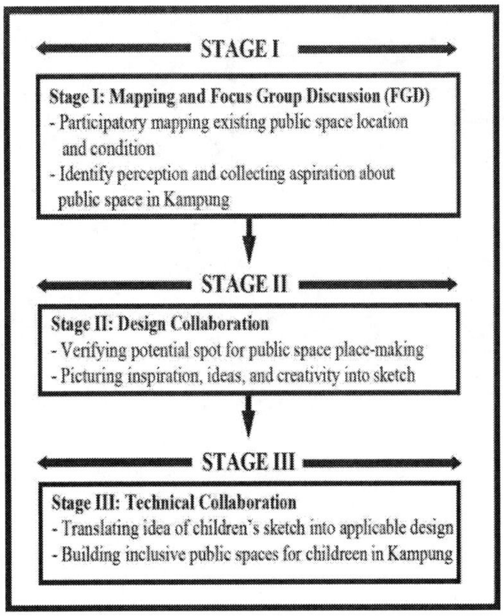

Fig. 2. Study Area of Child Inclusive *Kampung*

The Place Making Method

The place making method is one of the community participation methods in design. This project departs from understanding the fact that both of the sites are characterized as slum settlement, vacant spaces to build a new public spaces are not available. Therefore, this project takes **place-making** approach in public space provision. The place-making movement is soared over the concerns of several issues, one of them is lack of public space especially in informal settlements that increase tension and stress among people who live in crowded and inadequate conditions (PPS UN Habitat, 2012). Scheekloth and Sibley (1995) defined place-making as "the way in which all of us as human beings transform the places in which we live". Place-making offers cross-cutting perspective on activities which are often categorized as either formal or informal (Lombard, 2014).

At the context of children and young people, Sibley (1995; 1996) explains that place-making for children could be seen as an act to fight for a right, as the life-worlds of young people are bounded by rules created by adults, marking boundaries (re)establishing and naturalizing a moral division between adults and children.

Fig. 3. Activity of Place Making Project at Stage In *Sangkrah* and *Gandekan.*

This project highly relied on the Project Public Spaces (PPS)'s practices endorsed by The UN Habitat. The consideration to use this method is that it relies on the place-led, community-based process that will raise sociability values of the children (PPS UN Habitat, 2012). Therefore, this project was started by a place-making workshop, in collaboration with RUJAK Center for Urban Studies. It was equipped with an in-site simulation to conduct place-making method for children. (see Figure 3). Photography method was chosen to map the existing children's public space to ease children imagining their places. During the process of photography-based place-making ARKOM Jogja assist URDC Team by providing facilitators to lead the activity. The facilitators were also asked children several information related to the existing public spaces that include; the accessibility, the frequency of utilization, perceptions, and aspirations of the existing public spaces. This information was then aggregated into quantitative data listed below.

Result and Discussion

Actors Involved in *Kampung Layak Anak* Place Making Project

There are various actors / stakeholders involved in *Kampung Layak Anak* project. They also hold different roles that support the implementation of the project. This framework of where of implementation in both *Kampung Sangkrah* and *Kampung Gandekan* depart are part of research will explain the actors, their roles, and relationship with each other to put into context. The
In *Kampung Sangkrah*, the activities were focused on the SD Negeri Sawahan 2 (elementary school) located within the *kampung*. This was caused by the utilization of its playground space as the most frequently used public space in the *kampung*. There are several stakeholders involved, URDC UNS acted as the expert academician supported with architecture student volunteer. They play a key role as the main initiator and facilitator during the process. From the Non-Governmental Organization, there are (1) Bernard Van Leer Foundation as the main financial supporter for the whole *Kampung Layak Anak* project, (2) RUJAK Center for Urban Studies as financial supporter and knowledge-building collaborator, (3) ARKOM Jogja as a partner to train architecture student-volunteer to be a facilitator, and (4) Local Artists Community as a technical-support partner in design implementation process. From the government, (1) the authority of SDN Sawahan 2 supported the project by giving a permission to conduct activities in the school with its students (children participants) and ensuring the sustainability of the implemented public space and (2) BAPPEDA (Development Planning Agency) supported the project by giving permission and initial data as the project's objectives are in line with Surakarta's child-friendly objectives. From the local community, the only actor involved is the school student (4th grade) who also live in the *kampung* where the school is located.

Differs from the Kampung Sangkrah, Kampung Gandekan's implementation did not go through school authorities to gather children, it chose existing children local community instead. Similar to *Kampung Sangkrah*, URDC UNS acted as the expert academician; the main initiator and facilitator during the process. Bernard Van Leer Foundation, RUJAK Center for Urban Studies, and ARKOM Jogja, Local Artist Community, has also the same role. From the governmental side, it has the same support from BAPPEDA (Development Planning Agency of Surakarta). From the local community, the project was supported by (1) *Ketua RT* (administrative local leader) by granted permission to conduct activities in Kampung Gandekan, (2) *Taman Pendidikan Al-Quran*, a children local community that teaches Islamic religious teachings, that helped facilitator to gather children, (3) PKK, a local women community, to also granted permission to conduct activities, and (4) children as participants of this project.

The differences caused by different actors involved in both *kampungs* affects the process of project implementation. In Kampung Sangkrah, school authority played important role in advising and permitting the project implementation. As the result, the "mandatory" feeling could be felt during the process as data gathering and design collaboration process were held in the classroom. Limited time also offered during the process. By contrast, Kampung Gandekan relies the In this case,

 https://doi.org/10.1051/shsconf/20184107007

it offers more flexible time and less-mandatory feeling as children would be also able to do the activities outside.

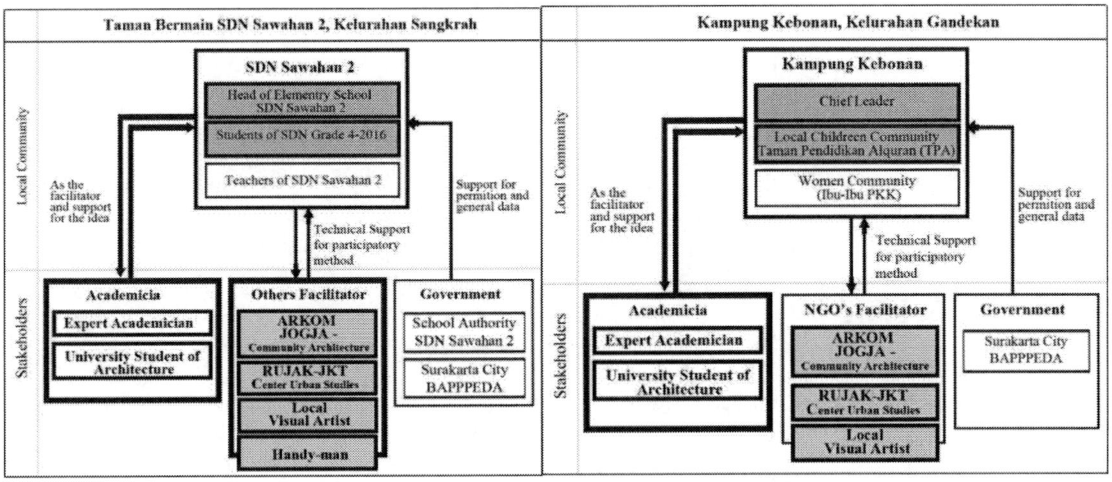

Fig. 4. Diverse Actors Involved in The *Kampung Layak Anak* Project.

Designers (facilitators) also have more free space to communicate with the children. As the results, more creative ideas from the children could be gathered in Kampung Gandekan compared to Kampung Sangkrah.

The Level of Children Participation in Kampung Layak Anak Place-Making Project

This part of analysis is intended to show the level of children participation in *Kampung Layak Anak* place-making project. The importance is that the project had already putting itself with the notion of child participation as its main value as well as its tagline. Therefore, to evaluate it becomes necessary. Hart (1992; 1997; 2013) in his extensive researches had adopted Arnstein's (1969) participation-ladder theory into children and young people context. In this research, his theory becomes the baseline to evaluate children participation in each stage of the projects, because the level of participation may vary.

There are three stages of the project, similarly implemented in both *Kampung* Sangkrah and *Kampung* Gandekan (See Figure 5 below). The first stage is the series of focus group discussions in which children were gathered to map public spaces utilization within their *kampungs* and to gather their aspirations towards better public space. The participation of children in the first stage is categorized into the sixth (6) rung "adult initiated but shared decisions with children". The reason is that the designers, in this case Urban Rural-Design and Conservation Laboratory and the architecture students who volunteer initiated the activity. But the activity included children and treated them and their voices as priority, thus the decisions is on the children's hand.

The second stage is the design collaboration process, in which children were asked to picture the public space they want to have in their *kampungs*. In this stage, the children draw the public space, that include elements and amenities of it, into the paper given by the architecture students that took a role as facilitators. The participation of children in this second stage is categorized into the sixth (6) rung of the ladder; "*adult initiated but shared decisions with children*". The reason is similar with the previous stage, because the adults initiated this activity and the children became the participants. But still, the main aim of this stage is to collect ideas and the needs of the children. Therefore, the children hold the decisions.

Both first and the second stage are categorized into sixth rung of the ladder due to the situation at the beginning of the project that the community and the children themselves in both *kampung* are not fully aware of the public space inadequateness issue. Therefore, to manage participation into the highest level (initiated by children) is not possible. In those phases, the initiative and initial information comes from the facilitators. As the result, the whole activities were dependent to the facilitators, children were continuously needed to be invited to those activities. Information about the adequate versus non-adequate is being "infiltrated" by the facilitators, in this case with strong academic background. Nevertheless, the children's knowledge still could be the baseline for developing the design. The benefit to be situated in this level is that the knowledge-exchange could occurs between the facilitators and the children. A lesson learned from these phases is that pre-project socialization activities could help children initiated initiatives on their own by triggering awareness to understand the issues that they faced earlier, so that the project could be in the highest participatory level possible.

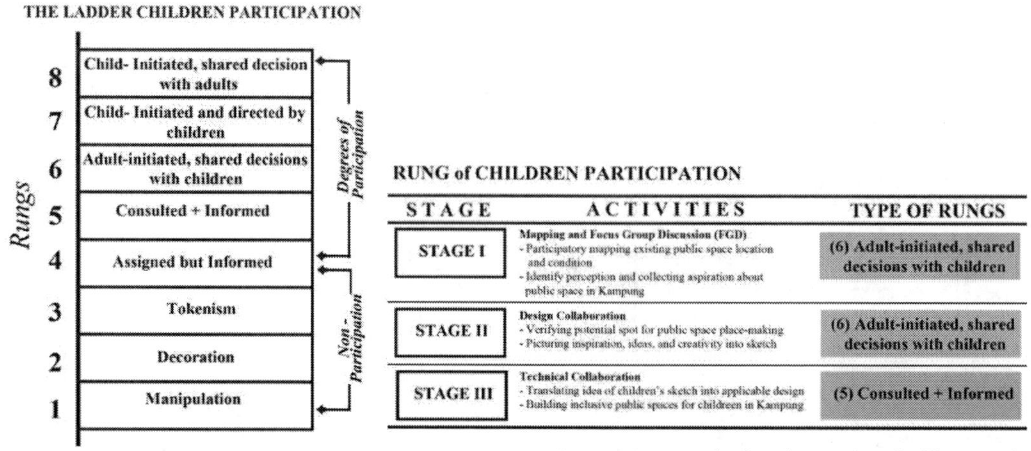

Fig. 5. The Ladder Children Participation in The *Kampung Layak Anak* Project.

The last stage is the design implementation process, in which the architecture students took a role as designer, translate children's ideas from the previous stage and realize them into a real public space. There are two activities within this stage, the first one is the design translation where the designers discuss and translate children's pictures into real and applicable design and the second one is the design realization with additional technical collaboration with local artist and construction workers (handy-man). This stage falls into the fifth (5) rung of the ladder; "consulted and informed". In this stage, children are not intended to collaborate in hands-on experience of building the public spaces but their opinions and ideas are treated seriously, even becomes the initial ideas of the design. During the activities, the designers do not close the chance of accidental involvement of children who live near the construction site. As a result, during the realization there are several children watching the process, even lending small help of watering plantation of the public space, even though they are not intended to participate. Nevertheless, this stage is not classified as true participation. Therefore, better method to conduct design implementation with children included is needed to be outline in order to improve this process to become a true participation.

Participation Techniques Analysis

Participation Techniques in Data Gathering Process (Stage I)

The first participatory stage of the project in general was "*public space mapping*" to collect data about existing public space, that was equipped with focus group discussion to verify the results (see figure 1). Nevertheless, even that the framework of activity was the same, the method to conduct is different. At *Kampung Sangkrah* implementation, the method used to gather information was **only** the focus group discussion that was led by architecture student volunteer as facilitators. At the context its typology, it falls into *group interaction method* where the facilitators led the discussion of relevant public space issues. The facilitators asked series of questions (See Table 1) and the children answered the questions in public-voting manner, then the facilitators recapped all of the information gathered during the process and put them into aggregated data.

Table 1. Questions Asked During Data Gathering Process.

No.	Question Asked	Sub-Questions
1.	Profiling	Age
		Gender
		Address
2.	The Location of Public Spaces They Frequently Use	-
3.	The Reason They Choose The Current Public Space	-
4.	Characteristic of Utilization	Time Spent
		Time Preferences
		Type of Users
		Frequency of Utilization
		Type of Activity
5.	Perceptions of Children Towards The Existing Public Space (Scale)	Comfortability
		Safety
		Accessibility
		Cleanliness
		Friendliness
6.	Aspirations	-

By contrast, the implementation at *Kampung Gandekan* chose different method to conduct the mapping (data-gathering) stage. Influenced by the previous workshop held by URDC UNS in collaboration with RUJAK Center for Urban Studies, the ***Gandekan*** team chose mainly place-making photography-based **transect walk**, equipped with individual questionnaire to gather information. The facilitators (1) gathered 30 children who lived in the *kampung*, (2) divided them into 10 small-groups consisted of three children that each were accompanied by a mentor (architecture student), (3) equipped each small-groups with a pocket/cell-phone camera, and (4) accompanied children wandering their environmental surroundings to show existing public spaces that they frequently use. At the end of the process, the facilitators asked each children with an online questionnaire the same questions asked in *Sangkrah* (see Table 2). Therefore, this method is a mix between ***"awareness methods"*** in which the facilitators initiated walking tour to facilitate children's awareness of their environment—especially the public spaces, and ***"indirect methods"*** as in the end of the process the facilitators used questionnaire to gather more information.

Table 2. Result of Frequency of Utilization, Time Preference, Scale of Accessibility (Perception), and Time Spent.

Frequency of Utilization	Time Preferences
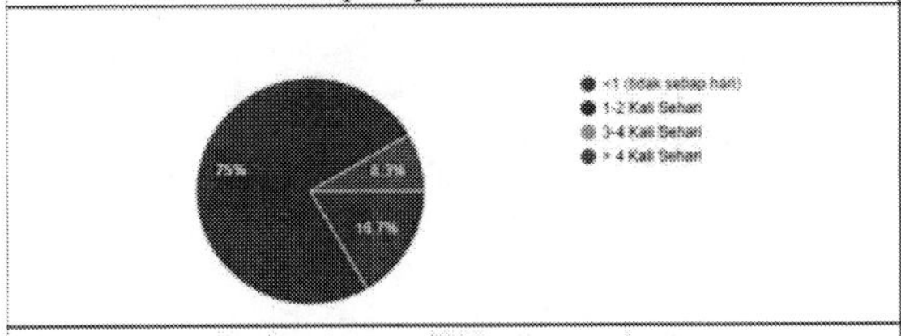	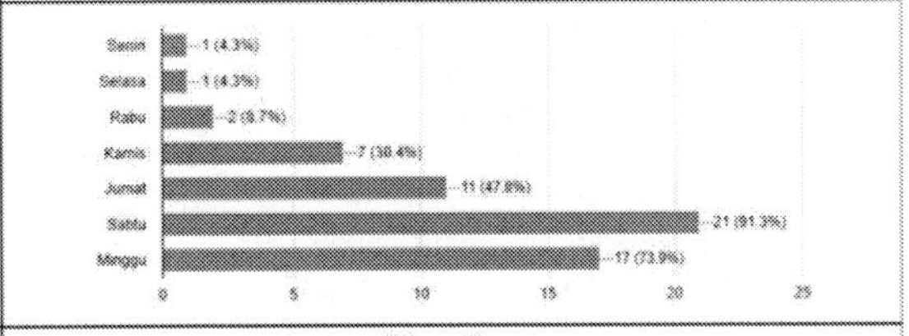
Scale of Accessibility (Perception)	**Time Spent**
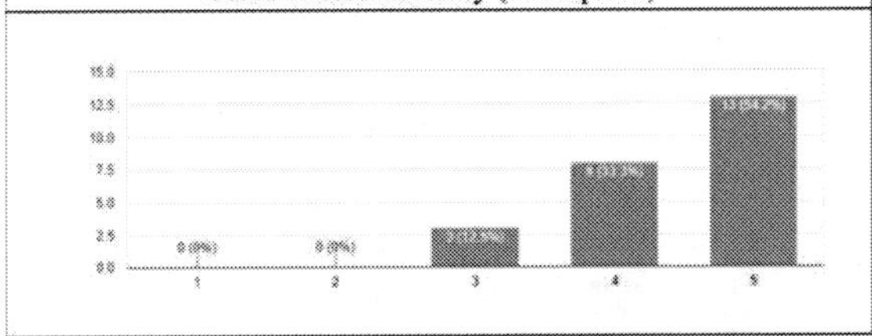	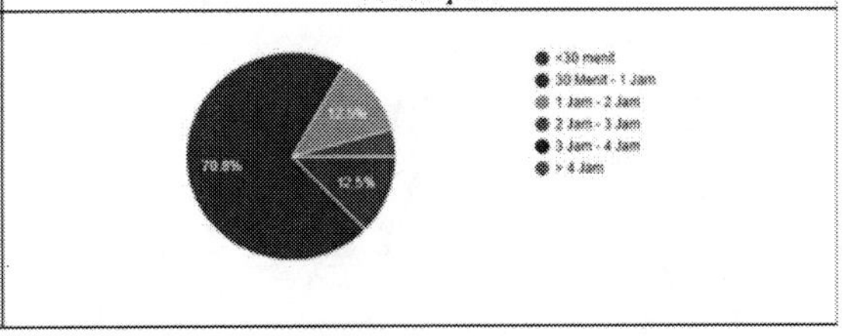

current conditions of public spaces from the perspective of children, only the methods are different. The data was also used to select the most frequent-used public space for the next stage—design collaboration. But different additional yet important information of the exact locations of public space and the pictures of each public space taken by children are be able to gathered through *Kampung Gandekan's* method. In *Sangkrah*, they only plotted the location of public space through spoken information from children, but not the visualized ones. In this case, children tended to echo other children's voices in mapping the public space. By contrast, the children in Kampung Gandekan that used transect walk method dominatingly utilize their sense of place and individual memory to spot the public spaces they use. From this stage, this research finds that the transect walk method is more effective to gather comprehensive initial data for design participation with children.

The results from both *kampung* are different to some degrees, as the methods used in each *kampung* was different (see Figure 7).

Fig. 7. Map of Public Space Location Produced Through Transect Walk and FGD.

The same aggregated information gathered are the same, as they used the same questions to identify the

Participation Techniques in Design Collaboration Process (Stage II)

The general framework for the second stage for both *kampungs* was the same, it was design collaboration focus group discussion (FGD); in which children were asked to outline their design ideas into selected public spaces that were selected from the previous stage. In both *Kampung Sangkrah* and *Kampung Gandekan*, the facilitators (1) gathered the same children-participants from the previous stage, (2) handed them a piece of blank paper and marker/pen, and (3) asked them directly to draw the public space they want that include the design ("*how the public space will look like*?') and its element *("what amenities and additional decorations (murals, etc.) do you want to have?")*. Therefore, the method falls into "*brainstorming method*" close to its ***gallery*** type, as it facilitates wild idea generations with no criticism by giving out media (paper) for the children participant to use. Result of The "Gallery" Technique Conducted During The Design Collaboration Phase (See Figure 8).

additional elements of the public space such as the picture of murals (See figure 9). This finding indicate that the different previous method taken in public space mapping affect the children's ability to visualize their aspiration. Children in *Kampung Gandekan* were previously equipped with real vision of their existing condition of public spaces, as *Kampung Sangkrah* could only imagining the place they use.

Participation Techniques in Technical Collaboration Process (Stage III)

Technical collaboration in this stage is in both *kampungs* intended to be implemented without direct participation from children. Therefore, in this stage the facilitators firstly (1) collected children design ideas, (2) conducted discussion with fellow facilitators and the academic experts (URDC UNS), (3) translated children's idea into applicable scaled design for each public spaces selected (see figure x), and (4) collaborated with local artist community and handy-man for implementation.

Fig. 8. Result of gallery type in *Kampung Gandekan.*

Fig. 9. Result of gallery type in *Kampung Gandekan.*

Nevertheless, the results produced from the focus group discussion in each *kampungs* are different. From the result, it could be identified that the drawings of public space that the children aspired were more precisely pictured in *Kampung Gandekan*, compared to *Kampung Sangkrah* (see the comparison on figure 8). The pictures produced in *Kampung Gandekan* appear to be more applicable and mimics the real condition of the existing public space in their *kampungs*. Some of the children were also careful enough to put detailed amenities with specific use that they want into the drawings. By contrast, the drawings produced in *Kampung Sangkrah* were less detailed and applicable. Some of them even had difficulties to picture the public space that they want, and prefer to write down their aspirations instead. Some of the children that were successful to picture the idea were only focused on the

Local artists collaborated in creating murals for public spaces, while handy-men helped the physical development process.

Conclusion

This study has examined the technique of participatory design for inclusive public space provision in *Kampung Kota* of Surakarta. The importance of this study to encourage better technique community participation in the future. Based on the results of these analyses several issues have been clarified:

1. There are **two important key player** engaged in *Kampung Layak Anak* Place-Making project. They are **stakeholders** (collaboration of multi-sectoral stakeholders such as academician, NGO's or others facilitator (visual artisan, handy man), and

government. The **local community actor** represents by the head of students, teachers of SDN Sawahan at *Kelurahan Sangkrah*. Meanwhile in the local community represent by the chief leader and local children community in *Kelurahan Gandekan*. Facilitators hold the main role in project implementation. A lesson learnt from this analysis is that involving non-formal local community as intermediate partner to involve children is better than formal organization such as school for flexible time-span and communication with the children.

2. *Kampung Layak Anak* place-making project has involved multi-level children participation. The level of children participation in Stage I and Stage II were categorized into the sixth (6) rung of the ladder; "adult initiated but shared decisions with children". In the last Stage III categorized in fifth (5) rung of the ladder; "consulted and informed". Therefore, in the further projects the children participation should be encouraged in the upper level in eight (8) rung of the ladder; "child initiated, shared decision with adults" by initiating series of pre-project socialization to raise children's awareness of issues they faced related to their environmental surroundings. This step would trigger children's own initiatives to intervene such issues, thus maximizing their participation level.
3. It is necessary to use various participation techniques in design collaboration process. Transect walk is more effective for data gathering process, as this method enables children using their sense of place and memory to map public spaces. This method will result in better design proposals created by the children as they would save memory of the existing condition of the public spaces they want to improve. There is also an importance to develop a new participation technique that would be able to include children in technical collaboration.

References

Arnstein, Sherry R. (1969). *A Ladder Of Citizen Participation*. Journal of the American Planning Association, 35: 4, 216 — 224.

Hart, Roger. (1992). *Children's Participation Form Tokenism to Citizenship*. UNICEF

Hart, Roger. (1997). *Environments for Children: Understanding and Acting on the Environmental Hazards That Threaten Infants, Children and Their Parents*. (with Satterthwaite, D., Levy C., Ross D., and Stevens, C.). New York: UNICEF, and London: Earthscan

Hart, Roger. (2013). *Undesigning for Children: Creating Space for Free Play and Informal Learning in Community Gardens*. New York: Design Place for Public Space.

Kusumaningdyah N.H, et all. (2011). *Participatory Mapping in Surakarta - Case Study of Mini Atlas Solo Kota Kita -*, Journal of Habitat Engineering 2011, Volume 3, Number 2, 189 – 200.

Lee, Y. (2006). *Design Participation Tactics: Redefining User Participation in Design*. 2006 Design Research Society International Conference in Lisbon (IADE).

Lee, Y. (2008). *Design participation tactics: the challenges and new roles for designers in the co-design process*. CoDesign Journal Vol 4, Issue 1, March 2008, 31-50

Lefebvre, H. (1972). *The Production of Space*. John Wiley & Sons Publication.

Neuman, W. Lawrence. (2013). *Social Research Methods : Qualitative and Quantitative Approach*. Pearson.

Putri, Rufia Andisetyana et. al. (2016). *A Spatial Analysis for Assessing The Suitability of Elementary School as Social Infrastructure at The Neighborhood Unit Scale in Supporting Child-Friendly Surakarta*. Geoplanning: Journal of Geomatics and Planning. [S.I], v. 3, n. 1, p. 33-52 apr. 2016.

Sanoff, Henry. (2000). *Community Participation Methods in Design and Planning*. John Wiley & Sons, inc.

WRAP UP SESSION

Human-Centered Architectural Education In The Contemporary Complexities In Asia

Chair of Scientific Committee: *Revianto* B. Santosa[1,*]
Moderators: *Abdul* R. Maghzaya[1], *Nensi* G. Yuli[1], *Jarwa* PS Handoko[1], *Baritoadi* BRG Rito[1], *Wisnu* H. Bayuaji[1], and *Maria* Adriani[1]

[1]Department of Architecture, Universitas Islam Indonesia, Indonesia.

The world of architecture in the global era, like many other field studies on professions, is facing enormous challenges, which at least come from two sources. First, its "old" theories which were put a requiem back in the 1970s are challenged by vast changes in today's world, in which technology and media responsible. Meaning of space and place, for example, need to be re-configure as time is moving not in linear way and not become the only constructor of nowadays world. Second, media and technology also tremendously shakes the legitimation of architects in their role of producing space, as they are also facing ethical-professional values to be constantly in confront. What Pinterest, Instagram and many other social media are pleasing consumers with visuals and narratives of imagined space, have resulted to the disturbance in the process of being architecture and becoming architects. Big Data and the Internet of Things currently direct of our understanding of the world and soon govern many aspect of our lifes. Although the world is running in dispersed directions, we, the academics, goalie in the knowledge of architecture production, the prerequisite to adapt to this interrupted changes.

Interconnectedness very much characterize contemporary situations. For millenia, the great land mass of mainland Asia as well as the vast oceans of its archipelagic regions have been the region to cultivate great interconnectivity and diversity producing enormous multi-culturalism. Currently, however, global movements and real time communications are accellerated in unprecedented speed. The ability of Asia to embrace and nurture diversity is challenged due to this extreme pace.

Architecture, as we know it, are experiencing fundamental changes. On the one hand, architecture is urged to deal with the increasing complexity and accelleration of changes by being adaptive, inclusive and progressive. On the other hand, architecture has to return to humanity being the fundamental values to uphold. Such situation is highlighted by a Pritzker laureate architect, Rem Koolhas in an interview, "... stands with one leg in a world that's 3,000 years old and another leg in the 21st century, . . .This almost ballet-like stretch makes our profession surprisingly deep."

Being the largest archipelago in the world, Indonesia develop heavily in the desire to improve with the high notion on the *Aufklarung* as architecture also follows, the pursuit on the fathom is reckoned to the product of knowledge, rather than steeped the process of producing knowledge. This situation shall not blame. However, there is also an urgency to move beyond merely continuously confirming "truth". A context of the study, whether it is a setting, methods, tools, or variations of participants, anything else is needed to be re-translated into a manner of broadening or knowledge. It supposes to be not hard, as our setting, the world of architecture in Indonesia is in the contestation between going into the future and holding up to the pasts, engaging inclusiveness or staying in exclusiveness, as well as going large and global or going small and local.

The conference, entitled, "Re-charting the knowledge of architecture", is aimed to pursuit of architectural knowledge in the position of global arena. As it took venue on Indonesia, the conference cannot be away from the dynamic inquires of decentring and diversifying of the architecture of local versus global. A day conference with a great numbers of participants, is prepared and carried out in a very limited time elucidate many aspects of the main theme. Prof. Junsuk Lee underlines the way we should manage the education of architects in a holistic way. The complexity of culture and some insightful ways of dealing with it are well stated by Prof. Johannes Widodo. Prof. Marcus Hackel presents the importance of first hand experience as an excellent way to enhance the ability of the students to deal with multi-culturalism; while Prof. Ibrahim Numan emphasizes on the ramification of socio-political context in the formulation of knowledge and method of education in architecture. Prof Yandi A. Yatmo provides a framework to understand the renewal of theories, methods and design practices. Dr. Ilya Maharika challenges the most fundamental notion on the transformation of the word "architecture" questioning the ontological basis of the field.

* Corresponding author: revianto@uii.ac.id

Presentations are arranged into a number of topics, namely:

1. Humanity and its Challenges, questioning the fundamental how architecture serving and being interrelated with humanity in challenging situations
2. Multitude of Cultural Complexity, investigating how architecture deals with the increasing complexity of culture
3. Uncertainty of Urbanism and Territoriality, thorougly discussing the fading the physical, social, and knowledge boundaries between architecture, urbanism and territoriality
4. Digital Reality and Digital Technology, trying to understand the based advencement of digital technology to affect life and architecture
5. Environmentally Appropriate Technology, reconsidering the ways to incoporate technoloy to create sustainable environment in the whole sense.

As many excellent presenters discuss those topics, I would like to take the liberty to propose different ways to wrap up by proposing a frame on the need of education, as follows:

1. Reconsidering Paradigm
   As architecture is one of the human-centered field study, while the world dramatically changing, a need for re-inventing architecture with considering dynamic of the interconnectedness of spatial fluidity and temporary arrangement is valuable to push a bit further our perception and definition about the study. We must not, however, drawn in the ocean of texts and information in the global age, and thus repositioning our architecture is a must in an attempt to put encounter of the wisdom of the past and promising future, of micro and macro scales. Architecture should be the meeting ground between variety of social and economic forces, between permanence and impermanence, between past, present and future; while deeply rooted in the humanity.
2. Reconfiguration of the Theory
   As architecture always signified genius loci rich with stories, backgrounds, and context of humans, the theory of architecture should not be seen, and never be an independent theory. As we must not forget this, production of knowledge in architecture should address the multi-level and multi-aspects interrelations. Thus, whatever research approaches we will use from technology to ecology, by positioning urbanism as a phenomenon, context, or setting may draw patterns of habitation which might be intervened with architectural design, and eventually leads to a new realm of knowledge.
3. Reformulation of Method
   As a process is important in the knowledge production; methods to do research and design suppose to expand knowledge and ability to deal with complexity. On the other hand, our students as actors in the global age potentially perceive architecture, space, and time quite differently. Thus methods to engage with them in various learning process will larger benefit, as same as to employ digital devices in dealing with variety of themes and interrelationship in the architectural world.

Humanity and its Challenges, questioning the fundamental how architecture serving and being interrelated with humanity in challenging situations

Community involvement on spatial issues is important to be considered and studied. People in Indonesia have an important role in shaping space even in an informal process that grows naturally in the midst of society. The process can be drawn to be part of a more formal design process which during this design process tends to focus on the architect. The patterns of informal space growth usually coincide with formal spaces that are interconnected with one another. As in the study of the growing relationship of informal patterns around the campus. The morphology, visual, and function dimensions around the campus are determined by social and temporal perceptions.

More broadly, the role of the community can also informally determine the formation of spatial patterns, as happened on the island of Karanrang. The spatial pattern in Karanrang is determined by the actors of activity, occupancy, territory that make up unique typology based on activity, child, mother, father, and combination of the three. Without comprehensive community involvement, a design engineering will function less than optimal.

The studies that have been conducted in the Cihapit traditional market of Bandung trying to include creative activities in the show that the existence of events and creative artifacts can indeed improve the physical image, but it does not give an economic share for the market. This is because the impetus of traditional market economic activity is stronger and requires more space. As one example is done in Kampung Kota Surakarta involving government and stakeholder to create public space for the child. The key point, the more varied the participants involved, the better the results will be. This understanding needs to be initiated in the education process for students.

The concept of co-design that is taught to students as a method of designing can introduce students to the society or community. The challenge is that this process needs to be done in a longer time. There are at least three stages to do is to determine the NOC (Number Of Clients), then determine the LOIC (Length of Initial Contact), and LOCE (Length of Community Engagement). The LOIC shorter may lead to the longer LOCE. The pattern of social space in Karanrang island shaped by a) Children activity, b) Father activity, c) Mother activity, and d) Combined activity.

Co-design is not something new but practicing it in academic setting is dificult because of time limitation. To understand informality we need to see not only physical aspect but also perceptual dimension and temporal dimension. When we face the uncertainty in informal space we need to unveil and unravel. Creative movement in traditional market only good for media but not improve the market economically. In participatory design , it is necessary to include various participator.

Multitude of Cultural Complexity, investigating how architecture deals with the increasing complexity of culture

There are three models in architecture studio learning: internship, supervised and simulation design studio who depends on supervised, senior architect and student disciplinary. Only internship and supervised studio are recommended for learning process to achieve the architect competence in professional program of architect. For bachelor program, problem based learning is important method in architecture pedagogy to have 3 important basic skills: complex problem solving, critical thinking and creativity. If this method could be combine to lecturer's responsible in three pillars of higher education, it would be a great potential to develop both students and lecturer's skill in development of architecture education.

To focus the cultural studies in architecture design studio, there are some design principles that delivered from traditional value, i.e. Sulappa Eppa concept from Bugis traditional community, the concept of Gowa traditional house which converted in the brick producers houses and the preservation of Islamic value that still use in Arabian community in Pekalongan and how the historical city still use their old spirit which transferred to the modern city concept but never loss its form and identity.

Cultural aspect should be transformed by its value, the spirit of space, basic fundamental, behaviour and habit of the community and the most important thing is the authenticity of its tradition of life. We should change our mind to not only see and transferred the form of traditional architecture but we should consider those aspects. All the architecture students have to be aware about how the cultural will be influence their design precisely if they could be use the tradition and the aspects.

Syam, et all find the concept of territory of social space in the context of the interaction of two ethnic Bajo tribe and Bugis tribe in the form of settlement, as we seen at kampung bajo, Sulawesi. Territory formed in Bajo settlement is inseparable from historical values, natural environment condition, sea as main orientation and privacy demand in the form of motivation of economic interest. Territory categories that have been unique to the uniqueness, social aspects and similarities of marine and economic cultural backgrounds can foster a strong sense of kinship, thus eliminating the boundaries of space and time in their daily interactions and activities.

Sari, et all. find the main elements of historical objects and examine their relationship with local communities and tourists perception. The research provide of place attachment towards all mainly historical objects as "spirits" of cultural heritage for people to gather and interact. The research indicates the significance of the tangible heritage and intangible heritage attachment in shaping the heritage city. Cultural heritage is a great asset and a proud to be preserved. The main elements of historical objects become of identity the city because of its has architectural style and historical background. As we seen in Malioboro, Yogyakarta.

Panjaitan,et all. suggest that The development of the area for tourism purposes should have good sensitivity related to the product to consumption aspects, where the interaction between the local community is seen as the supply (insider) produce the products, which can then be consumed by the visitor as demand (outsider).

Warakanyaka, et all. argue that the presence of time is a crucial aspect in the progression of the interior architectural education. The interior discipline that focuses on the aesthetic and constructional aspects, curently also about aspects of sociocultural conditions to enhance the wellbeing of its inhabitant. They also argue how the presence of time might affect, transform and even generate context-specific interior architectural design methods that enables several dynamic forms of inhabitation.

Harahap, et all discuss how *reading* is central to the architectural practice and how it carries an immense potential for architectural design method. It argues that an act of *reading* is important to be introduced and learned in the architectural education institution within which students learn and develop various design methods.

Siregar, et all. have studied focused on locating the genius loci of Medan City through tracing the historical meaning by adapting the method undertaken by Norberg-Schultz in tracing the spirit of place and genius loci. Finding fo this research is about the role of culture and economic background that plays a major role in the formation of the character of Medan City center. The city is formed from the history of the plantation industry as well as the diverse cultures that share the same attachment and goals in the economic field.

Some important points how architecture deals with the complexity of culture:

1. Cultural/ historical and economic background become the main influenced factor for the form of settlement or city.
2. Social culture, technology and legal standing is essential factor, how to conserve the historical architecture.
3. The presence of time is crutial aspect in the progression of the interior architectural education.
4. Reading as an approach of design culture method.

Digital Reality and Digital Technology, trying to understand the based advancement of digital technology to affect life and architecture

Three main themes are presented on the digital technology papers, which are : a) The theoritical approach of layered relation between digital technology and architecture design (Johanes et al) b) The practice and implementation of the digital technology such as parametric based design and Virtual reality tools within architecture educational and design for specific purposes, cases and context (Al Athas, Gunagama, Pamungkas et al) c) Examining the implementation of the digital technology within the architecture discourse (Indraprastha).

Johannes et al find digital realms layers are then layered into materialization, simulation, translation and generation , which then can be conceived in essential roles of two form of the development of the architectural education curricula: the layers of the medium which allow us to work in surpass the interface and work more tactically by understand its operating language, and the layers of methods which bridging the fundamental knowledge of digital medium with the more practical architectural design processes.

M Galieh finds Evolutionary Solver in the Galapagos could reach capabilities the dynamics of the algorithm process that was arranged for the acceleration of alternative design creations. Although the simulation is only able to propose an alternative approaching the ideal design conditions, due to the extent of the ideal definition and the difficulty of converting it into the mathematical notion, the simulation results show that each different initial condition yields a different design alternative.

Syarifah establishes a give-and-take relationship between top- down formal emphasis and a bottom-up material influence through intriguing case study of parametric design of Gridshell structures.

Pamungkas et al, suggest that design cognition as the essence of learning in the architectural design needs to be approached by the spatial experience honed by three-dimensional thinking from the medium diversity. By using Virtual Reality tools, they find VR simulation has a good impact on a spatial experience of office building design for 2nd-year architecture students by simulating good quality of visual and lighting, as well as brings up the ability to feel the scale of space.

Indraprastha suggests that rather than only providing them a set of computer skills for production and presentation techniques, the challenging issue mainly is to teach computational thinking at undergraduate level of the design. The computational-based course for undergraduate program needs to focuses on the know-how aspect of pedagogy, which focuses on to know how to use, and how to solve a certain problem using some set of tools or algorithms.

The based advancement of digital technology to affect life and architecture:

1. How to find pedadogical teaching method/approach to closing the gap between formal ways of using software vs informal and creative ways of design process.
2. Importance to know how to use, into how to understand the language and anatomy of the software
3. The dynamic in architecture education allow teacher to explore various approach in developing architecture student analytical skill of environmentally friendly design approach. Some utilise creative method whilst other employ computer based simulation. One of the example of creative approach is the use of Ecology to be utilised to learn about sustainability through a quantitative approach to achieve an efficient and environmental friendly problem solving. The idea of ecology as a study on interrelated systems could be transformed into an analytical tool that could reveal complex systems and issues. Its nature of having complexity and dynamicity makes ecology a complete analytical tool that could lead to a deep understanding through an explorative process as well as spatial understanding.
4. Another example of creative approach is Biophilic Design Method. It is an approach to architectural design to consider design elements (or human) as of a living biological organism with its mind-body systems indicator of health, productivity, and wellbeing. This method syntesizes the utilization of various factors i.e. resources both natural and man made with the human receptance and its implication of i.e. health, social, economy and well being.
5. The computer simulation and technology has evolved and result in the variety of approach and method in measuring variables in building planning and design. Each method has its own strengths and weaknesses as they are being developed using variables and elements that are unique to its origin, context and formula. Architect and both student and teacher should have a basic knowledge of various method to be able to justify which method is best suit their case of study and design.
6. Software technology allows architect and researcher to gather various information from climate to surrounding context, and other necessary design variables to perform and simulate building model to be able to both create and evaluate the most efficient building design. The use of natural resources for building materials as well as comfort generator has been an important aspect in achieving sustainable and energy efficient building. In a more specific example, the use of software through computational simulation may even help to identify and reveal the response of traditional architecture in Bangsal Srimanganti of Kraton Yogyakarta to acoustical aspect of Gamelan music. In other case, it may be utilise to study and evaluate a structural firmness of a construction against endangering disaster i.e. earthquake.
7. Daylighting calculation method, how relation of context affect the calculation result. New find. Challenges in implementing the ecology approach and strategies. Any other example of project using ecological approach? Creative trigger is important as a vessel to bring the process into intended goal. Architecture has evolved and so should the approach of teacher in delivering knowledge and insights to the students. Various creative ways such as biophilic and ecology as analytical tools is an example as well as the use of software technology and on site live observation. Those approaches should enrich the education system in preparing students as both a critical unique mind body system and part of a more complex and larger network of minds.

Environmentally Appropriate Technology, reconsidering the ways to incoporate technoloy to create sustainable environment in the whole sense.

The dynamic in architecture education allow teacher to explore various approach in developing architecture student analytical skill of environmentally friendly design approach. Some utilise creative method whilst other employ computer based simulation. One of the example of creative approach is the use of Ecology to be utilised to learn about sustainability through a quantitative approach to achieve an efficient and environmental friendly problem solving. The idea of ecology as a study on interrelated systems could be transformed into an analytical tool that could reveal complex systems and issues. Its nature of having complexity and dynamicity makes ecology a complete analytical tool that could lead to a deep understanding through an explorative process as well as spatial understanding.

Another example of creative approach is Biophilic Design Method. It is an approach to architectural design to consider design elements (or human) as of a living biological organism with its mind-body systems indicator of health, productivity, and wellbeing. This method syntesizes the utilization of various factors i.e. resources both natural and man made with the human receptance and its implication of i.e. health, social, economy and well being.

The computer simulation and technology has evolved and result in the variety of approach and method in measuring variables in building planning and design. Each method has its own strengths and weaknesses as they are being developed using variables and elements that are unique to its origin, context and formula. Architect and both student and teacher should have a basic knowledge of various method to be able to justify which method is best suit their case of study and design. Software technology allows architect and researcher to gather various information from climate to surrounding context, and other necessary design variables to perform and simulate building model to be able to both create and evaluate the most efficient building design. The use of natural resources for building materials as well as comfort generator has been an important aspect in achieving sustainable and energy efficient building. In a more specific example, the use of software through computational simulation may even help to identify and reveal the response of traditional architecture in Bangsal Srimanganti of Kraton Yogyakarta to acoustical aspect of Gamelan music. In other case, it may be utilise to study and evaluate a structural firmness of a construction against endangering disaster i.e. earthquake.

Daylighting calculation method, how relation of context affect the calculation result. New find. Challenges in implementing the ecology approach and strategies. Any other example of project using ecological approach? Creative trigger is important as a vessel to bring the process into intended goal. Architecture has evolved and so should the approach of teacher in delivering knowledge and insights to the students. Various creative ways such as biophilic and ecology as analytical tools is an example as well as the use of software technology and on site live observation. Those approaches should enrich the education system in preparing students as both a critical unique mind body system and part of a more complex and larger network of minds

Uncertainty of Urbanism and Territoriality, througly discussing the fading the physical, social, and knowledge boundaries between architecture, urbanism and territoriality

Continually and contextually changes. Thus, changes in *habitus* might open a creative cross-cut in reproducing knowledge about perspective of space. In a very least for example, changes in behavior in inhabiting space, such as from inhabiting water to land (Rosidin; Prayitno), from landed houses to high-rises (Ridwana; Adiyanti), from studio to real urban setting (Widodo and Atmodiwirjo), provides "facts" needed for architecture aims. Concerning on "facts" however, at some point might give undercurrent negative note on the knowledge reproductions.

However, as knowledge production needs to because of it is urbanism, a deep consideration in understanding contesting *field* invites even more engagement in human-related field of studies. Thus, narrative inquiry, ethnography, history, and many methods in revealing counter grand-narratives, are potential to be used as creative approaches in better understanding of now-world, as well as leading to re-chart our position, as source-full field, in the global knowledge world.

AUTHOR INDEX